PHARMACOCHEMISTRY LIBRARY – VOLUME 29

# TRENDS IN DRUG RESEARCH II
## Proceedings of the 11th Noordwijkerhout-Camerino Symposium

PHARMACOCHEMISTRY LIBRARY, edited by H. Timmerman

Other titles in this series

PHARMACOCHEMISTRY LIBRARY
Editor: H. Timmerman

Volume 29

# TRENDS IN DRUG RESEARCH II

## Proceedings of the 11th Noordwijkerhout-Camerino Symposium

11-15 MAY 1997, NOORDWIJKERHOUT, THE NETHERLANDS

Edited by:

HENK VAN DER GOOT

*Department of Pharmacochemistry, Free University Amsterdam, The Netherlands*

ELSEVIER

Amsterdam – Lausanne – New York – Oxford – Shannon – Singapore – Tokyo 1998

ELSEVIER SCIENCE B.V.
P.O. Box 1527
1000 BM Amsterdam, The Netherlands

ISBN 0-444-82633-5



This book is printed on acid-free paper.

Transferred to digital printing 2006
Printed and bound by CPI Antony Rowe, Eastbourne

PHARMACOCHEMISTRY LIBRARY

# PREFACE

This volume of Pharmacochemistry Library comprises the text of invited lectures presented at the 11th Noordwijkerhout-Camerino Symposium *Trends in Drug Research*, held in Noordwijkerhout, The Netherlands, from 11-15 May 1997.

The first medicinal chemistry symposium held in Noordwijkerhout took place in 1974. From 1987 on the Noordwijkerhout meetings were combined with the series organized in Camerino, Italy, and since then each second year a Trends symposium, either in Receptor Research (Camerino) or Drug Research (Noordwijkerhout) is organized. Inspecting the programmes of the previous symposia learns that "trends" develop not seldom into generally accepted subjects or even become routine aspects of the fields. In the 1974 programme it was very obvious that two groups of people were meeting, synthetic chemists and pharmacologists, whereas a small group of QSAR-ists tried to speak and understand the language of both fields. The 1997 programme shows that much has changed since then. Medicinal chemistry has developed into an independent subdiscipline of chemistry.

The 1997 trends are clearly dominated by the question: "How do we generate really new medicines?" The two major issues are the generation of new biological target systems and the generation of new lead molecules, respectively. In the programme important attention has therefore been given to those two aspects, with presentations on e.g. combinatorial chemistry, compound libraries, database search, high throughput screening and molecular biology.

Other topics discussed were the perspectives for new medicines for the gastro-intestinal tract, the major developments in the search for effective anti-HIV drugs and new aspects in synthetic approaches. In a special session three topics which currently draw much attention were discussed: How to deal with the major problem of resistance against antimirobial agents? Can the apoptosis mechanism be used as a drug target? Is the newly observed phenomenon of inverse agonism a general principle and has it consequences for drug development (and use)?

The organizers of the Noordwijkerhout-Camerino symposia express their sincere thanks to those who supported the 1997 symposium financially: Astra, Byk Gulden, Glaxo Wellcome, Janssen-Cilag, Lundbeck, Merck Darmstadt, Merck Sharp & Dohme, Organon, Pharmacia & Upjohn, Solvay Pharma and Synthélabo.

H. Timmerman,
Chairman Organizing Committee

# CONTENTS

H. van der Goot, (Editor)
TRENDS IN DRUG RESEARCH II
1998 Elsevier Science B.V.



# Insights into the structure and function of genetic disease genes from genome research and clues for drug therapy

P.L. Pearson

Department of Human Genetics, Utrecht University, PO Box 80030, 3508 TA Utrecht, the Netherlands

## 1. INTRODUCTION

Approximately 30% of all human morbidity has a (partial) genetic causation and over 60% of all persons in industrialized societies will suffer from a genetic disease during their life time. Although an obvious distinction can be made between infectious and genetic diseases in that susceptibility disease genes are inherited from one generation to the next in the case of genetic disease, the ability of the human body to mount a defense against infectious agents is itself under genetic control. Since World War II, the pharmaceutical industry has made major advances in reducing infectious disease, but it has had relatively little impact in alleviating diseases with a genetic causation. This effect is exemplified by changes in the incidence and causation of perinatal death, in which the large reduction in mortality during the last few decades has resulted mainly from decreases in infectious disease, while the genetic contribution has remained fairly constant. The net result has been that genetic diseases are playing an increasingly larger role in overall morbidity in the industrialized nations. This trend is further reinforced by the increasing age of the population and a concomitant increase in old-age related diseases with a genetic etiology, including Alzheimer's and NIDDM.

The majority of genetically determined morbidity arises not in single gene defects (Mendelian diseases) but in multifactorial (combination of multiple genetic and environmental factors) diseases and involves extremely common diseases, such as diabetes types I and II, arthritis, hypertension, obesity, hypercholesterolemia, atherosclerosis, cardiomyopathy, asthma, psoriasis, schizophrenia, autism, hyperactivity combined with learning disability, and various forms of depression. Generally, these disorders have proved relatively intractable to pharmaceutical treatments and it is arguable that most of the therapeutic strategies used to date have been based largely on empirical testing rather than on real insights into the causative molecular genetic mechanisms involved. For example, except for rare examples of mutations in the insulin receptor gene in NIDDM, the nature of insulin resistance and its relationship to factors such as obesity, aging, hypercholesterolemia and inheritance remains poorly understood. In the case of IDDM, genetic analysis has demonstrated the possible involvement of more than 20 loci (chromosome locations) which contribute to its inheritance. Although nearly all of the genes involved still have to be isolated and their structure and function determined, a genetic approach appears to offer realistic opportunities for determining the multiple components involved in IDDM.

A further strong argument for the use of genetic analysis in defining the pathophysiology of human disease is the underlying belief that the experiments of nature (mutations), which are transmitted within families and which are the primary determinants of genetic disease, also sign-post the molecular pathways most likely to be involved in "non-inherited" morbidity of the same type. Some of the best documented examples of inherited/non-inherited similarities in disease etiology involve tumors in which mutated genes occur constitutively and are transmitted from one generation to the next within families, with the production of "inherited tumors" in multiple family members: sporadic cases occur by de novo mutations arising in the same gene and confined to the tissue of origin for the tumor concerned. Thus, although the mechanism of origin differs between inherited and non-inherited forms, the molecular genetic changes invoked following initiation of the disease are similar for both forms.

## 2. EVIDENCE FOR GENETIC PREDISPOSITION

Observation of diseases occurring within families according to Mendelian expectations (i.e. segregating as an autosomal dominant or autosomal recessive disorder) yields indisputable evidence for a genetic etiology. However, there is no such clear distinction for many diseases and the first recognition of the involvement of a genetic component comes from the statistical demonstration of a significantly higher disease frequency within defined families than within the general population. Such a familial predisposition is normally expressed as the relative risk for a first-degree relative of an affected person within a family also being affected compared to the risk for a member of the general population. Table 1 gives some examples of relative risk estimates for known genetic diseases. Note the extremely high values for proven single-gene defects, also known as Mendelian (>500), and the relatively low values (<15) for polygenetic disorders. In practical terms, the use of modern genetic mapping methods employing highly informative DNA markers (polymorphisms) for detecting association between inheritance of a genetic disease and particular chromosome regions permits defining the genetic component of diseases with a relative risk as low as 2, as long as a sufficiently large number of patients and their family members are sampled.

*Table 1. Relative risk estimates for various genetic diseases*

| Disorder | Family risk | Population risk | Relative risk |
|---|---|---|---|
| Huntington's Chorea | 1/2 | 1/20,000 | 10,000 |
| Duchenne | 1/2 | 1/10,000 | 5000 |
| Cystic Fibrosis | 1/4 | 1/2200 | 555 |
| Breast Cancer | 1/2 (1/4) | 1/11 | 5 (3) |
| Diabetes I | 1/16 | 1/300 | 18 |
| Diabetes II | 1/16 ? | 1/30 | 2 |
| Schizophrenia | 1/3 | 1/50 | 16 |
| Hypertension | 1/3 | 1/7 | 2 |

Another classical approach, which was originally defined more than 100 years ago by Francis Galton, is that of the twin concordancy test. This test permits a clear determination of the heritability of diseases by comparing the frequency with which diseases occur in both members of identical twin pairs (100% genetic similarity) to the frequency in non-identical pairs (50% genetic similarity). For genuine genetic diseases, the disease concordancy estimates are always higher for identical than for non-identical pairs and, conversely, disease discordancy is higher in non-identical twins. The twin-pair method has the added advantage that estimates of the environmental contribution to the disease expression (phenotype) can also be directly derived from the data. The main reason for mentioning the old-fashioned twin-pair approach is to remind non-geneticists that old methods are not necessarily out-dated and that the twin-pair method has provided, and will continue to provide, a powerful means of estimating the genetic contribution to a large number of disorders. This, combined with the current general interest in defining the genetic components of many common disorders, means that large, well-characterized collections of twin data are important starting points for genetic disease analysis.

Another tangible advantage to the pharmaceutical industry that will come from the genetic analysis of diseases is the ability to stratify common diseases, so far defined solely on clinical manifestation, into sub-groups based on genetic differences and hence into categories with potentially different molecular mechanisms of disease disfunction. Evidence for such genetic heterogeneity is already available for Mendelian disorders where up to 10% of the disorders investigated at the molecular level (~700) are caused by defects in more than one gene. The same is certain to be true for multifactorial disorders where the much greater numbers of patients involved are a more attractive proposition for the pharmaceutical industry. Genetic stratification of common diseases should permit therapeutic strategies to be designed to meet the specific needs of subgroups of patients rather than the current policy of "one size fits all".

## 3. HUMAN GENOME PROGRAM

The genetic dissection of human disorders has received an enormous impetus through the Human Genome Program. This exciting program, was initiated in 1990 and it is estimated that it will be completed within 15 years, at an annual expenditure of $200 million in the USA alone. It has been described in emotive terms as "searching for the holy grail" or "writing the book of life". Its principal aims are to determine the structure and function of all genes using gene mapping and DNA sequencing as the primary analytical tools. From the beginning, the US congress stipulated that continued funding would be dependent upon significant advances being made in the prevention and therapy of genetic diseases.

To date, the genes of over 1200 of the approximately 4000 known genetic diseases have been mapped and disease-specific mutations have been detected in some 600 genes. Complex patterns are emerging of different mutations in a single gene giving rise to several distinct, clinically recognized diseases or, conversely, of mutations in several different genes producing a single disease (genetic heterogeneity). Of the estimated 70,000 genes, only about 6000 have been analyzed in detail to date, including definition of their structure and function. However, short DNA sequences (<500 bp) derived from coding regions have been determined from more than 500,000 partial cDNA sequences, themselves originating from mRNAs isolated from at least 150 different types of tissue. This approach was heralded as the short and cheap(er) route to detecting all human genes without the sweat and blood of mapping and sequencing the entire genome. However, DNA homology searches with this vast number of sequences, known as Short Tandem Repeats (ESTs), have demonstrated that probably only some 40,000 unique genes are represented [1], a long way short of the 70,000 genes believed to occur in man. Further, the probity of the majority of these ESTs will remain in doubt until each gene sequence has been isolated in its entirety, mapped, and its structure and function determined. It appears likely that detection of all human genes will only be possible following complete DNA sequence analysis of the human genome analogous to the situation in yeast where more than 4000 new genes out of a total of 6000 [2] were detected only after the total yeast genome had been sequenced and analyzed.

Characteristically, some commercial interests, notably the company *Human Genome Sciences,* saw an opportunity to gain an upper hand in the development of genome applications by analyzing many ESTs superficially and by a combination of controlling access to the information and patenting (discussed later). A widely expressed sentiment was that the *Human Genome Sciences'* attitude was not in keeping with the spirit of free information exchange as generally practiced in the Genome Program, and also expected within a program financed for >95% from public funds or non-profit making foundations. In an unexpectedly bold and altruistic move, *Merck & Co.* in collaboration with Washington University, mounted a counter operation to rapidly generate more EST data and to make the information immediately and freely available via the Internet. This and much other genome-related data, ranging from DNA sequences, protein structure to gene mapping information, can be easily accessed using the World Wide Web, which permits information to be freely accessed and linked together.

The vast majority of the genome still remains to be discovered and the development of pharmaceuticals based on our current perception of the molecular mechanisms of disease is, by definition, based on hugely incomplete information. This is amply illustrated by our understanding of the mechanisms involved in oncogenesis where molecular discoveries are forcing us to constantly readjust our interpretation of the processes involved, almost without exception towards greater complexity. Although we are still in the Stone Age in our understanding of the ways in which genes contribute to disease, rapid progress is being made and will permit more efficient therapeutic strategies to be developed. The potential benefits will come from five directions: (1) understanding the underlying molecular genetic mechanisms of disease processes will permit design of pharmaceuticals finely tuned to the disease concerned; (2) an understanding of the pathophysiology of clearly defined genetic diseases will provide indicators of which processes may also malfunction in other related (non-genetic) diseases; (3) analysis of the approximately 40,000 human genes still to be detected and analyzed will permit recognition of entirely new molecular mechanisms and their contribution to disease; (4) given the high complexity and potential redundancy of the human genome, malfunctioning in one pathway may be therapeutically compensated by activation of alternate pathways; (5) genetic engineering will open up new possibilities for pharmaceutical modalities.

Examples of two diseases holding enormous potential for the pharmaceutical industry are migraine and obesity. In both cases breakthroughs in genome analysis have led to the isolation and characterization of genes [3,4,5] whose involvement could never have been predicted in the pre-genomics era. Important as they are, these discoveries are just starting points for further investigations to determine the function of the genes and how they fit into their respective physiological pathways. In the case of obesity, the isolation of genes for leptin and its receptor, which are involved in regulating lipid uptake in fat tissue, was widely heralded [6] as an important breakthrough towards designing pharmaceuticals for weight control. However, the situation is turning out to be extremely complex and appears to involve interactions of various transcriptional, metabolic, hormonal and storage pathways [7] determining how, when and where lipids are stored in, or released from, adipocytes. Further, the very recent detection of genes that probably regulate calorie expenditure [8] have added more complexities and emphasize the need for caution before deciding which pharmaceutical strategies will be the most efficient.

## 4. OWNERSHIP ISSUES

From the very beginning of the Human Genome Program there has been a vigorous ongoing debate over ownership rights. The opinions have varied from "the genome is the universal property of mankind and cannot be patented" - strongly supported by third world countries fearing the potential of pharmaceutical multinationals to mount search and grab operations - to one of "we have a right and obligation to our financial investors to patent everything we can". The gatekeepers, the US and European Patent Offices, do not appear to have formed a clear policy on what constitutes reasonable grounds for making a patent application. The majority of people recognize the need for commercial firms to be able to protect their investment by being able to patent genome products at some level, otherwise the firms are simply not going to get involved in this work. The main problem is deciding on an appropriate level of new information to justify patenting.

The dilemma has been highlighted by the attempts of *Human Genome Sciences* to patent ESTs. The original application was submitted by NIH and *Human Genome Sciences*, but NIH withdrew their application and the US Patent Office decided not to take a decision at the time. The community of human genome scientists represented by HUGO were totally in agreement. However, the US Patent Office has recently stated that it is prepared to issue "broad" patents on EST sequences, much to the concern of the constituents of HUGO and NIH. The major concern with issuing patents on partial sequences without a well-defined function is that substantive claims made on the basis of full sequence and function information may run into problems through the less well-defined patent granted earlier. A case in point involves the patenting of the leptin receptor gene (OBR) by *Millenium* and *Hoffmann LaRoche*, who were the first to recognize the function of this gene. However, *Progenitor Inc.* had already applied for a patent for what they claimed were haematopoietin receptor sequences and there was no mention of either an association with obesity or binding to leptin in their application. However, in retrospect, it is clear that the *Progenitor Inc.* sequences show a high homology to OBR and they are contesting the rights of *Millenium* and *Hoffmann LaRoche* to the sole patent rights of the OBR gene. This is an ugly situation that is likely to get stuck in the courts for a long time. Neither the research nor pharmaceutical community is well served by the current situation, which will only lead to the continuing hesitation of many pharmaceutical companies to become involved in genome research. Many such examples will emerge if patents are issued on the basis of limited sequences and ill-defined function information. For a review of the salient issues see [9].

## 5. PHARMACEUTICAL COMPANIES' INVOLVEMENT

Several multinationals, notably *Merck & Co.* and *SmithKline and Beecham*, have recently made large investments and appointed eminent genome scientists to head genome programs. However, the majority of the pharmaceutical industry is still waiting on the sidelines or making only minor commitments. Nearly all genome research is being carried out in universities or in biotechnology companies largely supported by venture capital. The current high stock market value of biotech companies is based largely on genome discoveries and the eagerness of major pharmaceutical corporations to take them over (buy their results) and not on their record of bringing viable pharmaceuticals to the market. It could be argued that this is a good deal both for biotech firms who, in the main, have neither the infrastructure nor resources to mount large drugs trials, and for the pharmaceutical industry, which lacks a background in genome research. However, most genetic studies require significant investments to collect the patient materials required to carry them out and the studies are, by definition, long term. Funding agencies are unwilling to make the long-term commitments required to carry out such research and there is a clear niche for pharmaceutical and biotech companies to collaborate with university research groups in initiating genetic studies on common diseases. It is evident that genetic research can offer realistic possibilities of providing the information needed to design and implement therapeutic strategies which will reduce the morbidity associated with many of the diseases currently being dissected at the genome level.

## REFERENCES

1. L. Hillier, Genome Res. 6 (1996) 807-828.
2. A. Goffeau, Science 274 (1996) 546-567.
3. R.A. Ophoff et al., Cell 87(3), (1996) 543-552.
4. Y. Zhang et al., Nature 372 (1994) 425-432.
5. C.A. Monroe and R.I. Tepper, Cell 83 (1995) 1263-1271.
6. Anon, Nature Genetics 11(1), (1995) 1-2.
7. B.M. Spiegelman and J.S. Flier, Cell 87 (1996) 377-389.
8. S. Enerbäck et al., Nature 387 (1997) 90-94.
9. R.S. Eisenberg, Nature Genetics 15 (1997) 125-130.

H. van der Goot, (Editor)
TRENDS IN DRUG RESEARCH II



# Structure-based design: fact or fiction?

P. D. J. Grootenhuis, R. M. A. Knegtel, J. C. Heikoop and C. A. A. van Boeckel

N.V. Organon, P.O. Box 20, 5340 BH Oss, The Netherlands

## 1. INTRODUCTION

The rationale behind structure-based drug design finds its origins in the notions put forward a century ago by Fischer [1] and Ehrlich [2] concerning the highly specific interaction between certain small molecules and proteins. Once it was established that protein function can be modulated by small molecule binding and detailed structural data on protein folds became available through X-ray crystallography, the stage was essentially set for the advent of structure-based drug design.

The original aim of structure-based drug design was the discovery of new drugs by application of computational methods. Structure-based design traditionally focused on molecular association (good hydrogen-bonding, electrostatic and hydrophobic complementarity between ligand and target) rather than on other issues essential to drugs, i.e. administration, distribution, metabolism, excretion ('ADME'). Therefore, it is not surprising that the *de novo* design of an actual, marketable drug is still far beyond the reach of currently available technology. A term such as structure-based *ligand* or *molecular* design would therefore be closer to the truth and perhaps would have generated less high expectations (and sometimes disappointment) within industry and academia.

Structure-based design methods find application in widely diverse projects ranging from the docking of small molecules from 3D-databases in enzyme active site clefts to the construction of 3D-models of protein mutants. There are, however, a number of requirements that need to be fulfilled in order to successfully apply design methods.

The most important requirement for structure-based design is the availability of at least one high resolution structure of the target protein. When experimentally determined structures are not accessible one can try to make use of the ever increasing number of protein structures that have been elucidated, some of which may be of proteins of the same structural family. In such cases, structural models of the target protein may be constructed on the basis of homology modelbuilding or threading tools [3,4]. However, one should realize that although homology based models may be generally correct, details in the predicted structure will be incorrect as has been confirmed by the recent CASP-trials [5,6,7]. When the errors are located in 'hot' parts of the protein structure, they may lower the predictive value of subsequent modelling approaches [8].

Once a suitable 3D structure of the target protein has been obtained, a next step may be the placement of small molecules, in realistic conformations, in regions of interest of the protein. Several docking tools have been developed over the years [9,10] and the current trend is to include multiple conformations or even full conformational flexibility for both the protein and potential ligands [10,11]. During and after orientation of a ligand in the binding site its interaction with the receptor needs to be evaluated. Several approaches exist towards predicting binding free energies in receptor-ligand complexes [12] but from our point of view the development of reliable and robust 'scoring' methods constitutes one of the most challenging problems in structure-based design [13,14].

If the target protein itself is a drug candidate, other types of structure-based design can be applied in order to guide site-directed mutagenesis. Various methods exist to increase the thermostability of proteins, e.g. by introducing (additional) disulfide bonds or salt-bridges [15,16]. One may also 'minimize' the protein in such a way that only the most essential regions are conserved while other parts are removed [17,18].

Structure-based design methods (for recent reviews see: [19-24]) work best when applied in a 'cyclic' fashion, i.e. the computational models lead to insights that are translated into new compounds or mutants, which after testing lead to verification and fine-tuning of the models, etc. Here, we will illustrate that information on 'traditional' structure-activity relationships for a particular system, can be of great value for the modelbuilding. In the remainder we will limit ourselves to discussing three practical examples that highlight the strengths and weaknesses of various design methodologies.

## 2. EXAMPLES OF STRUCTURE-BASED DESIGN

### 2.1 Structure-based design of antiparasitic agents

An interesting example of the combined use of homology model building of protein structures and molecular docking for lead identification is provided by the structure-based discovery of anti-malarial drugs [25]. In this case, no crystal structure was available for the cystine protease (falcipain) that is crucial to the drug-resistant malaria strain *Plasmodium falciparum* for the degradation of host hemoglobin. Successful inhibition of falcipain would deprive the parasite from its primary source of amino acids and has indeed been shown to arrest parasite growth in cell culture [26].

The enzyme has a 30% sequence identity with the plant cystine proteases papain and actinidin, of which crystal structures had been solved. Of the conserved amino acids, 60% are located near the active site which increased the accuracy of the homology building of this region considerably. These similarities allowed for the construction of a homology model using side chain rotamer and loop conformation libraries. The final model was energy minimized and subjected to Ramachandran map analysis and checks for correct side chain packing.

The resulting model was used for searching the Available Chemicals Directory of commercially available small molecules with the DOCK [27] computer program. The DOCK program places small molecules (in a single conformation) into a protein binding site. The complementarity between the enzyme active site and a putative inhibitor was evaluated on the basis of both molecular shape and the intermolecular force field score. Of 4400 compounds that were ranked best by the DOCK program, 31 compounds were selected for testing on the basis of visual inspection of the docked ligand orientations in the active site. The inspection of such a relatively large number of compounds was deemed necessary in this particular case due to inaccuracies likely to be present in the homology model. In cases where a high resolution crystal structure is available, usually only 100-500 compounds need to be inspected in order to compensate for the limitations of scoring on interaction enthalpy or shape complementarity alone.

Of the 31 compounds tested, four inhibited the enzyme at concentrations lower than 100 μM. One of these compounds, oxalic bis[2-hydroxy-1-naphthylmethylene)hydrazide], see Figure 1, had an $IC_{50}$ of 6 μM and inhibited parasite metabolism in culture at a similar concentration. Chemical modification (partly based on the enzyme homology model) of the original lead compound identified by DOCK yielded inhibitors with improved selectivity and binding properties ($IC_{50}$ = 150 nM) [28].

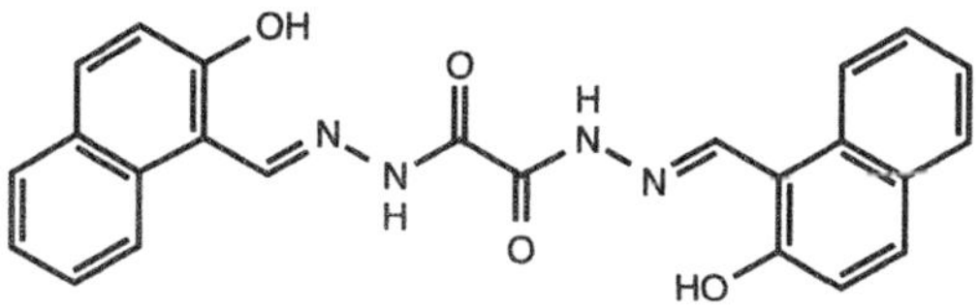

Figure 1. Oxalic bis[2-hydroxy-1-naphthylmethylene)hydrazide] which was identified by DOCK as a 6 μM inhibitor of falcipain and which served as a lead compound for further development.

The above mentioned example demonstrates that even in the absence of accurate structural data on a biomolecular target homology models can still be useful in structure-based drug design by serving as a rough pharmacophore model for database searches. On the other hand, the lack of a fast, accurate method for scoring of the enzyme-inhibitor complexes suggested by DOCK makes time-consuming visual inspection of large numbers of compounds still inevitable. In addition, the use of a single conformation for ligand molecules severely increases the number of false-negatives in such applications.

Finally, it should be noted that the binding mode of the anti-malarial lead compound suggested by DOCK has not (yet) been confirmed experimentally. The structure elucidation could still yield some surprises as was, for instance, the case in the application of DOCK to HIV-protease. Although this yielded a micromolar inhibitor (haloperidol, [29]), it was shown to bind approximately 3 Å away from its predicted location after determination of the crystal structure of the complex [30]. In this case, the unexpected replacement of the water molecule bound between the flaps of HIV-protease by a chlorine ion was found to be the cause of the discrepancy. This example stresses the importance of frequent experimental verification of modelling results. In addition, it illustrates that after synthesis and testing of analogues of the lead new insights may be obtained as to how to improve the model of the complex.

### 2.2 Structure-based design of a new class of antithrombotics

The natural product heparin has proven to be an important lead for the research on antithrombotics [31]. Heparin is able to activate the endogenous protease inhibitor antithrombin (AT) III. This important coagulation factor controls the blood coagulation by potent inhibition of several blood proteases including thrombin and factor Xa. In contrast to factor Xa, thrombin also contains a heparin binding domain, albeit of less affinity and selectivity than that of AT III. Inspired by traditional medicinal chemistry insights, systematic molecular modification of heparin fragments has resulted in synthetic derivatives with antithrombotic properties. One of them, a pentasaccharide which is the synthetic counterpart of the AT-III binding pentasaccharide, is now being tested clinically. The research towards pentasaccharides has yielded highly selective, synthetic antithrombotics which can be applied under *venous* thrombotic conditions. However, there also is a need for more selective and efficacious drugs against *arterial* thrombosis. There are strong indications that for prevention and treatment of this kind of thrombosis inhibition of both thrombin and factor Xa is most effective.

Heparin, which is able to inactivate both thrombin and factor Xa, has served as our lead although by itself it is insufficiently selective. Already in the 1980's it was known that heparin fragments that inactivate thrombin via AT-III, need to fulfill two basic requirements: the fragment needs to consist of a minimum of 18 carbohydrate moieties and should include the unique pentasaccharide domain. Thus, the second requirement renders the fragment to induce a conformational change in AT-III (also needed for factor Xa inactivation). The first requirement is often attributed to a template effect of heparin; heparin catalyzes complex formation by forming a "bridge" between AT-III and thrombin. In contrast to the very selective heparin binding domain of AT-III, the heparin-binding domain of thrombin is thought to be rather aselective and of low affinity. Since factor Xa seems to lack a heparin-binding domain, the template effect plays no role for its inactivation.

Structural models [32,33] of the interactions between AT-III and pentasaccharides have been generated in the past; at first on the basis of homology modelbuilding using the crystal structure of $\alpha$1-antitrypsin, and later on the basis of the crystal structure of AT-III [34]. It is noted that the orientation of the pentasaccharide on the complex was guided in an essential way by structure-activity relationships of many pentasaccharide derivatives. Our models of the AT-III - pentasaccharide complex in combination with the crystal structure of thrombin [35] and NMR-study of heparin [36] allowed us to construct a model (Figure 2) of the ternary AT-III - heparin - thrombin complex [37,38].

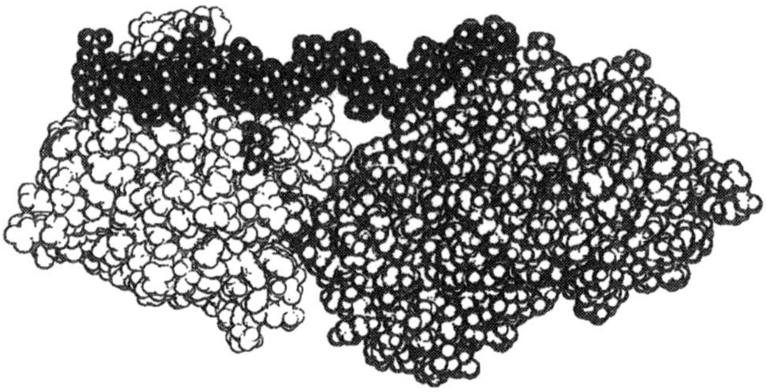

Figure 2. Space-filling representation of the ternary AT-III (right) - heparin (top) - thrombin (right) complex.

In contrast with the literature, the model of the ternary complex (Figure 2) clearly revealed that the thrombin-binding domain of heparin should be contiguous to the *non-reducing* terminus of the pentasaccharide. Furthermore, it was observed that the oligosaccharide spacer that connects the pentasaccharide and thrombin-binding domains has virtually no interactions with the two protein surfaces. We concluded that synthetic molecules with a mixed factor Xa/thrombin inhibitory profile could be obtained by elongating a synthetic pentasaccharide at its non-reducing end with a linear, neutral spacer bearing a thrombin-binding domain at the other terminus.

One of the first compounds (Figure 3) that were prepared based on this concept showed the desired, mixed profile; in addition to high anti-factor Xa activity, it also appeared to inactivate thrombin [38]. The conjugate and its derivatives present the first example of synthetic mixed profile oligosaccharides. Very interestingly, the mixed profile character can be tuned in a rational way; by modification of the spacer length, the anti-thrombin activity could be varied between 15 to 150 anti-thrombin units. Thus the ratio between the anti-factor Xa and anti-thrombin activity can be rationally tuned which is of paramount importance for the ultimate clinical application of drug candidates in various thrombotic disorders. Out of this new class of antithrombotics a candidate for pre-clinical studies has been selected.

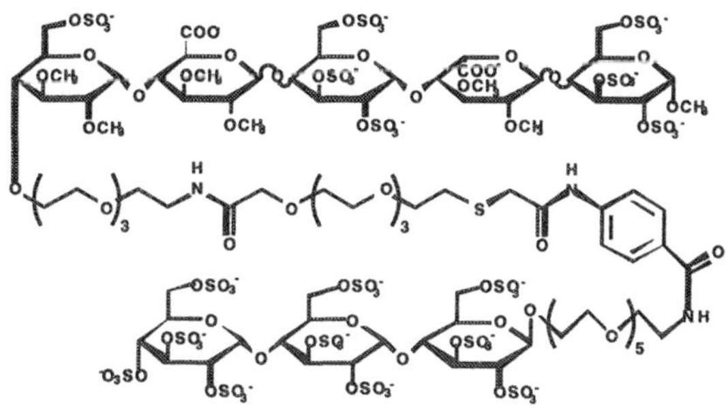

Figure 3 Antithrombotic glycoconjugate consisting of three elements: an AT-III binding (pentasaccharide) domain, a spacer, and a thrombin binding domain

### 2.3 Structure-based protein engineering of gonadotropins

The glycoprotein hormones [39] form a family of structurally related proteins. These so-called gonadotropins are heterodimers composed of two dissimilar subunits, named α and β, which are associated by non-covalent bonds. Within a species, the α subunit is the same for each member of the gonadotropin family. The β subunits are different for each member and confer receptor-binding specificity on the hormones. The human choriogonadotropin (hCG) is one of the members of the gonadotropin family and is used in infertility treatment. The hormone is secreted by the placenta and is involved in maintaining the pregnancy.

The association of the α and the β subunit is an important step in the biosynthesis of the glycoprotein hormones, since only the intact dimers are biologically active. Thus, the correct assembly and formation of a stable heterodimer is essential for efficient secretion, receptor binding and signal transduction of the hormone. Numerous mutation studies have shown that the dimerization is a highly specific process that can easily be prohibited by single point mutations. Structure-function analysis of the glycoprotein hormones is often hampered by unwanted secondary effects on the assembly of the subunits.

Therefore, we and others have tried to design and generate heterodimeric gonadotropins with covalent linkages between the two subunits. The structure-based design of such systems became possible after the recently published crystal structures of hCG [40,41]. Two strategies have been followed. In the first strategy, the two subunits were covalently linked to each other by a peptide spacer leading to so-called 'single chain' gonadotropins[42]. In the second strategy, intersubunit disulfide bonds were introduced in the gonadotropins by site-directed mutagenesis of cystine residues at appropriate positions [43].

In the case of the single chain gonadotropins we reasoned that the spacer should minimally disturb the observed heterodimeric hCG structure and should not interfere with regions associated with receptor binding and signal transduction. Taking this into account, we designed several single chains with Ser-Gly repeats between the (truncated) C-terminus of the β-subunit and the (truncated) N-terminus of the α-subunit. The length of the spacer peptides was determined by searching peptide loop fragments from a database consisting of high-resolution protein structures, followed by subsequent modelbuilding of the single chains after which the shortest possible Ser-Gly sequences were selected. The designed mutants were expressed in Chinese hamster ovary cells. All mutants were produced at levels comparable to wild-type hCG. It was found that the three mutants were bioactive, albeit with somewhat lower in vitro receptor binding and signal transduction than wild type hCG. The most interesting mutant probably was the one with truncated α and β subunits in which the essential β26-110 disulphide bond from the so-called 'seat-belt', a motif that plays an essential role in the subunit association of the wild-type hormone, can not be formed. Surprisingly, this mutant is still bioactive. Such a mutation could never have been evaluated in 'normal' heterodimeric hCG. In the near future we would like to reduce the structural complexity of the gonadotropin hormones by removing as many non-essential regions in the protein as possible while keeping the activity intact resulting in so-called 'mini-gonadotropins'.

In the second strategy a more subtle approach was followed. We tried to link the α- and β-subunit of gonadotropins via non-natural, *intersubunit* disulfide bonds. For the gonadotropins this may be complicated since already 11 native disulfide bonds are present and additional cystines could influence the formation of these disulfide bonds, thereby disturbing proper subunit folding and assembly. The crystal structures of CG were analyzed by using the SSBOND program of Hazes and Dijkstra [44] which assists the selection of sites in a protein where pairs of cystine residues may be introduced. The expression constructs for the mutants containing additional cystine residues were prepared and transfected into Chinese hamster ovary cells. The mutants, which were produced at varying levels, typically display wild-type receptor binding and bioactivity. In nearly all mutants the designed intersubunit disulfide bonds were indeed formed leading to a non-natural, covalent linkage between the α- and β-subunits. The concept of intersubunit disulfide bonds appeared to be translatable to other members of the gonadotropin family, including follicle stimulating hormone and luteinizing hormone.

It was found that the mutants with non-natural intersubunit disulfide bonds display significantly enhanced thermostabilities relative to the corresponding heterodimeric glycoprotein hormones. The in vivo properties of these mutants are currently under study.

## 3. CONCLUSIONS

From the examples described above it is clear that structure-based design methods may play an important role in drug discovery projects. The structural models of the target proteins mentioned in the three examples inspired the conception of new classes of inhibitors, conjugates and mutants. It also clear that in every example the experimental verification of computational models turned out to be absolutely necessary; in a number of cases it lead to refined or corrected models.

We strongly feel that with the current state-of-the-art a pragmatic attitude in drug discovery projects is more effective than the pursuit of highly sophisticated *de novo* design strategies. E.g. we took advantage of the fact that the plasma half-life of synthetic pentasaccharides is directly related to the half-life of the physiological plasma protein AT-III [45]. When we would chemically diverge too much from heparin, problems may be anticipated with respect to ADME. In our hands the rather qualitative tools for modelbuilding, docking and analysis had more predictive value than some of the quantitative design tools such as the quantitative prediction of binding free energies, etc.

Thus, structure-based design is most effective in the hands of scientists involved in drug discovery when it is used to inspire novel strategies and proposals at the residue and/or domain level. We note that this fits in completely with the current trend in medicinal chemistry to apply combinatorial chemistry and high-throughput screening in order to synthesize and test thousands of compounds. Such techniques may benefit greatly from the qualitative insights obtained by structure-based design; the need to predict properties of single compounds seems to be less prominent. In addition, the feed-back from synthesis and testing to modelbuilding can be done quicker and with more information. An effective integration of structure-based design and combinatorial chemistry / high throughput screening therefore holds great promise for the future [46].

## REFERENCES


1. E. Fischer, Chem. Ber. 27 (1894) 2985-2993.
2. P. Ehrlich, Chem. Ber. 42 (1907) 17.
3. R. Sanchez and A. Sali, Curr. Opin. Struct. Biol. 7 (1997) 206-214.
4. A.V. Finkelstein, Curr. Opin. Struct. Biol. 7 (1997) 60-71.
5. J. Moult, Curr. Opin. Biotechnol. 7 (1996) 422-427.
6. D. Eisenberg, Nature Struct. Biol. 4 (1997) 95-97.
7. R.L. Dunbrack, D.L. Gerloff, M. Bower, X. Chen, O. Lichtarge and F.E. Cohen, Folding & Design 1 (1997) R27-R42.
8. L.M.H. Koymans, P.D.J. Grootenhuis and C.A.G. Haasnoot, Recl. Trav. Chim. Pays-Bas 112 (1993) 161-168.
9. G. Jones and P. Willett, Curr. Opin. Biotechnol. 6 (1995) 652-656.
10. T. Lengauer and M. Rarey, Curr. Opin. Struct. Biol. 6 (1996) 402-406.
11 R.M.A. Knegtel, Kuntz, I.D. and C.M. Oshiro, J. Mol. Biol. 266 (1997) 424-440.
12. Ajay and M. M. Murcko, J. Med. Chem. 38 (1995) 4953-4967.
13. P.D.J. Grootenhuis and P.J.M. van Galen, Acta Cryst. D51 (1995) 560-566.
14. R.M.A. Knegtel and P.D.J. Grootenhuis, in H. Kubinyi, G. Folkers and Y.C. Martin (eds.), 3D-QSAR in Drug Design Vol 2, ESCOM Science Publishers BV, Leiden, 1997.
15. C. Vieille and J.G. Zeikus, TIBTECH 14 (1996) 183-190.
16. A. Shaw and R. Bott, Curr. Opin. Struct. Biol. 6 (1996) 546-550.
17. A.C. Braisted and J.A. Wells, Proc. Natl. Acad. Sci. USA 93 (1996) 5688-5692.
18. B. Li, J.Y. Tom, D. Oare, R. Yen, W.J. Fairbrother, J.A. Wells, and B.C. Cunningham, Science 270 (1995) 1657-1660.
19. I.D. Kuntz, E.C. Meng, B.K. Shoichet, Acc. Chem. Res. 27 (1994) 117-123.
20. C.L.M.J. Verlinde and W.G.J. Hol, Structure 2 (1994) 577-587.
21. R.C. Jackson, Curr. Opin. Biotechnol. 6 (1995) 646-651.
22. P. Bamborough and F.E. Cohen, Curr. Opin. Struct. Biol. 6 (1996) 236-241.
23. H.J. Böhm, Curr. Opin. Biotechnol. 7 (1996) 433-436.
24. H.J. Böhm and G. Klebe, Angew. Chem. Int. Ed. Engl. 35 (1996) 2588-2614.
25. C.S. Ring, E. Sun, J.H. McKerrow, G.K. Lee, P.J. Rosenthal, I.D. Kuntz, F.E. Cohen,

Proc. Natl. Acad. Sci. USA 90 (1993) 3583-3587.
26. Z. Li, X. Chen, E. Davidson, O. Zwang, C. Mendis, C.S. Ring, W.R. Roush, G.Fegley, R. Li, P.J. Rosenthal, G.K. Lee, G.L. Kenyon, I.D. Kuntz and F.E. Cohen, Chem. & Biol. 1 (1994) 31-37.
27. I.D. Kuntz, J.M. Blaney, S.T. Oatley, R. Langridge and T.S. Ferrin, J. Mol. Biol. (1982), 269-288.
28. R. Li, X. Chen, B.Gong, P.M. Selzer, Z. Li, E. Davidson, G. Kurzban, R.E. Miller, E. O. Nuzum, J.H. McKerrow, R.J. Fletterick, S.A. Gillmor, C.S. Craik, I.D. Kuntz, F.E. Cohen, G.L. Kenyon, Bioorg. & Med. Chem. 4 (1996) 1421-1427.
29. R.L. DesJarlais, G.L. Seibel, I.D. Kuntz, P. S. Furth, J.C. Alvarez, P.R. Ortiz de Montellano, D.L. DeCamp, L.M. Babé and C.S. Craik, Proc. Natl. Acad. Sci. USA 87 (1990) 6644-6648.
30. E. Rutenber, E.B. Fauman, R.J. Keenan, S. Fong, P.S. Furth, P.R. Ortiz de Montellano, E. Meng, I.D. Kuntz, D.L. DeCamp, R. Salto, J.R. Rosé, C.S. Craik and R.M. Stroud, J. Biol. Chem. 268 (1993) 15343-15346.
31. C.A.A. van Boeckel and M. Petitou, Angew. Chem. Int. Ed. Engl. 32 (1993) 1671-1690.
32. P.D.J. Grootenhuis and C.A.A. van Boeckel, J. Am. Chem. Soc. 113 (1991) 2743-2747.
33. C.A.A. van Boeckel, P.D.J. Grootenhuis and A.Visser, Nature Struct. Biol. 1 (1994) 423-425.
34. H.A. Schreuder, B. de Boer, R. Dijkema, J. Mulders, H.J.M. Theunissen, P.D.J. Grootenhuis and H.W.G. Hol, Nature Struct. Biol. 1 (1994) 48-54.
35. W. Bode, L. Mayr, U. Baumann, R. Huber, S.R. Stone and J. Hofsteenge, EMBO J. 8 (1989) 2467-3475.
36. B. Mulloy, M.J. Forster, C. Jones and D.B. Davies, Biochem. J. 293 (1993) 849-858.
37. P.D.J. Grootenhuis, P. Westerduin, D. Meuleman, M. Petitou, M. and C.A.A. van Boeckel, Nature Struct. Biol. 2 (1995) 736-739.
38. C.A.A. van Boeckel, P.D.J. Grootenhuis, D. Meuleman and P. Westerduin, Pure & Appl. Chem. 67 (1995) 1663-1672.
39. Y. Combarnous, Endocr. Rev. 13 (1992) 670-691.
40. A.J. Lapthorn, D.C. Harris, A. Littlejohn, J.W. Lustbader, R.E. Canfield, K.J. Machin, F.J. Morgan and N.W. Isaacs, Nature 369 (1994) 455-461.
41. H. Wu, J.W. Lustbader, Y. Liu, R.E. Canfield and W.A. Hendrickson, Structure 2 (1994) 545-558.
42. J.C. Heikoop, M. J. A. C. M. van Beuningen-de Vaan, P. van den Boogaart and. P.D.J. Grootenhuis, Eur. J. Biochem. 245 (1997) 656-662.
43. J.C. Heikoop, P. van den Boogaart, J.W.M. Mulders and P.D.J. Grootenhuis, Nature Biotechnol. 1997, in press.
44. B. Hazes and B. W. Dijkstra, B. W. Protein Eng. 2 (1988) 119-125.
45. R. G. M. van Amsterdam, G. M. T. Vogel, A. Visser, W. J. Kop, M. T. Buiting and D. G. Meuleman, Arterioscler. Tromb. Vasc. Biol. 15 (1995) 495-503.
46. F.R. Salemme, J. Spurlino and R. Bone, Structure 5 (1997) 319-324.

H. van der Goot, (Editor)
TRENDS IN DRUG RESEARCH II



# New developments in synthetic medicinal chemistry
# Introduction

F. Gualtieri

Dipt. Scienze Farmaceutiche, Università degli Studi di Firenze,
Via G. Caproni 9, Firenze 50121, Italy

Medicinal Chemistry, at its more basic level, remains an empirical science, but in the past decades the tools available to the medicinal chemist to identify, design and test drugs have increased dramatically in quantity and sophistication. Computational methods, combinatorial chemistry, biotechnologies, high-throughput screening, are among the many powerful techniques that have been harnessed to bring an element of rationality to the search for new drugs. The role of the ongoing research of human genoma on drug discovery is difficult to evaluate at the moment, but will very likely be crucial, at least for some diseases.

The dominant role of synthetic chemistry has been increasingly challenged by knowledge of the structure and functions of enzymes, receptors, channels, membrane pumps, nucleic acids and by the exponential growth of information on biology, genetics and pathology, giving paramount importance to the dialogue of chemists and biologists. It is a common practice in Medicinal Chemistry lectures and presentations to pass over synthetic procedures to comment structure-activity relationships and biological and pharmacological data.

According to G. Wess who discussed the problem in a recent perspective on the challenges for Medicinal Chemistry [1], "the pharmaceutical industry appears to be in transition from a chemistry-based industry to one based more on human biology and genetic information".

Nevertheless, as in the old days, the development of new chemical entities is still highly dependent on the ability of chemists to obtain, with simple, reliable, fast and possibly inexpensive methods, the molecules that have been designed. At least until someone is able to test a virtual molecule (existing only on the screen of the computer) by connecting the computer directly to a mouse.

Until a few years ago, pharmacological screening was the critical step and the compounds synthesized piled up waiting to be tested. The development of high-throughput screening techniques has reversed this situation and the possibility to evaluate tens and hundreds of compounds a day, has led to an epochal change in the philosophy of drug discovery and synthesis: it is better to spend time generating diversity rather than highly pure but inactive compounds.

Combinatorial chemistry represents one possible answer to this new strategy. It accelerates the drug discovery programs and increases the chemical diversity of the drugs available for biological tests. If in general, combinatorial chemistry appears to be an excellent tool to generate chemical diversity through the synthesis of large libraries, parallel synthesis, which affords a limited number of compounds, seems more suitable for lead optimization.

In both cases the best results are obtained by the use of solid phase synthesis that makes the isolation and purification of the products much easier. A dramatic limit to the extensive use of this technique is that, of the many reactions available to the synthetic organic chemist, only a few (less than two hundred) have been successfully adapted to solid phase synthesis [2]. As a consequence, more work to adapt the most useful organic reactions to solid phase synthesis is badly needed. It can be expected that much of the effort in this direction will regard chiral reactions, even if classical chiral organic synthesis will very likely maintain its fundamental role in drug research.

For a very long time humans have exploited the chemical skill of microorganisms to produce or modify organic molecules. From them and from other sources, many enzymes are now available that can cleanly, rapidly and inexpensively afford organic reactions on a variety of substrates.

A combination of the techniques of organic synthesis with the power of enzymes to asymmetrically catalyze reactions, now within reach, can be another major improvement toward the synthesis and optimization of drugs. This enzyme-organic approach has proven to be particularly useful in the synthesis of oligosaccharides and other carbohydrates [3] but can obviously be used also for other classes of compounds.

In any case, even if it is an undisputed fact that biology has become exceedingly important in drug research, it is reasonable to imagine that chemistry, in particular synthetic organic chemistry, will continue to play a major role in academic research and in the R&D departments of drug companies of the third millennium. As a consequence, medicinal chemists, while being receptive to the progress of knowledge in biology and pharmacology, must continue to excel in synthetic organic chemistry. It is interesting that nearly identical conclusions have been reached in a recent survey on the training of medicinal chemists for drug industries [4].

## References

1. G. Wess, "Challenges for Medicinal Chemistry", Drug Discovery Today, 1 (1996) 529.
2. I.C. Choong, J.A. Ellman, "Solid-Phase Synthesis: Application to Combinatorial Libraries", Ann. Rep. Med. Chem., 31 (1996) 309-317.
3. E.K. Wilson, "Scripps Research Institute Thrives at Interfaces of Chemistry and Biology", Chemical & Engineering News, 74 (1996) 39.
4. W.D. Busse, C.R.Ganellin, L.A. Mitscher, "Vocational Training for Medicinal Chemists: Views from Industry", Eur. J. Med. Chem., 31 (1996) 747.

H. van der Goot, (Editor)
TRENDS IN DRUG RESEARCH II



# NEW BIOCATALYTIC APPROACHES FOR THE SYNTHESIS OF CHIRAL DRUGS, INTERMEDIATES, AND SUBSTRATES

Kurt Laumen, André Brunella, Martin Graf, Matthias Kittelmann, Paula Walser, and Oreste Ghisalba
Novartis Pharma AG, Pharmaceuticals Division, Core Technology Area, CH-4002 Basel, Switzerland

## 1. INTRODUCTION AND SUMMARY

Chiral building blocks play an important role for the manufacture of modern pharmaceutical drugs. Due to the improved and increased application of modern target structure-based biorational techniques throughout the drug discovery process, it can easily be predicted that the role of chirality will become even more important for the drugs of the future. Therefore, new efficient and selective ways are needed to produce interesting chiral building blocks. Very attractive possibilities are offered by the use of biological reactions for some of the key synthesis steps.

Chiral amines represent a highly versatile and attractive group of chiral building blocks for drug R&D. Some of the companies dealing with speciality chemicals offer a limited selection of enantiomerically pure chiral amines. However, these compounds are often available in small quantities only and at prohibitive prices. Chiral amines are of interest e.g. as building blocks for $GABA_B$ antagonists or substance P antagonists. In the course of an extended screening programme in our laboratory, three novel types of highly enantioselective microbial amidohydrolases were discovered. The production of these enzymes from *Rhodococcus equi, Arthrobacter aurescens,* and *Rhodococcus globerulus* was optimized and the isolated enzymes were then broadly evaluated for synthetic applications. The three enzymes (showing different and opposite selectivities) proved to be highly suitable for the preparative scale synthesis of e.g. *(R)*- and *(S)*-1-(3-cyanophenyl)ethylamine, *(R)*- and *(S)*-1-phenylethylamine, or *(S)*-2-amino-1-phenyl-4-pentene, and many other chiral amines from the corresponding racemic N-acetyl derivatives.

Chemo-enzymatic routes are also of great interest for the preparation of research biochemicals. This was demonstrated in our laboratory e.g. for the preparation of D-*myo*-inositol-1-phosphate and of inositol polyphosphates.

## 2. CHIRAL AMINES AS BUILDING BLOCKS FOR NEW DRUGS

### 2.1. Chiral amines as pharmaceutical building blocks: target compounds

In our study, the major target compounds in demand for pharmaceutical applications and drug profiling were *(R)*- and *(S)*-1-(3-cyanophenyl)ethylamine, *(R)*- and *(S)*-1-(4-cyanophenyl)ethylamine, *(S)*-2-amino-1-phenyl-4-pentene, and *(S)*-2-amino-1-(4-chlorophenyl)-4-pentene.

*(R)*-1-(3-cyanophenyl)ethylamine was required as a building block for the $GABA_B$ receptor antagonists CGP 56999A (for the treatment of Alzheimer patients), CGP 62349 (PET-ligand), and related structures [1,2], as well as for radioligands[2]. The *S*-enantiomer and the two enantiomers of 1-(4-cyanophenyl)ethylamine were needed for comparative studies.

*(S)*-2-amino-1-phenyl-4-pentene was required as a building block in an alternative synthesis of the substance P antagonist CGP 49823 (anxiolytic, antidepressant), and related structures[3,4]. *(S)*-2-amino-1-(4-chlorophenyl)-4-pentene was needed for comparative studies.

The separation of racemic mixtures of these target compounds into their enantiomers via crystallisation of diastereomeric salts is a possible but rather costly solution, whereas the synthesis of the pure enantiomers by the use of novel microbial enzymes or even whole cells of microbes is easier and more economic to perform.

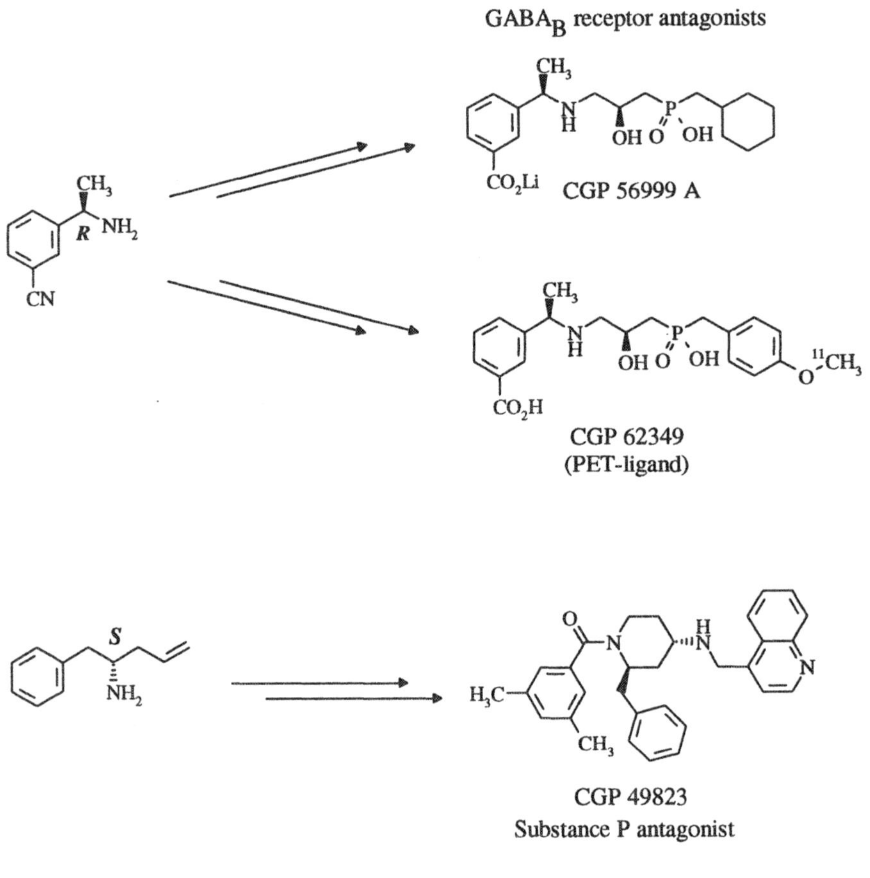


Figure 1. Chiral amines as building blocks for $GABA_B$ receptor antagonists and substance P antagonist

## 2.2. Biocatalytic approaches described in literature

Enantiomerically pure chiral amines are of high interest for both the pharmaceutical and the agrichemical industries. Besides classical approaches using separation of enantiomers, several biocatalytic ways for the production of such compounds have been proposed and developed more or less successfully. The most relevant examples are:

- The production of *D*-2-aminobutanol via microbial hydrolysis of the corresponding N-acyl derivatives[5];
- The enantiomeric separation of amines from racemates having the amino group on a secondary carbon atom through a process involving the action of ω-amino acid transaminases[6];
- The biological synthesis of *(S)*-1-phenylethylamine from L-alanine and acetophenone by *Acinetobacter sp.* MBA-15 or from alkyl-(1-phenylethyl)carbamate in the presence of bacteria of the genera *Rhodococcus* or *Arthrobacter*[7,8];
- The production of optically active 1-aryl-2-aminopropane by enantioselective transfer of the amino group in the racemate by means of the amino acid transaminase from *Bacillus megaterium*[9];
- The production of optically active primary and secondary amines by enantioselective acylation of a racemic amine with an activated ester as an acyl donor in the presence of the lipase from *Pseudomonas spp.*[10];
- The preparation of chiral amines by specific amidation of the *R*-enantiomer via acyl transfer reactions using lipase B from *Candida antartica*[11-14], and by *S*-specific amidation catalyzed by subtilisin[15,16];
- The hydrolysis of racemic N-acetyl-amines using lipase B from *Candida antartica* to prepare enantiomerically pure primary *(R)*-amines,[13,17].

## 2.3. *(S)*-N-acetyl-1-phenylethylamine amidohydrolase (acylase)

The biological syntheses of *(S)*-1-phenylethylamine from L-alanine and acetophenone by *Acinetobacter sp.* MBA-15 or from alkyl-(1-phenylethyl)carbamate in the presence of bacteria of the genera *Rhodococcus* or *Arthrobacter*, as described in literature[7,8], are unfortunately characterized by low yields and long incubation times.

Therefore, a more effective biocatalytic way to produce the enantiomerically pure 1-phenylethylamine derivatives was requested. Our main target compounds were *(R)*- and *(S)*-1-(3-cyanophenyl)ethylamine and *(R)*- and *(S)*-1-(4-cyanophenyl)ethylamine.

An intensive screening programme was performed[18]). In order to provide a new, simple, and generally applicable route to chiral amines, the resolution of the enantiomers from the corresponding racemic N-acyl, preferably N-acetyl amides via enzymatic enantioselective deacylation was investigated.

1-Phenylethylamine was chosen as a model compound for the screening. This had the advantage that both enantiomers are commercially available at a reasonable price to be used as reference compounds for analytics and for the synthesis of the enantiomerically pure acylamides as the enzyme substrates. In order to find enantioselective microbial acylases (amidohydrolases) enrichment cultures from habitat samples were prepared using racemic N-acetyl-1-phenylethylamine as the sole carbon source (four propagations). The isolation of suitable pure microorganisms was performed with agar plates containing the individual pure enantiomers of the acetylamide as the sole carbon source. 26 bacterial strains were isolated that grew only on the plates containing the *(S)*-N-acetyl-1-phenylethylamine. Crude cell extracts of these strains hydrolysed only the *(S)*-N-acetyl-1-phenylethylamine, but not the *(R)*-enantiomer. The strain Ac6 with the highest specific amidohydrolase activity and >95% ee for the *(S)*-amine was selected for further investigations. This strain was later identified as *Rhodococcus equi* (deposited as DSM 10278).

The production of the acylase has to be induced by the *(S)*-substrate, whereas the *(R)*-substrate is not active as an inducer. In optimized growth medium enzyme yields up to 56 U/l culture broth were obtained.

The enzyme from *Rhodococcus equi* Ac6 was purified and characterized as follows[18)]: specific activity: 0.87 U/mg (for *(S)*-N-acetyl-1-phenylethylamine); molecular weight: 94,000 Da, homodimeric substructure; pH-optimum: 6.5-7.0; half life at 30°C: 350 days; strong inhibition by PMSF but not by chelating agents (this suggests a serine protease like reaction mechanism). The enantioselectivity E of the enzyme is 350 for *(S)*-N-acetyl-1-phenylethylamine, the $K_m$-value is 0.6 mM.

Substrate specificity and enantioselectivity: The enzyme from *Rhodococcus equi* Ac6 deacetylates with good to very high enantioselectivities the *S*-enantiomers of a broad range of N-acetylamines with aromatic side-chains (see Table 1) but does not accept N-methyl-N-acetylamines.

With this new enzyme it became very easy to economically prepare (with very high yields) larger amounts of the desired pure *(S)*-enantiomers of 1-(3-cyano-phenyl)ethylamine and 1-(4-cyanophenyl)ethyl amine, as well as structurally related compounds.

## 2.4. *(R)*-N-acetyl-1-phenylethylamine amidohydrolase (acylase)

As described in literature, lipase B from *Candida antarctica* (e.g. Novozym SP-435) can catalyze the enantioselective hydrolysis of racemic N-acetylamines to primary *(R)*-amines. However, this lipase shows extremely low specific activities for the reported amide substrates[17], including N-acetyl-1-phenylethylamine ($3.3 \times 10^{-6}$ U/mg). Lipases in general do not act effectively on amide bonds and are therefore not really suitable biocatalysts for amide cleavage on industrial scale. Therefore, we searched for a more effective biocatalytic way to produce the enantiomerically pure *(R)*-1-phenylethylamine derivatives.

*(R)*-1-(3-cyanophenyl)ethylamine and *(R)*-1-(4-cyanophenyl)ethylamine can not be prepared by acidic or alkaline hydrolysis of the corresponding *(R)*-N-acetylamides (the left over enantiomer from the racemate under catalysis of the acylase from *Rhodococcus equi* Ac6) because this treatment results in the degradation of the cyano group. Therefore, an *(R)*-specific enzyme was needed in addition.

Since the first screening on racemic N-acetyl-1-phenylethylamine exclusively yielded *(S)*-specific microorganisms, additional enrichment cultures were performed with *(R)*-N-acetyl-1-phenylethylamine as the sole carbon source. However, no strain with an *(R)*-specific enzyme could be obtained by this second screening strategy. In a third screening programme sodium acetate was added to the first two propagates of the enrichment cultures. Due to this simple trick, the enriched microorganisms could adapt to acetate utilization and partly deactivated cells had the possibility to regenerate before they were forced (in the third propagation) to utilize *(R)*-N-acetyl-1-phenylethylamine. In this screening two bacterial strains possessing a highly *(R)*-specific enzyme could be isolated from 45 tested habitat samples[19]. The strain

AcR5b with the higher specific amidohydrolase activity (0.98 U/mg of protein in the crude cell extract) and >98% ee for the *(R)*-amine was selected for further investigations. This strain was later identified as ***Arthrobacter aurescens*** (deposited as DSM 10280)

In contrast to the *(S)*-specific enzyme of ***Rhodococcus equi*** Ac6, this *(R)*-specific amidohydrolase of ***Arthrobacter aurescens*** AcR5b is produced constitutively (i.e. no inducer is necessary in this case). In optimized growth medium very high enzyme yields up to 6420 U/l culture broth were obtained in laboratory fermentor scale. The enzyme amounts up to 20% of the total protein content of the cells!

The enzyme from *Arthrobacter aurescens* AcR5b was purified to homogeneity and characterized as follows[19]: specific activity: 16.0 U/mg (for *(R)*-N-acetyl-1-phenylethyl amine); molecular weight: 220,000 Da, heterotetrameric substructure $\alpha_2\beta_2$ with subunit sizes of 89,000 and 16,000 Da; pH-optimum: 7.5-9.5; half life at 30°C: 25 days (at 23°C: 75 days); strong inhibition by several metal cations which can be reversed by chelating agents and only weak inhibition by PMSF (this indicates that the enzyme is neither metal cation dependent nor acting with a serine protease like reaction mechanism). The enantioselectivity E of the enzyme is >500 for *(R)*-N-acetyl-1-phenylethylamine, the $K_m$-value is 7.5 mM.

Substrate specificity and enantioselectivity: The enzyme from *Arthrobacter aurescens* AcR5b deacetylates with good to very high enantioselectivities the *R*-enantiomers of a broad range of N-acetylamines with aromatic side-chains (see Table 2) but does not accept N-methyl-N-acetylamines. The substrate range of this enzyme is comparable to the one of the *Rhodococcus equi*Ac6 enzyme but the enantioselectivity is in the opposite direction.

With this new enzyme it became very easy to economically prepare (with very high yields) larger amounts of the desired pure *(R)*-enantiomers of 1-(3-cyanophenyl)ethylamine and 1-(4-cyanophenyl)ethyl amine, as well as structurally related compounds.

### 2.5. *(S)*-N-acetyl-2-amino-1-phenyl-4-pentene amidohydrolase (acylase)

Since the amidohydrolases from *Rhodococcus equi* Ac6 and *Arthrobacter aurescens* AcR5 did not hydrolyze racemic N-acetyl-2-amino-1-phenyl-4-pentene enantioselectively, a third type of amidohydrolase was demanded.

Therefore, an additional screening programme was performed via enrichment cultures starting from 74 habitat samples supplying racemic N-acetyl-2-amino-1-phenyl-4-pentene as the selective carbon source and acetate as a cosubstrate in the first culture propagates. 18 bacterial strains were isolated which preferentially hydrolyze one enantiomer of 2-acetylamino-1-phenyl-4-pentene, but only strain K1/1 showed high selectivity for the cleavage of the *(S)*-enantiomer. This strain was later identified as ***Rhodococcus globerulus*** (deposited as DSM 10337)

As in the case of the ***Rhodococcus equi Ac6*** enzyme, the production of the acylase from ***Rhodococcus globerulus*** K1/1 must be induced e.g. with N-acetyl-1-phenylethylamine whereby the *(R)*-enantiomer is a more effective inducer than the *(S)*-enantiomer. In optimized growth medium enzyme yields up to 47 U/l culture broth were obtained.

The enzyme from ***Rhodococcus globerulus*** K1/1 was purified and characterized as follows: specific activity: 3.17 U/mg (for *(S)*-N-acetyl-2-amino-1-phenyl-4-pentene); molecular weight: 92,300 Da, homodimeric substructure; pH-optimum: 7.0-7.5; stability at 30°C: no activity loss after 30 days; strong inhibition by PMSF but not by chelating agents (this suggests a serine protease like reaction mechanism). The enantioselectivity E of the enzyme is 75 for *(S)*-N-acetyl-2-amino-1-phenyl 4-pentene, the $K_m$-value is 1.24 mM.

Substrate specificity and enantioselectivity: The enzyme from *Rhodococcus globerulus* K1/1 deacetylates with good to very high enantioselectivities the *S*-enantiomers of a broad range of N-acetylamines with aromatic side-chains (see Table 3) but does not accept N-methyl-N-acetylamines. However, the substrate range of this amidohydrolase is clearly different from that of the *Rhodococcus equi*Ac6 enzyme.

With this new enzyme it became very easy to economically prepare (with very high yields) larger amounts of the desired *(S)*-2-amino-1-phenyl-4-pentene, or *(S)*-2-amino-1-(4-chlorophenyl)-4-pentene, as well as structurally related compounds. The enantioselective hydrolysis of the racemate can also be performed with resting cells instead of isolated enzyme (see Table 3). The resting cell system was used to perform the preparation of *(S)*-2-amino-1-(4-chlorophenyl)-4-pentene in a laboratory pilot scale. In this case 10% methanol were added for solubilization of the poorly soluble substrate.

## 2.6. Comparison of the three novel amidohydrolases

The comparison of the inhibition patterns, the subunit structures, and the storage stabilities indicates that the two *(S)*-specific acylases from *Rhodococcus equi* Ac6 and *Rhodococcus globerulus* K1/1 are biochemically very similar (but show different substrate spectra) and clearly differ from the *(R)*-specific acylase from *Arthrobacter aurescens* AcR5b. N-acetylamino acids are not hydrolyzed at significant rates by these three novel enzymes.

The N-terminal amino acid sequences of the three amidohydrolases were determined. The two *S*-specific *Rhodococcus*-enzymes do not show any significant N-terminal sequence homology and are also very different from the *R*-specific *Arthrobacter*-enzyme. Searches in the SWISS PROT sequence data bank did not lead to the identification of identical or highly homologous sequences.

## Substrate range and enantioselectivity

For the enzymatic hydrolysis of the racemic amides **1-8** (20mM, pH7, 30°C) with acylase from ***Rhodococcus equi*** **Ac6** (see Table 1) a partially purified enzyme preparation (2U/ml) was used. When the conversion reached about 50% (checked HPLC) the reaction was terminated and solutions extracted with at different pH values to separate the amine from the unreacted amide. The amides were used directly for optical purity determinations (chiral HPLC), whereas the formed amines were converted to the acetamides by adding triethylamine and acetic anhydride and then analyzed for optical purity.

For the enzymatic hydrolysis of the racemic amides **1-5**, **7**, and **9-11** (20mM, pH7, 30°C) with acylase from ***Arthrobacter aurescens*** **AcR5b** (see Table 2) a highly purified enzyme preparation (2U/ml) was used. The experiments were performed in the same manner as indicated for the Ac6-acylase.

For the enzyme catalyzed hydrolysis of the racemic amides **1**, **4**, **and 9-13** (10mM, pH7, 30°C) whole cells of ***Rhodococcus globerulus*** **K1/1** (1.5mg/ml) were used as biocatalyst (see Table 3). The analytical data were obtained in the same manner as indicated for the Ac6-acylase.

Table 1: Enantioselectivity of the *(S)*-specific amidohydrolase from *Rhodococcus equi* Ac6 acting on racemic amide substrates

| Substrate | Incubation time [h] | Conversion [%] | e.e. (*R*)-Amide [%] | e.e. (*S*)-Amine [%] | Selectivity E |
|---|---|---|---|---|---|
| 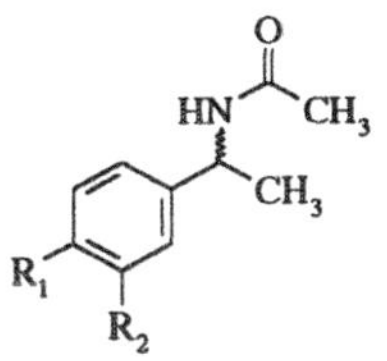 | | | | | |
| **1** $R_1$, $R_2$= -H | 18 | 50.6 | 99.3 | 96.8 | 350 |
| **2** $R_1$= -H, $R_2$= -CN | 19 | 46.1 | 85.5 | >99.9 | >500 |
| **3** $R_1$= -CN, $R_2$= -H | 19 | 49.4 | 97.4 | >99.9 | >500 |
| 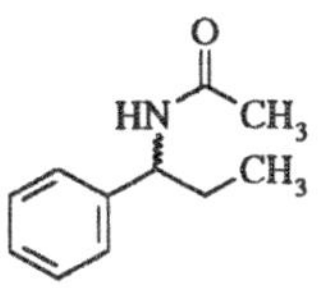 **4** | 19 | 50.4 | >99.9 | 98.5 | >500 |
| 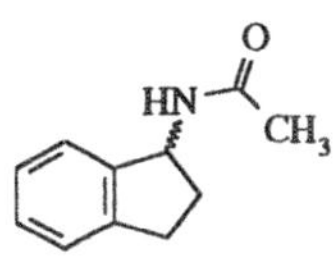 **5** | 20 | 52.3 | 96.7 | 88.2 | 65 |
| 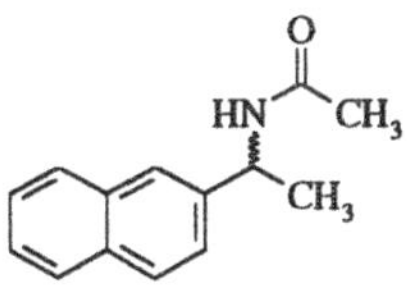 **6** | 19 | 14.4 | 16.4 | 97.4 | 88 |
| 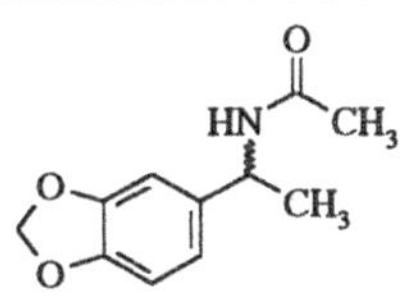 **7** | 20 | 47.9 | 90.1 | 97.8 | 280 |
| 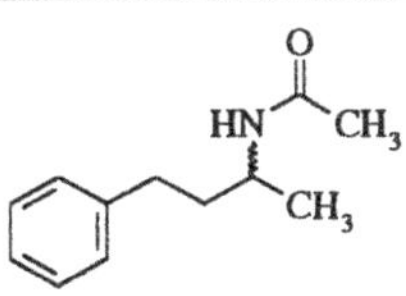 **8** | 20 | 51.7 | >99.9 | 92.1 | 230 |

Table 2. Enantioselectivity of the *(R)*-specific amidohydrolase from *Arthrobacter aurescens* AcR5b acting on racemic amide substrates

| Substrate | Incubation time [h] | Conversion [%] | e.e. (*S*)-Amide [%] | e.e. (*R*)-Amine [%] | Selectivity E |
|---|---|---|---|---|---|
| 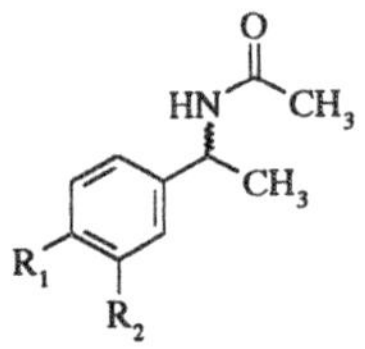 | | | | | |
| **1** $R_1$, $R_2$= -H | 22 | 50.3 | >99.9 | 98.2 | >500 |
| **2** $R_1$= -H, $R_2$= -CN | 75 | 47.4 | 82.4 | 91.5 | 60 |
| **3** $R_1$= -CN, $R_2$= -H | 75 | 51.3 | >99.9 | 95.0 | 320 |
| 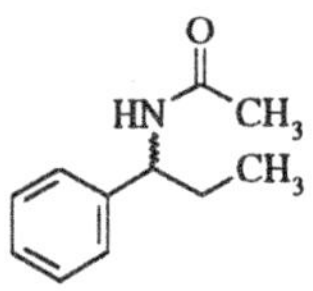 **4** | 139 | 46.0 | 87.4 | >99.9 | >500 |
| 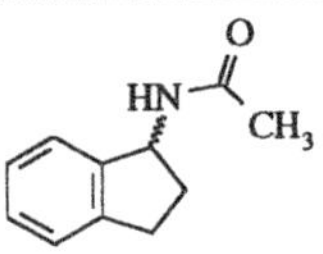 **5** | 46 | 50.1 | 96.1 | 95.7 | 180 |
| 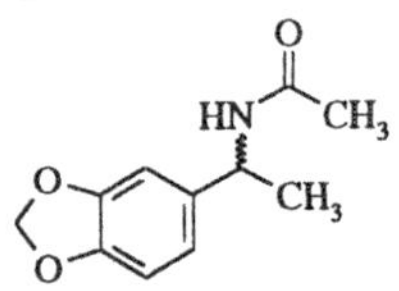 **7** | 22 | 47.5 | 82.4 | 91.5 | 60 |
| 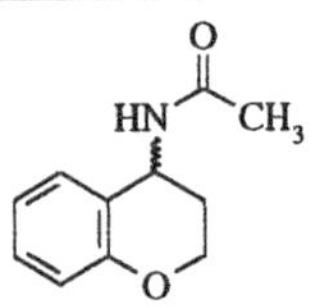 **9** | 194 | 35,2 | 46.6 | 90.5 | 32 |
| 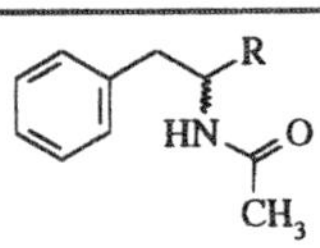 | | | | | |
| **10** R= $-CH_2CH=CH_2$ | 20 | 31.0 | 23.0 | 52.0 | 4 |
| **11** R= $-CH_2CH_3$ | 20 | 60.0 | 67.0 | 44.0 | 5 |

Table 3: Enantioselectivity of the *(S)*-specific amidohydrolase from *Rhodococcus globerulus* K1/1 acting on racemic amide substrates

| Substrate | Incubation time [h] | Conversion [%] | e.e. (*R*)-Amide [%] | e.e. (*S*)-Amine [%] | Selectivity E |
|---|---|---|---|---|---|
| 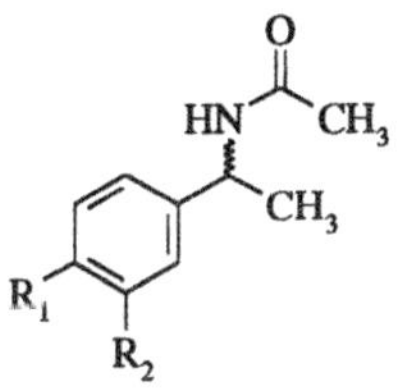 **1** $R_1, R_2$= -H | 3.0 | 50.6 | >99.9 | 97.5 | >500 |
| 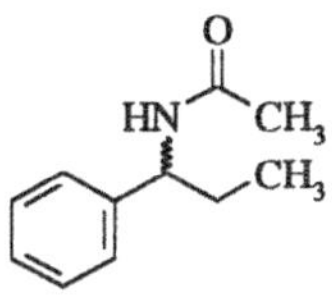 **4** | 3.5 | 52.6 | >99.9 | 90.0 | 140 |
| 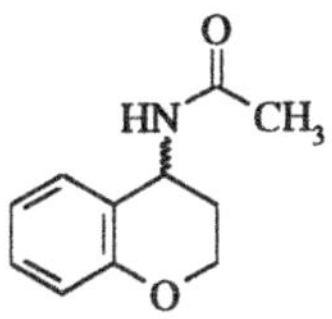 **9** | 3.5 | 51.7 | 99.9 | 93.5 | 290 |
| 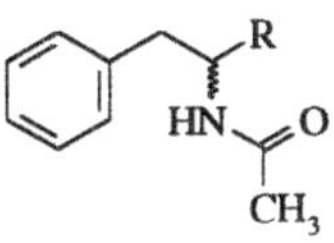 | | | | | |
| **10** R= $-CH_2CH{=}CH_2$ | 3.5 | 51.7 | 95.7 | 90.0 | 70 |
| **11** R= $-CH_2CH_3$ | 2.0 | 49.4 | 90.7 | 93.0 | 87 |
| **12** R= $-CH_2CH_2CH_3$ | 4.0 | 55.7 | 97.8 | 77.8 | 35 |
| 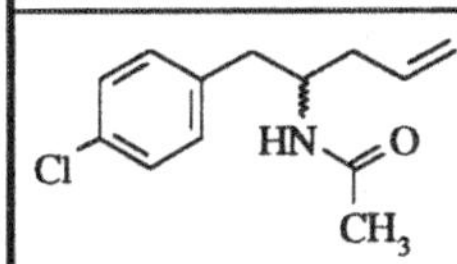 **13** solubilized with 10% MeOH | 5.0 | 49.6 | 98.5 | >99.9 | >500 |

## 3. INOSITOL PHOSPHATES AS TOOLS FOR PHARMA RESEARCH

### 3.1. Importance of inositol phosphates as a tool for research

A significant number of physiological processes in differentiated higher cells are closely linked with inositol metabolism. Important examples are e.g.: the activation of thrombocytes in the blood clotting process; hormonal signal transduction; signal transformation; contraction of muscles; transmission and processing of neural signals; control of cell proliferation; bone biosynthesis and calcium metabolism; anchoring of proteins in membranes, etc. In order to gain a deeper insight into such cellular processes inositol phosphates, as important substrates for many of the involved enzymes, would provide a very elegant tool. However, only very few types of inositol phosphates are commercially available, in minute quantities and at very high costs. Most of the commercially available material is derived from organ preparations.

An easier access was searched which would open the way to large quantities of a broad variety of enantiomerically pure inositol phosphates. The chosen strategy was a chemo-enzymatic approach starting from *myo*-inositol and involving commercially available enzyme preparations for the key synthesis steps.

### 3.2. *D-myo*-inositol-1-phosphate (*D*-1-$IP_1$)

*D-myo*-inositol-1-phosphate (*D*-1-$IP_1$), our first target compound. was synthesized by a short and facile route from optically pure *1D*-1-acetoxy-4,6-di-*O*-benzyl-*myo*-inositol, which was easily obtained by a highly regio- and enantioselective acylation of 4,6-di-*O*-benzyl-*myo*-inositol catalyzed by lipase PS from *Pseudomonas* sp. (Amano). The key intermediate 4,6-di-*O*-benzyl-*myo*-inositol can easily be prepared from *myo*-inositol by four chemical steps[20].

### 3.3. *D*- and *L*-1,3,4,5-*myo*-Inositol tetraphosphate (*D*- and *L*-1,3,4,5-$IP_4$)

*D*-1,3,4,5-$IP_4$ and the unnatural enantiomer *L*-1,3,4,5-$IP_4$ were prepared from *D*- and *L*-2,6-dibenzyl-*myo*-inositol by a chemical phosphorylation and deprotection step in high yields and purities without extensive purification.

The diprotected racemic 2,6-dibenzyl-*myo*-inositol **(±)-14** is an excellent precursor for the synthesis of racemic 1,3,4,5-$IP_4$ and can easily be prepared from *myo*-Inositol by four chemical steps[21]. The benzyl protecting group allows the final deprotection under mild and neutral conditions without any P-migration[20].

However the enantioselective synthesis of *D*- or *L*-1,3,4,5-$IP_4$ requires the optically pure building blocks **(+)-14** and **(-)-14**. A classical chemical resolution procedure for racemic **14** is so far unknown and seems to be very laborious.

An initial screening of several commercially available lipases for their acyltransfer activity towards **(±)-14** in vinyl acetate as an acyl donor indicates that only the lipase from *candida antarctica* (Novo SP 435) offers reasonable conversion. HPLC- and TLC-analysis showed that only one product was formed and the rate of the reaction decreased dramatically after the conversion reached 50%, indicating a very high enantiospecificity. A preparative experiment was worked up after approximately 50% conversion yielding 49% (98% theor. yield) of the unconverted inositol derivative **(-)-14** with an optical purity of >99% ($[\alpha]_D^{20}$= -29.9$^o$, c=1 EtOH) and 49% of the mono acetate **(-)-15** ($[\alpha]_D^{20}$= -6.1$^o$, c=1 EtOH). Based upon the two-dimensional COSY and the $^{13}C$-NMR-spectra, the mono acetate **(-)-15** was isomerically pure and the hydroxyl group at C5 was esterified. Chemical hydrolysis of the mono acetate **(-)-15** leads quantitatively to **(+)-14** ($[\alpha]_D^{20}$= +29.8°, c=1 EtOH, e.e.>99%)[22].

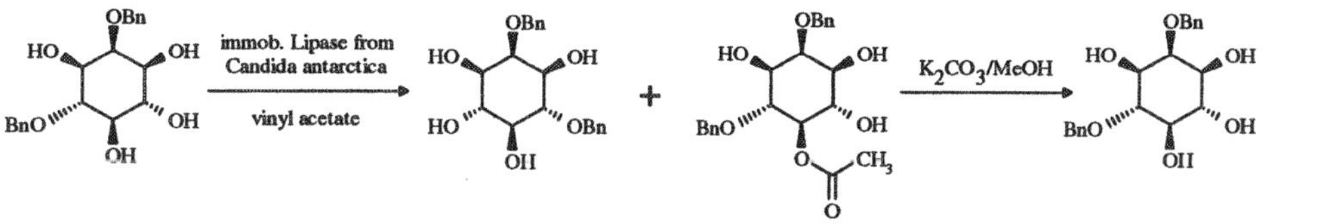


**(+/-)-14** **(-)-14 >99% e.e.** **(-)-15** **(+)-14 >99% e.e.**

Figure 2. The enzymatic key step in the synthesis of *D*- and *L*-1,3,4,5-$IP_4$

Both enantiomers of **14** were finally phosporylated using 2-di-ethylamino-1,3,2-benzo-dioxaphosphepane/tetrazole, followed by successive oxidation with $H_2O_2$.
Deprotection with $H_2$-Pd/C removes all benzylic protecting groups leading to *D*-1,3,4,5-$IP_4$ ($[\alpha]_D^{20}$= -4.6°, c=1 $H_2O$, pH=10.0) resp *L*-1,3,4,5-$IP_4$ ($[\alpha]_D^{20}$= +4.7°, c=1.1 $H_2O$, pH=10.3).

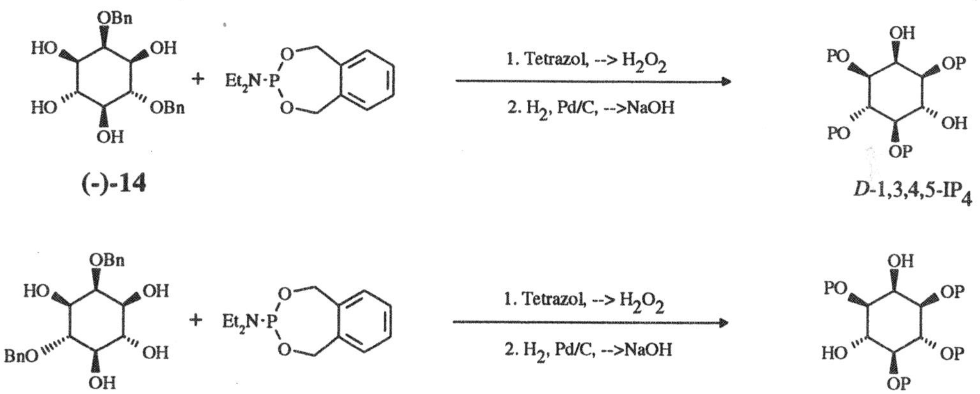


**(-)-14** *D*-1,3,4,5-$IP_4$

**(+)-14** *L*-1,3,4,5-$IP_4$

P = $PO_3Na_2$

Figure 3. The phosphorylation and deprotection of the chiral key intermediates **(-)-14** and **(+)-14**

Using the synthesis route described above 3.5g of each *D*-1,3,4,5-$IP_4$ and *L*-1,3,4,5-$IP_4$ as sodium salts (storage form) were synthesized.

By extending the strategies outlined above, a whole set of twelve different types of inositol phosphates could be made available in preparative scale[23].

## REFERENCES


1. W. Fröstl, S.J. Mickel, M. Schmutz, and H. Bittiger, Pharmacol. Rev. Commun. 8 (1996) 127-133
2. K. Kaupmann, K. Huggel, J. Held, P.J. Flor, S. Bishoff, S.J. Mickel, G. McMaster, C. Angst, H. Bittiger, W. Fröstl, and B. Bettler, Nature 386 (1997) 239-246
3. S. Ofner, K. Hauser, W, Schilling, A. Vassout, and S.J. Veenstra, Bioorg. Med. Chem. Lett. 6 (1996) 1623-1628
4. S.J. Veenstra, K. Hauser, W. Schilling, C. Betschart, and S. Ofner, Bioorg. Med. Chem. Lett. 6 (1996) 3029-3034
5. Chisso Corp., Japan, Japanese Patent (1997), JP 58-198296
6. D. Stirling, G. Branchburg, W. Matcham, A. Bridgewater, and L. Zeitlin, Celgene Corporation (1994), US Patent No. 5,300,437
7. Baiohru Co. Ltd., Japanese Patent (1994), JP 06253-875
8. Baiohru Co. Ltd., Japanese Patent (1994), JP 06253-876
9. G.W. Matcham and S. Lee, , Celgene Corporation (1994), US Patent No, 5,360,724
10. F. Balkenhohl, B. Haauer, W. Landner, and U. Pressler, (1993), German Patent Application DE 4332738 A1
11. V. Gotor, E. Menendez, Z. Mouloungui, and A. Gaset, J. Chem Soc. Perkin Trans. (1993) 2453-2456
12. S. Puertas, R. Brieva, F. Rebolledo, and V. Gotor, Tetrahedron 49 (1993) 4007-4014
13. M.T. Reetz and C. Dreisbach, Chimia 48 (1994) 570
14. N. Öhrner, C. Orrenius, A. Mattson, T. Norin, K. Hult, Enzyme Microb. Technol. 19 (1996) 328-331
15. H. Kitaguchi, P.A. Fitzpatrick, J.E. Huber, and A.M. Klibanov, J. Am. Chem. Soc. 111 (1989) 3094-3095
16. A.L. Gutmann, E. Meyer, E. Kalerin, F. Polyak, and J. Sterling, Biotechnol. Bioeng. 40 (1992) 760-767
17. H. Smidt, A. Fischer, P. Fischer, and R.D. Schmid, Biotechnology Techniques 10 (1996) 335-338
18. A. Brunella, M. Graf, M. Kittelmann, K. Laumen, and O. Ghisalba, Appl. Microbiol. Biotechnol. 47 (1997) in press
19. M. Graf, A. Brunella, M. Kittelmann, K. Laumen, and O. Ghisalba, Appl. Microbiol. Biotechnol, 47 (1997) in press
20. K. Laumen and O. Ghisalba, Biosci. Biotech. Biochem. 58 (1994) 2046-2049
21. Billington D. and Baker R., J. Chem. Soc., Chem. Commun. **1987**, 1011-1013
22. K. Laumen and O. Ghisalba, unpublished results
23. Details will be published elsewhere.

H. van der Goot, (Editor)
TRENDS IN DRUG RESEARCH II



# α-Diazocarbonyl Chemistry - Target Driven Applications

R. Pellicciari, G. Costantino, M. Marinozzi, L. Mattoli and B. Natalini

Istituto di Chimica e Tecnologia del Farmaco, Università di Perugia,
Via del Liceo, 1 - 06123 Perugia, Italy

In a time in which organic synthesis moves beyond its traditional objectives and major changes are introduced in the choice of synthetic targets, with the emphasis shifting from 'compounds' to 'properties', the area of 'Synthetic Methods Developments' continues to be an issue of great impact. Having the role to define the strategies and to provide the tools within which new target molecules can be reached, organic synthesis is an obliged crossway for disciplines such as medicinal chemistry, and an effective means for the discovery of new reactions or the uncovering of new aspects of previously described ones.

Our subject, α-diazocarbonyl chemistry, is illustrative of the creativity, courage, and patience needed to acquire chemical control on reagents as tremendously versatile as the diazocarbonyl one. An examination of the historical developments of this field reveals in how many directions the progress has been made. It is not the purpose of this chapter to cover exhaustively the area as this has been done in several excellent reviews [1-17]. Rather, we will highlight a few aspects of two major areas of diazocarbonyl chemistry:

a. α-Diazocarbonyl compounds as a source of carbenoids.
b. - Reactions of α-diazocarbonyl compounds with aldehydes and ketones.
   - α-Diazocarbonyl compounds as a source of vinyl cations.

Some of the examples chosen to illustrate these two areas are taken from research activities of our lab. Although predominantly engaged in medicinal chemistry projects, a part of our efforts has been devoted through the years to the study of the reactivity of diazocarbonyl compounds, being often able to capitalize the 'in house' experience thus acquired for the synthesis of biologically relevant target molecules. One of us (R.P.) is grateful to his mentor Professor Ernest Wenkert for having introduced him in the early 70's to the fascinating world of α–diazocarbonyl chemistry.

## 1. α-Diazocarbonyl Compounds as a Source of Carbenoids

α-Diazocarbonyl compounds can be applied in an impressive array of strategy-level reactions such as carbon- and heteroatom-hydrogen bond insertion, cyclopropanation, cyclopropenation, azyridation, ylide generation, dimerization, Wolff rearrangement (Fig. 1), which are increasingly integrated

in multistep processes (tandem, cascade, domino sequences, etc.) allowing structural changes of great complexity in single operations [18].
The use of acyl carbenes in the encoding process of combinatorial libraries is another revealing example of the synthetic versatility of this class of compounds [19].

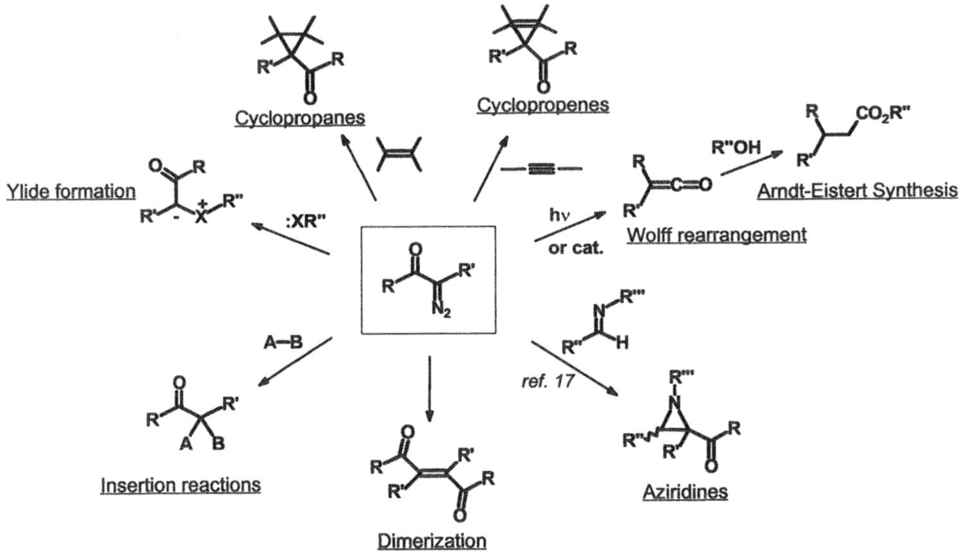


Fig. 1

It is worthwhile to recall, however, that the acquisition of chemical control on the reactivity of α-diazocarbonyl compounds after the first report by Curtius in 1883 of the preparation of ethyl diazoacetate from glycine [20], has been slow and hesitant. A major reason was that 'free' carbenes, generated by thermal (ca 150 °C) or photochemical splitting of nitrogen were highly reactive, unstable intermediates leading to highly complex reaction mixtures with products often difficult to characterize. Also, the behavior of diazocarbonyl compounds was poorly understood. It may be interesting to recall that the Arndt-Eistert synthesis, discovered by Wolff in 1902 [21], did not become established until 1935 when Arndt and Eistert stated that silver oxide, copper or platinum could be used as catalysts for the decomposition *with rearrangement* of diazoketones [22].

In the early 50's Reichstein [23] and Yates [24] independently discovered that diazocarbonyl compounds could be decomposed in the presence of copper bronze (i.e. very finely divided copper) or cupric oxide, to give metal-carbene complexes - (carbenoids) - leading to the formation of products *without rearrangement*. The introduction of transition metals as insoluble, heterogeneous catalysts was certainly an improvement in terms of increased chemo-control, and opened the way to the development of the new class of homogeneous transition metal catalysts and, with them, to a new phase of increased exploitation of carbenoid reactions.

### 1.1. Homogeneous Catalysts for Carbenoid Reactions

The first example of the application of soluble homogenous transition metal catalyst was provided in 1966 by Nozaki and Noyori [25] who reported the cyclopropanation of styrene with ethyl diazoacetate (EDA) by using copper(II)acetylacetonate as catalyst. Subsequently, many other homogeneous catalysts have been prepared and applied to a variety of carbenoid reactions such as (trialkyl- and triaryl-phosphite)copper(I) described by Moser [26], and copper(II)triflate first reported by Solomon and Kochi [27].

#### 1.1.1. Dirhodium(II)tetraacetate as Catalyst for Carbenoid Transformations

A major advance in the metal-catalyzed carbenoid transformations has occurred with the introduction, in 1973, of the dirhodium (II) tetracetate (**1**) [28]. Over the last two decades, the design and the preparation of a large number of dirhodium(II) complexes with ligands other than acetate has given a major impulse to the application of carbenoid reactions. Dirhodium (II) catalysts usually require much milder conditions than other homogeneous catalysts and an impressive chemo-, regio-, and enantiocontrol over a variety of substrates can be achieved in inter- and intramolecular reactions by modifying the electronic characteristic and the steric disposition of the bridging ligands around the carbenoid center.

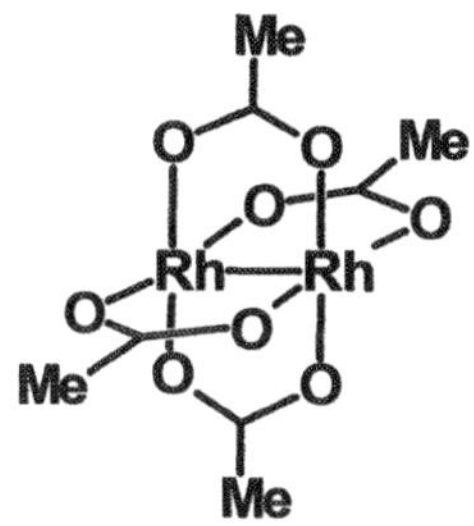


**1**

In the late 80's, it became apparent that much of the properties of dirhodium(II)-based catalysts could be understood on the basis of the existence of a dirhodium(II)-carbenoid complex which can be represented by two limiting resonance structures [11]:

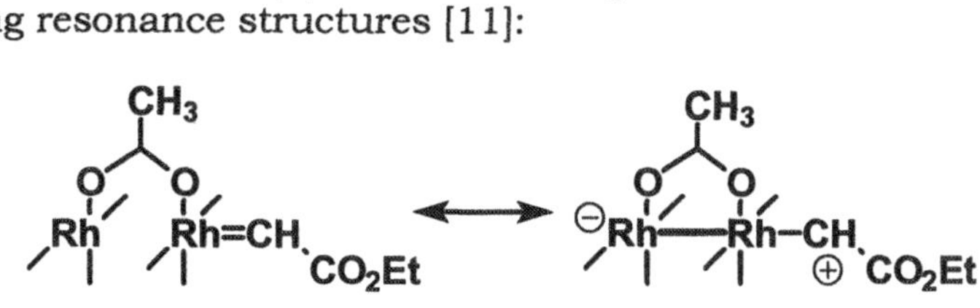


**1a** **1b**

While a large number of theoretical studies have appeared on the structure of dirhodium(II)tetracarboxylate complexes [29], the nature of the dirhodium(II)-carbenoid intermediate has been investigated at semi-empirical level by Doyle et al. only for the simplest case of $Rh_2(OAc)_4\text{-}CH_2$ [30]. We have therefore decided to undertake an extensive *ab initio* study on the electronic structure of different $L_nRh_2$-$CHCO_2Et$ intermediates in order to quantitatively analyze the effect of different ligands on the Frontier Molecular Orbital, on the electronic density and on the bond order of the metal carbene carbon atom. We report here some preliminary results on the structure of $Rh_2(OAc)_4$-$CHCO_2Et$ intermediate complex.

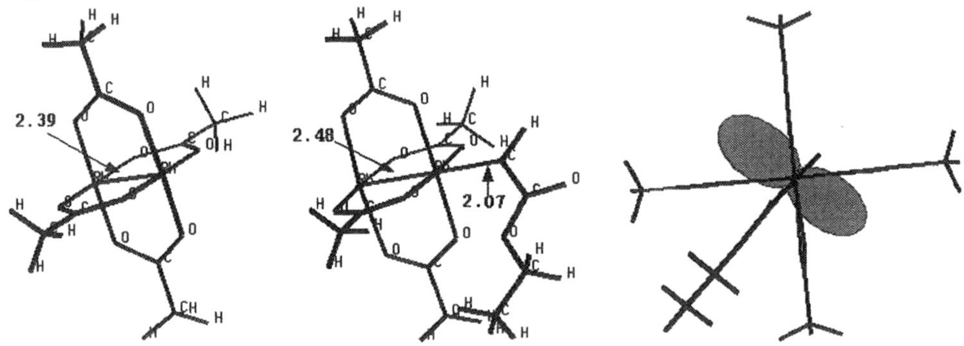


Fig.2 Structure of dirhodium(II)tetraacetate (left) and dirhodium(II)tetraacetate-carbenoid complex (middle) . The distances (Å) of Rh-Rh and Rh-carbenoid carbon atom are reported. On the right, a fron-view of the complex showing the spatial disposition of the empty p-orbital of the carbenoid carbon atom.

In the optimized structure (Fig. 2), the carbenoid portion of the complex forms an angle of 45° with respect to the Rh-O bonds defining one of the four (equivalent) quadrants of $Rh_2(OAc)_4$ **(1)**. Because of the trigonal hybridization of the carbenoid carbon atom bearing a partial positive charge, its formal atomic empty p-orbital also makes an angle of 45° with the other quadrant of $Rh_2(OAc)_4$, a disposition which certainly minimizes the steric requirement for the incoming nucleophile.

Since much of the discussion on the reactivity of dirhodium(II)-mediated carbenoid reactions is based on the 'philicity' of the carbenoid carbon atom, we are carrying out a detailed MCSCF (Multiconfigurational Self Consistent Field) analysis of the electronic structure of $Rh_2(OAc)_4$-$CHCO_2Et$ which will be reported elsewhere [31]. Preliminary results seems to indicate that a more complex picture of the resonance structures than that above reported (**1a**, **1b**) is required for understanding of the reactivity of the carbenoid species, especially when the reaction with unactivated olefins is studied.

## 1.2 Preparation of Carboxy-Buckminsterfullerenes on route to biologically active derivatives.

The accessibility of $C_{60}$ Buckminsterfullerene (**2**) in macroscopic quantities [32] and the development of synthetic strategies for its transformation into functionalized, often water-soluble derivatives has opened new promising perspectives for the exploitation of this large sphere-

shaped molecule for the preparation of new pharmaceutical candidates [33-35] and as scaffold for the preparation of combinatorial libraries.
Recently, a report by Choi et al. has attracted our attention in that a hexakiscarboxylate derivative [$C_{63}(COOH)_6$, **3**] of $C_{60}$ fullerene showed very promising neuroprotective properties when tested in mouse cortical neurons against excitotoxic necrosis and apoptosis. EPR experiments have suggested that this activity is likely to be due to the radical scavenger properties of **3** [36].

**2**

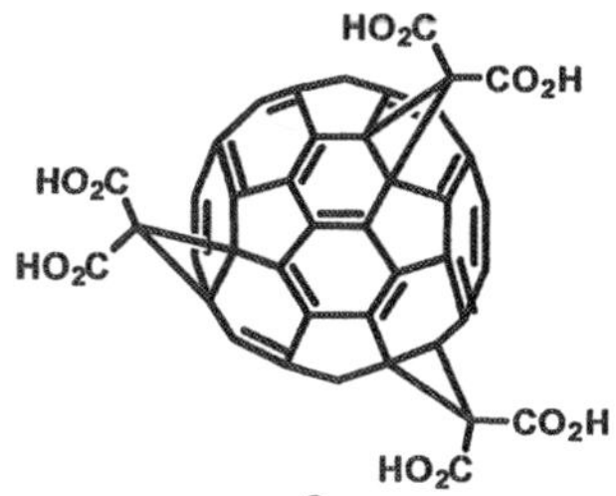


**3**

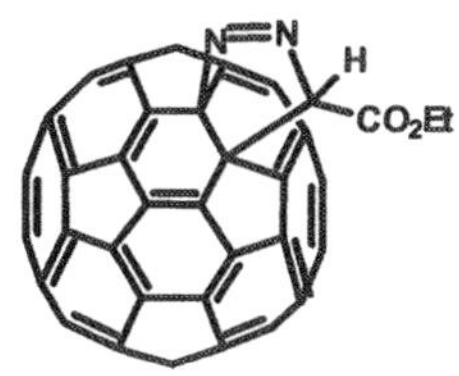


**4**

These preliminary observations have prompted us to undertake a research program devoted to the development of synthetic strategies and their application to the preparation of new fullerene derivatives to be evaluated for their neuroprotective properties. Functionalization of $C_{60}$ fullerene has largely relied on the exploitation of its moderate electrophilic character as an electron poor polyolefine which readily undergoes addition reaction with a number of nucleophilic species. Owing to the 1-3-dipolar character, diazocompounds, in particular, have been shown to react with the 1-3-dipolarophile $C_{60}$ to give a variety of substituted methanofullerenes and fulleroids. In 1991, Suzuki and Wudl first reported on the thermal addiction of diphenyldiazomethane to $C_{60}$ fullerene [37] to give a mixture of three products, identified as a [6,6]-closed methanofullerene, and two isomeric [6,5]-open fulleroids, respectively [38]. The thermal addition of diazoalkenes to $C_{60}$ does not involve the formation of free carbenes but, rather, proceeds through the formation of a pyrazoline intermediate (**4**) resulting from the addition of the diazocompound to the (6-6) double bond [39]. The extrusion of nitrogen from the pyrazoline ring then leads to the kinetically favored [6,5]-open fulleroid which, under reflux conditions, undergoes valence isomerism to give the thermodynamically favoured [6,6]-closed methanofullerene.
The same methodology has subsequently been extended to α-diazocarbonyl compounds, including α-diazoesters [40], α-diazoamides [41], and α-diazoketones [42]. As a general statement, the thermal addition of diazocarbonyl compounds to $C_{60}$ usually requires stronger conditions and gives lower yields than the more reactive diazoalkanes. A particular attention has been given to the reaction with fullerene of ethyl diazoacetate (EDA). In 1993, Diederich et al., have reported that EDA and $C_{60}$ react in toluene for 7 h at 110 °C to afford a mixture of the three products (overall yield of 35%)

that have been identified as [6,6]-closed methanofullerene (**5**), [6,5]-open (**6**) and [6-5]-open (**7**) fulleroids in the ratio 1:4:2, respectively (Table 1, Entry 1) [40b].

Table 1. $Rh_2(OAc)_4$-Mediated Addition of Diazocarbonyl Compounds to $C_{60}$ Fullerene

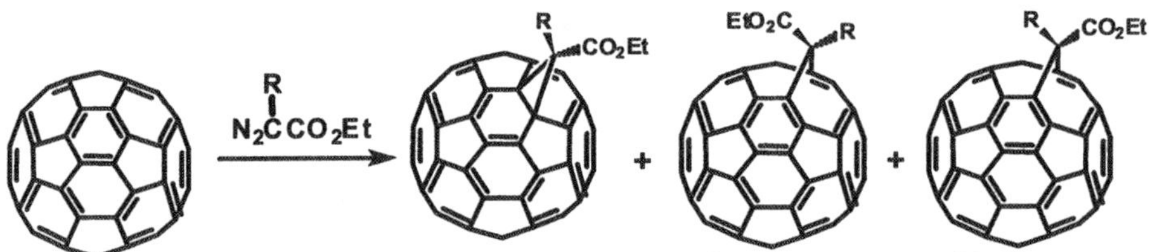


[6,6]-closed [6,5]-open [6,5]-open

**2** **5** **6** **7**

| Entry | Conditions | R | Solvent | T (°C) | t (h) | Yield (%) | Ratio of Formation 5 | 6 | 7 |
|---|---|---|---|---|---|---|---|---|---|
| 1. | Thermal | H | PhMe | 110 | 7 | 35 | 1 | 4 | 2 |
| 2. | Rh-catalytic | H | PhCl | rt | 20 | 15 | 9 | 1 | 1 |
| 3. | Rh-stoichiom. | H | 1-Me-Naphth. | rt | 8 | 42 | 52 | 1 | - |
| 4. | Thermal | $CO_2Et$ | PhMe | 110 | 20 | 10 | - | - | - |
| 5. | Rh-stoichiom. | $CO_2Et$ | 1-Me-Naphth. | 80 | 32 | 33 | 9 | 1 | - |

When the mixture was heated for additional 24 h in refluxing toluene, the conversion of the [6,5]-open fulleroids into the thermodynamically more stable [6,6]-closed isomer (**5**) was observed. We have investigated the still unexplored reactivity towards fullerene of transient carbenoids formed by reaction of α-diazoesters with dirhodium (II) tetraacetate with the aim of verifying the possibility of functionalizing $C_{60}$ with diazocarbonyl compounds by using methodologies different than the thermal decomposition. Thus, several reactions were carried out, giving a particular attention to the ratio between $Rh_2(OAc)_4$ and substrate, and to the solvent. The best results (see Table 1, Entry 3) were obtained when the reaction was carried out at *room temperature* in α-methylnaphthalene as solvent in the presence of a *stoichiometric quantity* of catalyst. Most interestingly, the $Rh_2(OAc)_4$-catalyzed reactions lead almost exclusively to the formation of the (6,6)-closed isomer (**5**) (Table 1, Entries 2, 3, 5), with the [6,5]-open isomers (**6**, **7**) present only in minor amounts or in traces (Table 1, Entry 3), a behaviour in contrast to that observed in the thermal addition of α-diazoesters, in agreement with the hypothesis that the reaction proceeds *via* a carbenoid intermediate and not by 1,3-cycloaddition. The possibility given by the use of $Rh_2(OAc)_4$ as catalyst to obtain directly the most stable [6,6]-closed isomer at room temperature represents a clear advantage with respect to the thermal addition.

By using $Rh_2(OAc)_4$ in the same conditions, the less reactive ethyl diazomalonate (R=$CO_2$Et) failed to react with fullerene at room temperature. When the reaction was carried out at 80 °C, however, bis-carboalkoxymethanofullerene was obtained in a yield (32%) significantly higher with respect to that obtained under thermal conditions [43] (Table 1, Entry 5 and 4, respectively).

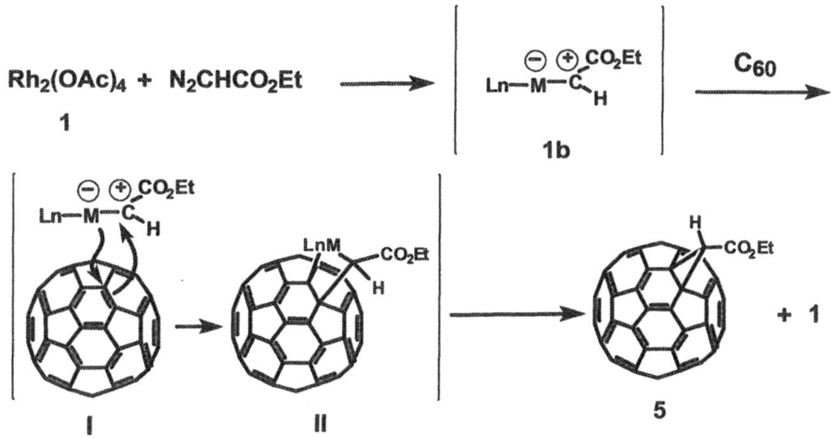


Scheme 1

The mechanism by which the metal-carbenoid intermediate reacts with the electrophilic fullerene is intriguing (Scheme 1). The rhodium atom can be assigned as the nucleophile center while the carbenoid carbon atom is the electrophile center. Nucleophilic addition of the rhodium atom to the electrophilic (6-6) double bond of fullerene (I), followed by electrophilic attack of the carbenium carbon atom, generates the metal-cyclobutane **II** that is the direct precursor of the expected carboethoxymethanofullerene (**5**). [44]. The above results clearly indicate that the stoichiometric $Rh_2(OAc)_4$-catalyzed reaction of ethyl diazoacetate with fullerene provides an efficient way to its functionalization. Evident advantages of our methodology over the previously reported thermal one are the much milder required conditions and the almost exclusive formation of the [6,6]-closed methanofullerene (**5**), being the overall yields of conversion close to 50% in the best cases.

### 1.3. Chiral Catalysts for Carbenoid Reactions

In 1966, Nozaki and Noyori [25] reported the first example of a catalytic asymmetric reaction of prochiral compounds induced by soluble chiral complexes. A chiral copper(II)-Schiff base complex was shown to catalyze the reaction of styrene with ethyl diazoacetate to give *cis*- and *trans*-2-phenylcyclopropylcarboxylates in <10 % ee, a degree of enantioselectivity that was unsatisfying in itself, but leading to an extensive screening of other chiral Schiff bases [45].

As a consequence of this work, R-7644, an highly efficient Cu(II)-Schiff base catalyst was developed and employed for the commercial scale preparation of (*S*)-2,2-dimethylcyclopropanecarboxylate, a constituent of cylastatin, an *in vivo* stabilizer of imipenem [46].

After the seminal work of the Japanese group, a significant advance in the field of soluble, homogeneous catalysts was achieved in 1986 with the introduction by Pfaltz et al. of chiral semicorrin copper (II) catalysts [47], such as (**8-10**), followed by the independent discovery in 1990 by Masamune et al. [48] and in 1991 by Evans et al. [49], of the bis-oxazoline-copper (II) catalysts (**11-20**).

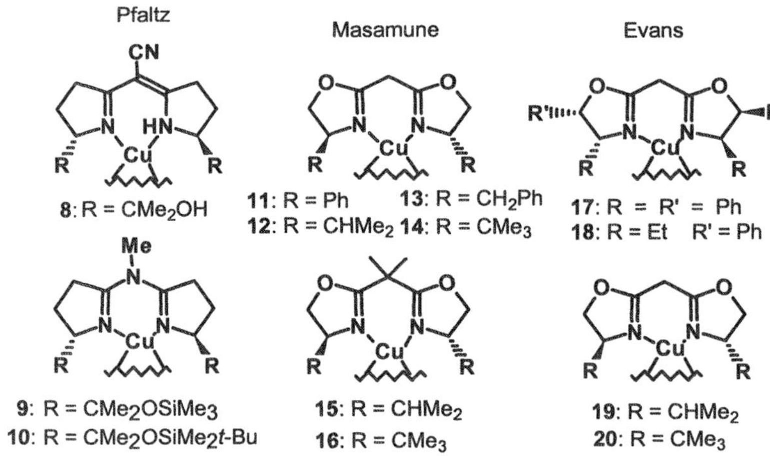


Parallel to the development of copper-based homogeneous catalysts, the late 80's have seen a growing interest in the design of dirhodium(II)-based chiral catalysts. In 1988, Brunner et al. [50] has reported the first chiral dirhodium(II)tetrakiscarboxylates, a work followed by the preparation in 1990 by McKervey of mandelate and proline chiral derivatives of dirhodium (II) [51]. Significative advance in this class of catalysts was the preparation in 1990 by Doyle et al. [52], of the chiral dirhodium(II)-4-alkyloxazolidines [$Rh_2$(4S-IPOX)$_4$ (**21**), $Rh_2$(4S-BNOX)$_4$(**22**), $Rh_2$(4S-MPOX)$_4$ (**23**)], in which the carboxamide ligands place the chiral environment in the proximity of the carbenoid center, thus allowing an increase in the enantiocontrol. Based on the notation that polar substituents on the oxazolidine ring could favorably interact with the p-orbitals of the transient carbene, thus providing an adequate orientation of the incoming nucleophile, a number of dirhodium(II)-2-pyrrolidone-5-carboxylate chiral catalysts such as [$Rh_2$(5*S*-MEPY)$_4$ (**24**) and $Rh_2$(4*S*-MEOX)$_4$ (**25**)] [52] were then developed. The field of dirhodium (II) catalysts has continuously evolved over the last decades and more and more sophisticated catalysts have been reported. A list of these dirhodium(II)-based catalysts is reported in Chart 1 [53].

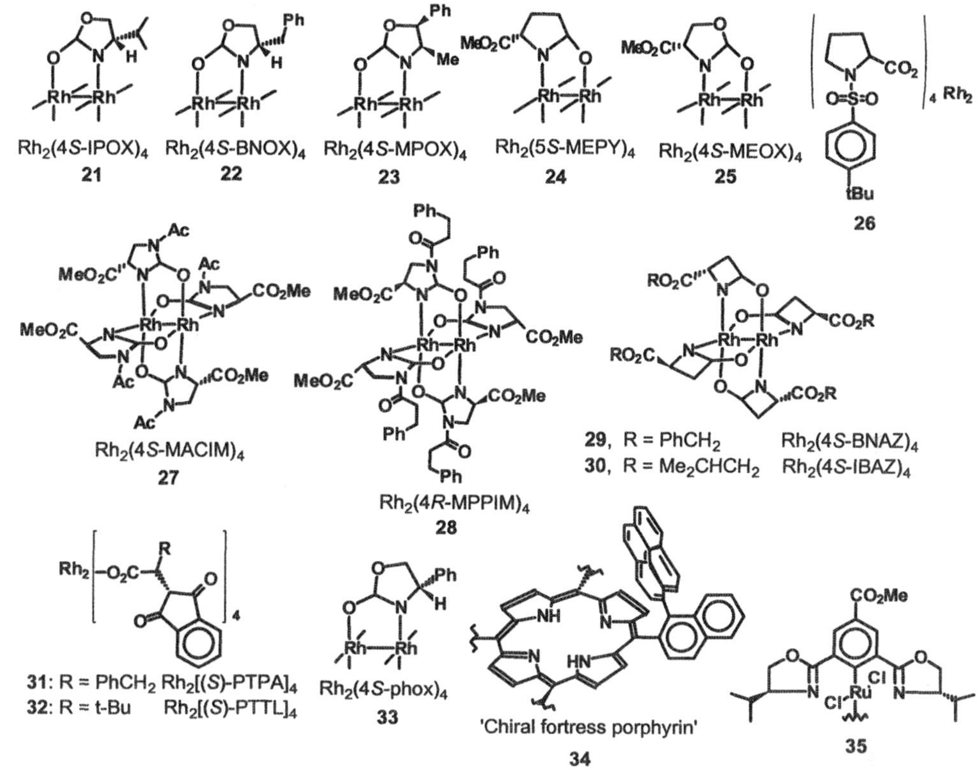


Chart 1

Cobalt- and ruthenium-based catalysts have also been reported. Among them, the Nishiyama's catalyst (**35**) has recently been found of particular interest [54]. The continuous development of transition metal-based catalysts such as those above reported has been motivated by the need of achieving a high degree of either chemo-, regio-, diastereo-, or enantiocontrol in carbenoid transformations. The problem of the chemoselectivity (i.e., intra-molecular cyclopropanation vs. C-H insertion) associated with the use of transition metal-based catalysts can be addressed by a careful choice of bridging ligands when rhodium(II) is used as catalyst. High chemocontrol can be achieved for example by moving from electron withdrawing to electron donating bridging ligands, as exemplified by the $Rh_2(cap)_4$ and $Rh_2(pbf)_4$ catalysts [30]. Diastereo- and enantiocontrol in intra- and intermolecular cyclopropanations are also a major issue. An interesting example of the application of chiral dirhodium catalysts to the synthesis of biologically active agents is the recent Corey's enantioselective synthesis of the antidepressant sertraline (**38**) [55] achieved by reacting stirene (**36**) with Davies's vinyl diazoacetate (**37**) [56] in the presence of McKervey's catalyst (**26**) [51].

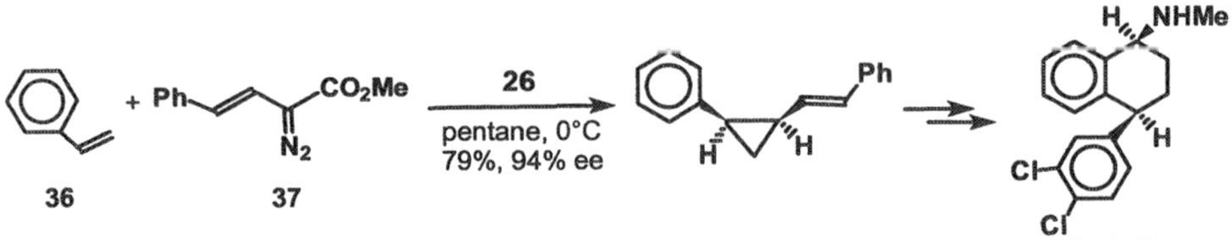


Sertraline
38

### 1.3.1. Chiral Catalysts for Intramolecular Cyclopropanations: Enantionselective Synthesis of PCCG-4, a Potent and Selective mGluR Antagonist.

The use of $Rh_2$(5*S*-MEPY)$_4$ and $Rh_2$(5*R*-MEPY)$_4$ as catalysts affords enantioselecyively the products of intramolecular cyclopropanation of allylic diazoacetates [57]. Dirhodium(II)-complexes, in particular, have been applied by Martin to the enantioselective synthesis of a renin inhibitor [58] and a collagenase inhibitor [59], by Rogers to the preparation of (+)-presqualene diphosphate [60] and by Taber to the synthesis of the (±)-8-*epi*-$PGF_{2\alpha}$ prostaglandin ethyl ester [61]. As an example taken from our work, we report here the application of the $Rh_2$(5*S*-MEPY)$_4$(**24**)-catalyzed intramolecular cyclopropanation to the synthesis of PCCG-4 (**45**), a potent and selective antagonist of the type II of the metabotropic glutamate receptors. Carboxycyclopropylglycines, conformationally constrained analogs of L-glutamic acid, have played a fundamental role in the pharmacological characterization of either ionotropic or metabotropic classes of glutamate receptors [62]. As a continuation of our work in this field, we have reported in 1996 the synthesis of all the sixteen isomers of 2-(2'-carboxy,3-phenylcyclopropyl)glycine by an enantiodivergent procedure based on the Cu(TBS)$_2$-catalyzed cyclization of suitable allylic diazoacetates [63]. The screening towards metabotropic and ionotropic glutamate receptors of this stereolibrary has revealed that some of the isomers possess interesting pharmacological properties and, among them, (2S,1'S,2'S,3'R)-2-(2'-carboxy-3'-phenylcyclopropyl)glycine (PCCG-4, **45**) was shown to be a potent and selective antagonist of the mGluR2 metabotropic glutamate receptor subtype. In the light of the importance of PCCG-4 (**45**) as a promising pharmacological tool, we have involved ourselves in its enantioselective synthesis (Scheme 2) [64]. Key step in this preparation is the intramolecular cyclopropanation of *cis*-3-phenyl-2-propen-1-yl diazoacetate (**39**) catalyzed by the chiral dirhodium (II) catalyst $Rh_2$(5S-MEPY)$_4$ (**24**) which affords the lactone (**40**) with 92% ee. The opening of the lactone (**40**) followed by selective epimerization of the morpholinocarbonyl moiety of the amide (**41**) to give the isomeric amide (**42**) and oxidation of its primary alcholic function affords the aldehyde (**43**) which is endowed with the correct stereochemistry at the three cyclopropane chiral centers. Diastereoselective Streker synthesis, involving the condensation of aldehyde (**43**) with optically active *R*-α-phenylglycinol followed by nucleophilic addition of cyanide ion to the

Schiff base affords almost exclusively the aminonitrile (**44**) which, after purification, was oxidized and treated with 6N HCl to give PCCG-4 (**45**).

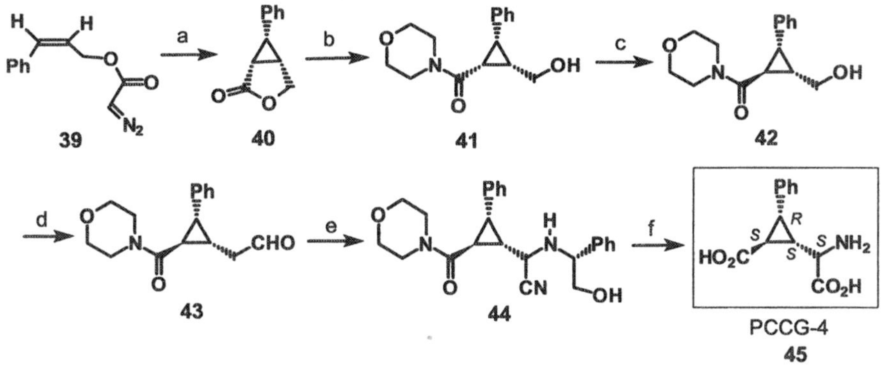


a) $Rh_2(5S\text{-MEPY})_4$, $CH_2Cl_2$, reflux; b) Morpholine, $AlMe_3$, $CH_2Cl_2$, reflux; c) Li-HMDS, THF, rt; d) PCC, $CH_2Cl_2$, rt; e) i. *R*-α-Phenylglycinol, MeOH, rt; ii. TMSCN, 0 °C, then rt; iii. mpc; f) i. $Pb(OAc)_4$, $CH_2Cl_2$/MeOH (1:1), rt; ii. 6N HCl, reflux; iii. Dowex 50WX2-200, 10% Py; iv. rp8-mpc.

Scheme 2

**1.4. Chiral Catalysts for Intramolecular C-H Insertion.**

The intramolecular insertions of carbenoids are synthetically important. In a paper of 1982 [65], the decomposition of cyclohexylmethyl-diazomethyl ketone (**46**) was studied and it was shown that when heated in cyclohexane in the presence of $CuSO_4$ as catalyst, it was transformed in different yields to the corresponding cyclopentanone (**47**), cyclobutanone (**48**) and in a solvent insertion product (**49**); whereas cyclobutanone (**48**) was only a minor side product, the formation of cyclopentanone (**47**) was highly dependent on the ring size of the cycloalkane residue, thus indicating the importance of ring conformation for supplying the required proximities of the functional group involved in the reaction process.

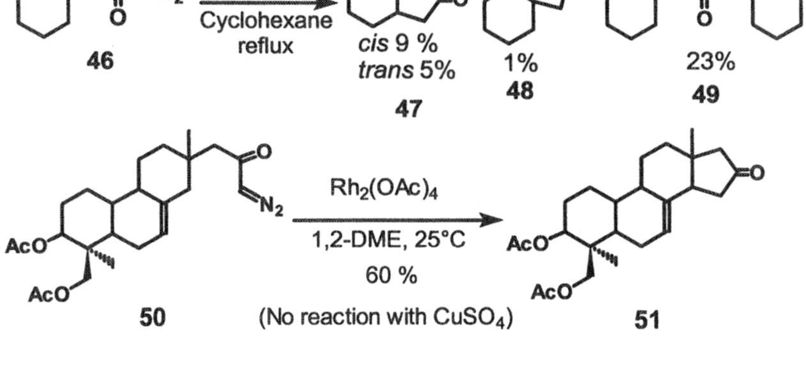

It was shown, for the first time, that the intramolecular C-H insertion could greatly benefit from the use of $Rh_2(OAc)_4$. As an example, it has been reported that the conversion of the isopimaradiene system (**50**) into the corresponding steroid skeleton (**51**) could be achieved in approx. 60% yield with $Rh_2(OAc)_4$ while no reaction was observed with $CuSO_4$ [65]. The opportunity offered by suitable catalysts such as $Rh_2(OAc)_4$ (**1**) in intramolecular C-H insertion was subsequently exploited with the introduction of more sophisticated catalysts, including chiral ones, in many

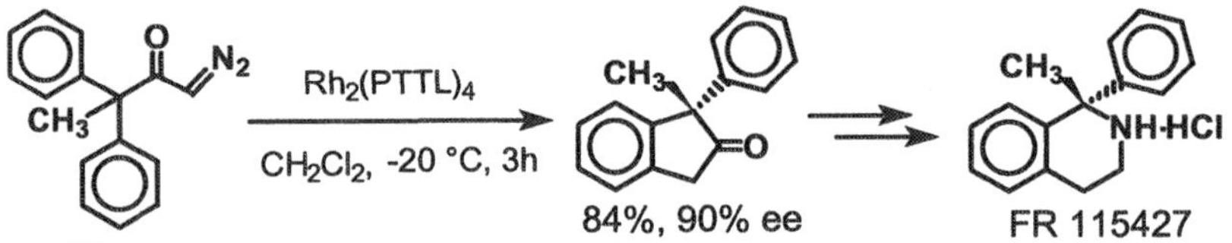


52

53

cases directed towards the synthesis of biologically active compounds. As a recent example, Watanabe et al [66] have described the enantioselective synthesis of FR-115427 (**53**) a non-competitive NMDA antagonist, where the key step is the enantioselective intramolecular aromatic C-H insertion of the diazoketone (**52**) catalyzed by the dirhodium (II) catalyst $Rh_2(PLLT)_4$(**32**).

## 2. Reactions of Diazocarbonyl Compounds with Aldehydes and Ketones

The thermal reaction of diazoacylmethanes acting as nucleophiles with aldehydes and ketones was first reported by Buchner in 1885 [67], intermittently studied and reported to be of limited synthetic value in view of its unselectivity [68,69]. In the 1970's a renewed interest in this reaction led to the discovery that α-diazocarbonyl compounds could undergo both base and Lewis acid catalyzed aldol type addition with ketonic substrates in an efficient and synthetically valuable way.

### 2.1. Utilization of the Lewis Acid Catalyzed Aldol-type Addition of α-Diazocarbonyl Compounds in Key Steps of the Synthesis of New NMDA Receptor Complex Modulators

An important contribution to the development of synthetically useful Lewis acid-promoted reactions came by Liu et al. [70] that in 1975 reported that the addition of ethyl diazoacetate to cyclic ketones in the presence of $BF_3{\cdot}Et_2O$ affords in good yields the corresponding homologated β-ketoesters. This reaction, which proceeds with loss of nitrogen, has been employed by our group for the preparation in three steps of (2*S*)- and (2*R*)-4-ketopipecolates (**56a**) and (**56b**) starting from protected (2*S*)- and (2*R*)-4-ketoprolines (**55a**) and (**55b**), respectively. Reduction and deprotection of **56a** affords (2*S*,4*S*)-pipecolic acid, an amino acid isolated from Acacia and Armenia species, that was submitted to sulfation to give *trans*-4-hydroxy-*S*-pipecolic acid-4-sulfate (*t*-HPIS, **57**) an amino acid extracted from the seeds of *Peltophorum Africanum*, representing the first example of a naturally occurring sulfate ester of a non-proteinogenic amino acid [71].

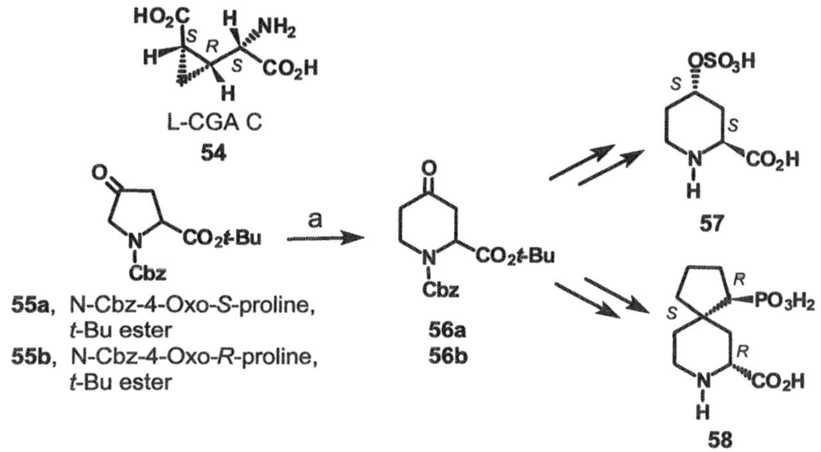


a) i. $N_2CHCO_2Et$, $BF_3.Et_2O$; ii. NaCl, DMSO, 140 °C; iii. flash chrom.

When tested as an acidic amino acid ligand, *t*-HPIS (**57**) was shown to be a selective and potent NMDA receptor site agonist (Ki 0.21 μM) against specific L-($^3$H)glutamate binding. In a subsequent study [72], the activity of *t*-HPIS was confronted with that of (2*S*,3*R*,4*S*)-cyclopropylglutamate (L-CGA C, **54**), another potent NMDA agonist first reported by us in 1986 [73] and 1988 [74]. The most significant finding of this study was that these two agonists behave differently when tested on the contraction of the longitudinal muscle-myenteric plexus preparation or on the depolarization of the mouse cortical wedges, with L-CGA C being almost three orders of magnitude more potent than NMDA on the first assay, and not significantly different from NMDA on the second. On the contrary, *t*-HPIS is 10 times more potent than NMDA on the depolarization of mouse cortical wedges. On the basis of these results, NMDA receptors have been hypotesized to exist into at least two functional groups: the first one preferentially stimulated by *t*-HPIS and particularly abundant in the mouse cortex and the second one, selectively stimulated by L-CGA C and particularly abundant in the guinea pig myenteric plexus.

In another work [75], we have utilized the (2*R*)-4-oxopipecolate [**56b**) to prepare two new spirobicyclophosphonate isomers as conformationally constrained analogs of the potent competitive NMDA antagonist CGS 19755. The preliminary biological evaluation of these compounds has revealed that only the (1*R*,5*S*,7*R*)-isomer (**58**) characterized by the (-)-gauche disposition around the (C1-C5)-bond is active as NMDA receptor competitive antagonist, a result that confirm previously reported pharmacophoric models.

## 2.2. Base-catalyzed Formation of α-Diazo-β-hydroxyacylmethanes by Aldol-type Condensation with Carbonyl Substrates. Synthesis of β-Dicarbonyl Compounds and β-Hydroxy Esters

The report in 1970 [76] of the successful preparation at –78 °C of ethyl diazo(lithio)acetate (LiEDA) and the subsequent report of its aldol-type condensation with aldehydes and ketones with retention of the diazo function to give the corresponding α-diazo-β-hydroxy ester has given new

interest to the study of these reactions. In our first work in this area [77a-i], treatment with LiEDA of thiochroman-4-one and isothiochroman-4-one afforded the corresponding α-diazo-β-hydroxy esters readily trasformed into the corresponding tetrahydrobenzothiepin-β-oxo esters in moderate yield by treatment with hydrochloric acid [77a]. Analogously, starting from chroman-4-one and isochroman-4-one this two steps sequence afforded the two corresponding ethyl-4-oxo-tetrahydrobenzoxepin-5-carboxylates [77c].
In 1979, as a major improvement over the rearrangement in the presence of HCl or by vacuun thermolysis, we reported that α-diazo-β-hydroxyacyl derivatives (Scheme 3, **59**) are rearranged to the corresponding β-ketoacyl derivatives (**60**) with catalysis by dirhodium(II)tetraacetate (Scheme 3, route A) at room temperature, in nearly quantitative yield.

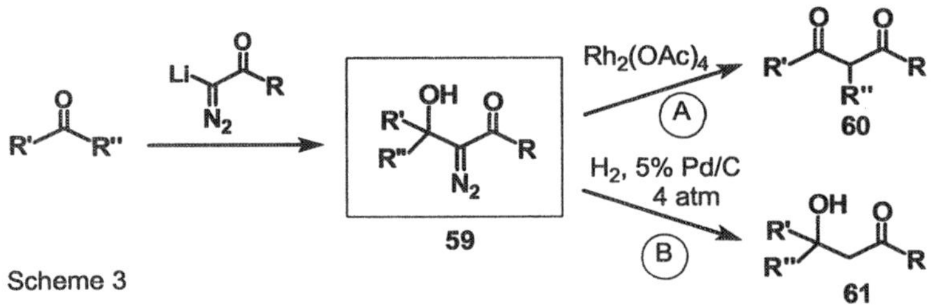


Scheme 3

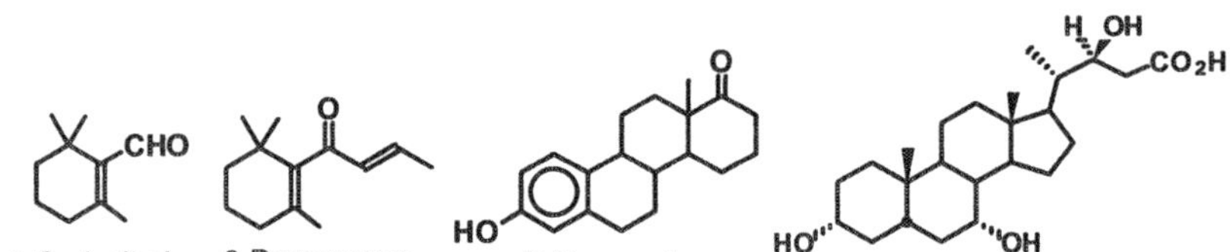

β-Cyclocitral **62** β-Damascone **63** D-Homoestrone **64** Haemulcholic Acid **65**

This methodology was subsequently utilized by us for the synthesis of β-damascone (**63**) starting from β-cyclocitral (**62**) [77f] and for the preparation of the D-homoestrone (**64**) [77i]. As an extension of the range of useful applications of α-diazo-β-hydroxy esters (**59**, R=OEt), we reported in 1985 that these compounds can be reduced with hydrogen (4 atm.) in the presence of 5% Pd/C to give the corresponding β-hydroxy esters (Scheme 3, route B) [77h]. This sequence was applied to a new synthesis of haemulcholic acid (**65**), a natural (22*S*)-22-hydroxy bile acid isolated from marine species, starting from chenodeoxycholic acid.

### 2.3. Reactions of α-Diazo-β-hydroxyacylmethanes with $BF_3{\cdot}Et_2O$

Treatment of β-hydroxyacylmethanes (**59**) derived from aldehydes (Scheme 4, R"=H) with $BF_3{\cdot}Et_2O$ in a polar solvent such as acetonitrile, afforded the corresponding acylacetylenic derivatives (**66**) in good yield [77d].

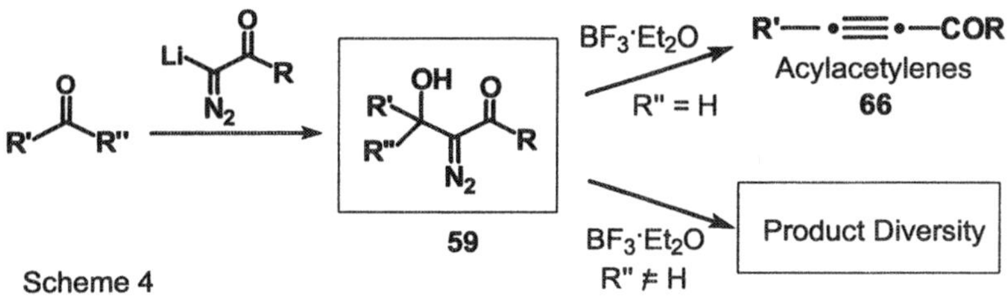


Scheme 4

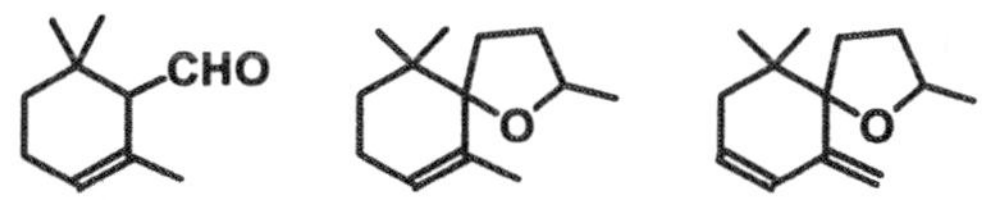


α-Cyclocitral **67** Theaspiranes **68** Vetispiranes **69**

This reaction, which involves the evolution of molecular nitrogen with neighbouring group participation, was applied to the synthesis of physiologically active terpenes such as theaspiranes (**68**) and vetispiranes (**69**) starting from α-cyclocitral (**67**).

## 2.4. Reaction of α-Diazo-β-hydroxyesters Derived from Ketones with $BF_3{\cdot}Et_2O$. Generation and Rearrangement of Vinyl Cations

The interesting results obtained in the $BF_3{\cdot}Et_2O$ catalyzed rearrangement of α-diazo-β-hydroxyacyl methanes obtained from aldehydes (see above) has prompted us to investigate the behavior towards $BF_3{\cdot}Et_2O$ of α-diazo-β-hydroxycarbonyl compounds (**59**) obtained from ketonic substrates (Scheme 4, R"≠H), and therefore missing the hydrogen atom α to the diazo moiety, crucial for the conversion **59** —> **66**. The reaction, first performed on α-diazoester (**70**) derived by treating cyclohexanone with ethyl lithio-diazoacetate, was shown to afford an unusual array of products (see Scheme 5) strongly dependent on ring size and solvent used [78].

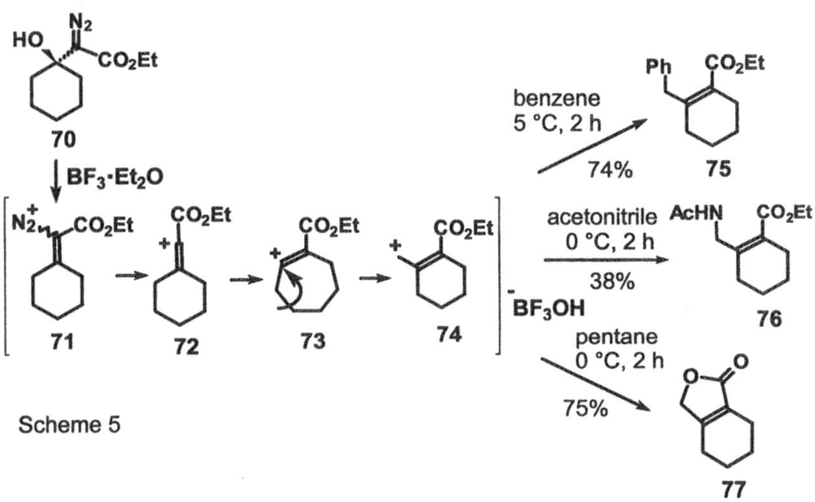


Scheme 5

The reaction proceeds by Lewis acid complexation of the alcohol functionality of the diazoester with $BF_3{\cdot}Et_2O$ followed by neighboring group participation of the diazo moiety to generate a cycloalkylidene diazonium salt (**71**) which is then transformed into a highly reactive, destabilized linear vinyl cation (**72**). Ring expansion *via* a 1,2-methylene shift leads then to the formation of an unstable, bent cycloalkenyl vinyl cation (**73**) which undergoes again a 1,2-methylene shift to give a more stable allylic cation (**74**) [78b]. This cation is either trapped by the solvent to give compounds such as (**75**) or (**76**) or else undergoes cyclization with the adjacent ester group to give a lactone (**77**). The diversity of products arising from the $BF_3{\cdot}Et_2O$ promoted transformations of α-diazo-β-hydroxy esters in various solvents can be of synthetic value. An expansion of the scope and applications of this reaction is currently under way.

## REFERENCES


1. M.P. Doyle, Aldrichimica Acta, 29(1) (1996) 3.
2. T. Ye and M.A. McKervey, Chem. Rev., 94 (4) (1994) 1091.
3. A. Padwa and D.J. Austin, Angew. Chem. Int. Ed. Engl., 33 (1994) 1797.
4. H.M.L. Davies, Tetrahedron, 44(24) (1993) 5203.
5. A. Padwa and K.E. Krumpe, Tetrahedron, 48(26) (1992) 5385.
6. J. Adams and D.M. Spero, Tetrahedron, 47 (10/11) (1991) 1765.
7. M.P. Doyle, Rec. Trav. Pays Bas, 110 (1991) 305.
8. A. Padwa and S.F. Hornbuckle, Chem. Rev., 91(3) (1991) 263.
9. A. Padwa, Acc. Chem. Res., 24(1) (1991) 22.
10. G. Maas, Topics in Current Chemistry, 137 (1986) 77.
11. M.P. Doyle, Chem. Rev., 86 (1986) 919.
12. E. Wenkert, Heterocycles, 14(10) (1980) 1703.
13. E. Wenkert, Acc. Chem. Res., 13 (1980) 27.
14. W. Ando, Acc. Chem. Res., 10 (1977) 179.
15. A.P. Marchand and N. MacBrockway, Chem. Rev., 74(4) (1974) 431.
16. F. Weygand and N.J. Bestmann, Syntheses Using Diazoketones, in W. Foerst, Newer Methods of Preparative Organic Chemistry, Academic Press, N.Y., London, (1964) 451.
17. K.G. Rasmussen and K.A. Jørgensen, J. Chem. Soc., Chem. Commun., (1995) 1401.
18. For recent application to the total synthesis of Natural Products, see: a) A. Padwa and A.T. Price, J. Org: Chem., 60 (1995) 6258; b) H.L. Davies, J.J. Matasi, L.M. Hodges, N.J.S. Huby, C. Thornley, N. Kong and J.H. Houser, J. Org. Chem., 62 (1997) 1095; c) A. Padwa, M.A.Brodney, J.P. Marino and S.M. Sheeban, J. Org. Chem., 62 (1997) 78; d) W.G. Dauben, J. Dinges and T.C. Smith, J. Org. Chem., 58 (1993) 7635.
19. H.P. Nestler, P.A. Bartlett and W. Clark Still, J. Org. Chem., 59 (1994) 4723.
20. T. Curtius, Berichte, 16 (1883) 2230.
21. L. Wolff, Liebigs Ann. Chem., 325 (1902) 144.

22. F. Arndt and B. Eistert, Ber., 68 (1935) 200.
23. R. Casanova and T. Reichstein, Helv. Chim. Acta, 33 (1950) 417.
24. P. Yates, J. Am. Chem. Soc., 74 (1952) 5376.
25. H. Nozaki, S. Moriuti, H. Takaya and R. Nogori, Tetrahedron Lett. (1966) 5239
26. W. R. Moser, J. Am. Chem. Soc., 91 (1969) 1141.
27. R.G. Solomon and J. K. Kochi. J. Am. Chem. Soc., 95 (1973) 3300.
28. R. Paulissen, E. Reimlinger, E. Hayez, A.J. Hubert and P. Teyssié. Tetrahedron Lett. (1973) 2233.
29. a) J.G. Norman, Jr. and H.J. Kolari, Inorg. Chem., 100 (1978) 791; b) J.G. Norman, G.E. Renzoni and D.A. Case, J. Am. Chem. Soc., 101 (1979) 5256; c) B.E. Bursten and F.A. Cotton, Inorg. Chem., 20 (1981) 3042.
30. A. Padwa, D.J. Austin, A.T. Price, M.A. Semones, M.P. Doyle, M.N. Protopopova, W.R. Winchester and A. Tran, J. Am. Chem. Soc., 115 (1993) 8669.
31. R. Pellicciari, unpublished results.
32. a) H.W. Kroto, J.R. Heat, S.C. O'Brien, R.F. Curl and R.E. Smalley, Nature, 318 (1985) 162; b) W. Krätschmer, L.D. Lamb, K. Fostiropoulus and D.R. Huffman, Nature, 347 (1990) 354.
33. A.W. Jensen, S.R. Wilson and D.I. Schuster, Bioorg. Med. Chem., 6 (1996) 767.
34. (a) S.H. Friedman, D.L. DeCamp, R.P. Sijbesma, G. Srdanov, F. Wudl and G.L. Kenyon, J. Am. Chem. Soc., 115 (1993) 6506; (b) R.P. Sijbesma, G. Srdanov, F. Wudl, J.A. Castoro, C. Wilkins, S.H. Friedman, D.L. DeCamp and G.L. Kenyon, J. Am. Chem. Soc., 115 (1993) 6510.
35. H. Tokuyama, S. Yamago, E. Nakamura, T. Shiraki and Y. Sugiura, J. Am. Chem. Soc., 115 (1993) 7918.
36. L.L. Dugan, D.M. Turetsky, C. Du, T.T. Lin, D. Lobner, R. Almli, M. Wheeler, D.W. Choi, Abstract of 26th ACS Annual Meeting; Washinghton, DC, 3 (1996) 2142.
37. T. Suzuki, Q. Li, K.C. Khemani, F. Wudl and Ö. Almarsson, Science, 254 (1991) 1186.
38. A.B. Smith, III, R.M. Strongin, L. Brard, G.T. Furst, W.J. Romanow, K.G. Owens and R.C. King, J. Am. Chem. Soc., 115 (1993) 5829.
39. T. Suzuki, Q. Li, K.C. Khemani and F. Wudl, J. Am. Chem. Soc., 114 (1992) 7301.
40. a) F. Wudl, Acc. Chem. Res., 25 (1992) 157; b) L. Isaacs, A. Whersig and F. Diederich, Helv. Chim. Acta, 76 (1993) 1231.
41. A. Skiebe and A. Hirsch, J. Chem. Soc., Chem. Commun., (1994) 334.
42. H.J. Bestmann, C. Moll and C. Bingel, Synlett, (1996) 729.
43. F. Diederich, J. Chem. Soc., Perkin Trans. I, 2 (1992) 391.
44. For related examples on the formation of a metal-cyclobutane see: (a) E. Wenkert, M. Guo, F. Pizzo and K. Ramachandran, Helv. Chim. Acta, 70 (1987) 1429. (b) C. A. Merlic and H. D. Bendorf, Tetrahedron Lett. 35 (1994) 9529
45. T. Aretani, Y. Yoneyoshi and T.Nagase, Tetrahedron Lett., (1977) 2599.

46. T. Aretani, Pure Appl. Chem., 57 (1985) 1839.
47. F. Hugo, U. Leutnegger, K. Siegman, A.Pfaltz, W. Keller and C. Kratky, Helv. Chim. Acta, 75 (1988) 1541.
48. R.E. Lowenthal, A. Abiko and S. Masamune, Tetrahedron Lett., 31 (1990) 6005.
49. D.A. Evans, K.A. Woerpel and M.M. Himman, J. Am. Chem. Soc. 113 (1991) 726.
50. A. Brunner, H. Kluschanozof and K. Wutz., Bull. Soc. Chim. Belg., 98 (1989) 63.
51. (a) M. Kennedy and M.A. McKervey. J. Chem. Soc., Chem. Commun., (1988) 1028; (b) M. Kennedy, M.A. McKervey, A.R. Maguire and G.H.P. Roos, J. Chem. Soc., Chem. Commun., (1990) 361.
52. M. P. Doyle, B. D. Brandes, A. P. Kazala, R.J. Pieters, M.B. Jarstfer, L. M. Watkins and C. T. Eagle, Tetrahedron Lett., 31 (1990) 6613.
53. For compounds **27**-**30**, see: M.P. Doyle, M.N. Protopopova, Q-L. Zu and J.W. Bode, J. Org. Chem., 60 (1995) 6654; for cpd. **31**-**32**, see ref 67; for cpd. **33**, see: P. Müller, C. Baud, D. Ené and S. Notallebi, Helv. Chim. Acta, 78 (1995) 459; for cpd. **34**, see: S. O'Malley T. Kodadek, Organometallics, (1992) 2299.
54. H. Nishiyama, Y. Itoh, Y. Suguwara, H. Matsumoto, K. Aoki and K. Itoh, Bull. Chem. Soc. Jpn., 68 (1995) 1247.
55. E.J. Corey and T.G. Grant, Tetrahedron Lett., 35 (1994) 5373.
56. H.M.L. Davies and D.K. Hutcheson, Tetrahedron Lett., (1993) 7243.
57. M.P. Doyle, R.J. Pieters, S.F. Martin, R.E. Austin, C.J. Oalmann and P. Müller, J. Am. Chem. Soc., 113 (1991) 1423.
58. S.F. Martin, R.E. Austin, C.J. Oalmann, W.R. Baker, S.L. Condon, E. Delava, S.H. Rosenberg, K.P. Spina, H.H. Stein, J. Cohen and H.D. Kleinert, J. Med. Chem., 35 (1992) 1710.
59. S.F. Martin, C.J. Oalmann and S. Liras, Tetrahedron, 49 (1993) 3521.
60. D.H. Rogers, E.C.Yi andC.D. Poulter, J. Org. Chem., 60 (1995) 941.
61. D. Taber, J. Org. Chem., 62 (1997) 194.
62. J.-P. Pin and R. Duvoisin, Neuropharmacology 34 (1995) 1.
63. R. Pellicciari, M. Marinozzi, B. Natalini, G. Costantino, R. Luneia, G. Giorgi, F. Moroni, C. Thomsen. J. Med. Chem. 39 (1996) 2259.
64. M. Marinozzi, B. Natalini, G. Costantino, P. Tijskens, C. Thomsen, R. Pellicciari. Bioorg. Med. Chem. Lett. 6 (1996) 2243.
65. E. Wenkert, L. L. Davis, B. L. Mylari, M. F. Solomon, R, R, da Silva, S. Shulman, R. J. Warnet, P. Ceccherelli, M. Curini, R. Pellicciari. J. Org. Chem. 47 (1982) 3242.
66. N. Watanabe, T. Ogawa, Y. Ohtake, S. Igekami, S. Hashimoto. Synlett (1996) 85.
67. E. Buchner and T. Curtius, Berichte, 18 (1885) 2371.
68. F. Schlotterbeck, Berichte, 42 (1909) 2565.
69. C.D. Gutsche and M. Hillman, J. Am. Chem. Soc., 76 (1945) 2236.
70. H.J. Liu and S.P. Majumdar, Synth. Commun., 5 (1975) 125.
71. R. Pellicciari, B, Natalini, R. Luneia, M. Marinozzi, M. Roberti, G. Rosato, B.Sadeghpour, J. Snyder , J. Monahan and F. Moroni, Med. Chem. Res., 2 (1992) 491.

72. F. Moroni, A Galli, G. Mannaioni, V. Carlà, A. Cozzi, F. Mori, M. Marinozzi and R. Pellicciari, Naunym-Schmildeberg's Arch. Pharmacol., 351 (1995) 371.
73. R. Pellicciari, M. Curini, B. Natalini and P. Ceccherelli, "Preparation and Properties of Cyclopropyl Glutamate Analogs". IX International Symposium on Medicinal Chemistry, Berlin (West Germany), (1986) 118
74. R. Pellicciari, B. Natalini, M. Marinozzi, L. Selvi, A. Cordi, J.B. Monahan, T.H. Lanthorn and J.P. Snyder, "Conformationally Restricted Glutamates As New Potent Agonists of the NMDA Receptor", Abstracts of The International Symposium Excitatory Amino Acids '88, Manaus (Brazil), (1988) 26.
75. R. Pellicciari, M. Marinozzi, B. Natalini, G. Costantino, D. Lankin, J.P. Snyder and J.B. Monahan, Il Farmaco, in press.
76. U. Schöllkopf and H. Frasnelli, Hangew. Chem. Int. Ed., 82 (1970) 291.
77. For our works in the area see: a) R. Pellicciari and B. Natalini, J. Chem. Soc., Perkin Trans. I, (1977) 1822; b) R. Pellicciari, E. Castagnino and S. Corsano, J. Chem. Res., (1979) 76; c) R. Pellicciari, B. Natalini, M.Taddei, A. Ricci. G. Alunni Bistocchi and G. DeMeo, J.Chem. Res., (1979), 142; d) R. Pellicciari, E. Castagnino, R. Fringuelli and S. Corsano, Tetrahedron Lett., 5 (1979) 481; e) R. Pellicciari, R. Fringuelli, P. Ceccherelli and E. Sisani, J. Chem. Soc., Chem. Commun., (1979) 959; f) R. Pellicciari, R. Fringuelli and E. Sisani, Tetrahedron Lett., 21(41) (1980) 4039; g) R. Pellicciari, R. Fringuelli, M. Curini and E. Sisani, J. Chem. Soc., Perkin Trans. I, (1981) 2566; h) R. Pellicciari, B. Natalini, S. Cecchetti and R. Fringuelli, J. Chem. Soc., Perkin Trans. I, (1985) 493; i) R. Pellicciari, B. Natalini and R. Fringuelli, Steroids, 49 (1987) 433.
78. a) R. Pellicciari, B. Natalini, B.M. Sadeghpour, G.C. Rosato and A. Ursini, J. Chem. Soc., Chem. Commun., (1993) 1798; b) R. Pellicciari, B. Natalini, B.M. Sadeghpour, M. Marinozzi, J.P. Snyder, B.L. Williamson, J.T. Kuethe, A. Padwa, J. Am. Chem. Soc., 118 (1996) 1.

H. van der Goot, (Editor)
TRENDS IN DRUG RESEARCH II



# Ligands for the 5-$HT_{2C}$ Receptor as potential Antidepressants and Anxiolytics

D. Leysen[a] and J. Kelder[b]

[a]Department of Medicinal Chemistry,
[b]Department of Computational Medicinal Chemistry,
Scientific Development Group, N.V. Organon, P.O. Box 20, 5340 BH Oss, The Netherlands

## 1. RATIONALE

The first generation antidepressants are plagued by serious side effects and the risk for suicide by depressed patients by overdosage. The second generation antidepressants have increased the likelihood of a clinical response with a reduction in unwanted toxicity. However, they brought their own particular pattern of adverse reactions. New agents with less side effects are needed which will act more rapidly and effectively [1,2].

A possible role for 5-hydroxytryptamine (5-HT, **1**) in the pathophysiology and treatment of depression is well documented in the literature [1-3]. By means of microdialysis studies, it can be clearly shown that the SSRI fluoxetine (**2**) and our recently introduced NaSSA mirtazapine (**3**), increase the synaptic levels of 5-HT. This increase induced by mirtazapine is mediated by blockade of the presynaptic $\alpha_2$-heteroreceptors located on the 5-HT neurones [4]. As a result of the increased synaptic levels of 5-HT by for instance the SSRIs, all 5-HT receptors present can be activated. Therefore all 5-HT receptors, especially those located post-synaptically, are potential candidates for the antidepressant effects of compounds such as **2**. As a consequence, agonists for these receptors are potential new antidepressants.

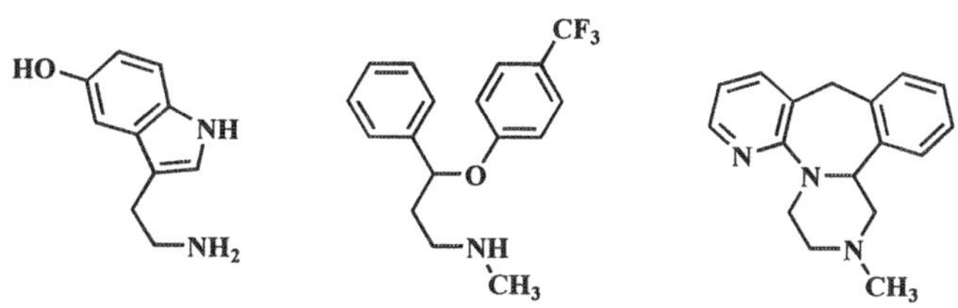


5-HT (**1**) Fluoxetine (**2**) Mirtazapine (**3**)

Which 5-HT receptor should be targeted? Progress in receptor pharmacology has reached a maximum in the field of 5-HT receptors. From the initial concept of just two functional sites [5], we have now reached a point where at least 14 distinct 5-HT receptor subtypes have been characterised : 5-$HT_{1A,\ 1D\alpha,\ 1D\beta,\ 1E,\ 1F}$, 5-$HT_{2A,\ 2B,\ 2C}$, 5-$HT_3$, 5-$HT_4$, 5-$HT_{5A,\ 5B}$, 5-

$HT_6$ and 5-$HT_7$. With the exception of the 5-$HT_3$ receptor, all these are G-protein coupled receptors.

We believe that the indirect activation of the 5-$HT_{2C}$ receptor by, for instance, SSRIs is the underlying cause for their AD-like effects, whereas the indirect activation of other 5-HT receptors leads towards potential side-effects such as nausea, headache and GI effects. Indeed, when we compare the *in vivo* effects of SSRIs with the effects of 5-HT receptor agonists, the uptake inhibitors and especially 5-$HT_{2C}$ agonists have a lot of effects in common. As indicated in Table 1, they have similar activities in mechanistic and other tests, whereas they have similar AD-like effects in animal models for depression [6-8].

Table 1
Similarities between the *in vivo* effects of SSRIs and 5-$HT_{2C}$ agonists

| AD-like effects | Mechanistic effects |
|---|---|
| • Restoration of bulbectomy induced passive avoidance acquisition deficit | • Induction of penile erections |
| • Reversal of chronic mild stress induced self-stimulation reduction | • Stimulus properties in conditioned taste aversion |
| • Inhibition of aversive brain stimulation escape | Others |
| • Selective inhibition of burying behaviour | • Appetite suppression |
| • Influence on sleep | • Induction of female rat lordosis |

## 2. TEST BATTERY

In order to identify and select selective 5-$HT_{2C}$ agonist, the test battery depicted in Table 2 was used. By using these tests we were able to measure the affinity, agonistic properties *in vitro* and *in vivo*, selectivity and *in vivo* activity in animal models for depression and anxiety in order to get some indication on the potential medical use of the compounds prepared.

Table 2
Test battery for selective 5-$HT_{2C}$ agonists

| | | |
|---|---|---|
| Mechanistic tests | | |
| • NIH-3T3 cells transfected with the human 5-$HT_{2C}$ receptor | | |
| • Displacement with [$^3$H]-5HT | ⇒ | Ki |
| • $IP_3$ formation | ⇒ | $EC_{50}$ and α |
| • Induction of penile erections in rats | ⇒ | *in vivo* activity |
| Selectivity profile | | |
| • Panel of approximately 100 different binding-sites | ⇒ | Selectivity |
| Behavioural models | ⇒ | Medical use |
| • Depression | | |
| • Anxiety | | |

## 3. TOPICS IN MED CHEM

### 3.1. Selectivity

It is not difficult to find 5-$HT_{2C}$ ligands, screening of existing chemical libraries will afford several hits. The challenge however is to obtain selectivity, especially 5-$HT_{2C}$ / 5-$HT_{2A}$. The reason is very obvious : there is a high degree of homology between these two receptors. Of all amino acids within the transmembrane regions 78% are identical. This is also reflected in the pharmacology of 5-$HT_2$ ligands. By looking at the affinities of compounds from various chemical and pharmacological classes with nM affinity for thc 5-$HT_2$ receptors, as listed for instance by J. Leysen [9], it is clear that several selective 5-$HT_{2A}$ ligands exist, whereas no selective 5-$HT_{2C}$ ligands were known in 1992.

Why do we explicitly want to get rid of the 5-$HT_{2A}$ affinity, when this activity is so difficult to avoid with 5-$HT_{2C}$ ligands? At least for the agonists there is a very clear reason. It is well documented in the literature that a high correlation exists between 5-$HT_{2A}$ receptor affinity and human hallucinogenic doses. No such correlation exists for the 5-$HT_{2C}$ receptor affinity. Therefore the critical issue for medicinal chemists is to find potent 5-$HT_{2C}$ agonists with low 5-$HT_{2A}$ activity.

The existence of 5-$HT_{2C}$ binding sites, at that time termed 5-$HT_{1C}$, has only recently been detected by using [$^3$H]-5-HT in pig choroid plexus and cortex [10]. Our knowledge about structure activity relationships concerning 5-$HT_{2C}$ agonists is very limited.

### 3.2. Molecular modifications

The major metabolite of the antidepressant trazodone (4), e.g. mCPP (5), is a known non-selective 5-$HT_{2C}$ agonist. Modification of the aromatic moiety afforded a series of 5-$HT_{2C}$ agonists, with varying selectivities and potencies. For the arylpiperazines a correlation was found between the lipofilicity, measured as logP, and the concentrations reached in the brain. From this series of compounds, Org 12962 (6) had an optimal logP value. It has been selected for further development based upon its potent *in vivo* activity.

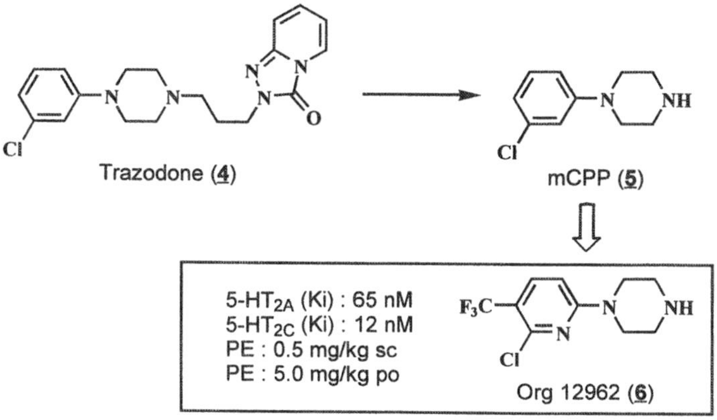

### 3.3. Substructure database searches / Molecular modifications

Between 1992 and 1995 we had a research collaboration with Hoffmann-La Roche on 5-$HT_{2C}$ ligands. The phenylpiperazine moiety present in mCPP (**5**) was used as a template to perform 2-D database searches. The tricyclic derivatives **7** were identified as weak 5-$HT_{2C}$ agonists. Ring opening afforded compounds **8** with increased potencies, whereas the 5-fluoro-6-chloro substitution pattern in compound **9** was optimal for *in vitro* potency and selectivity. Although being very active *in vitro*, these compounds lacked *in vivo* activity. It was found that this was due to metabolism of the primary amine by mono-amine oxidases. As it is known that α-branching disrupts the interaction with these enzymes, the α-methyl derivatives were prepared. Ro 60-0175 (Org 38491) will soon enter phase 2 clinical trials.

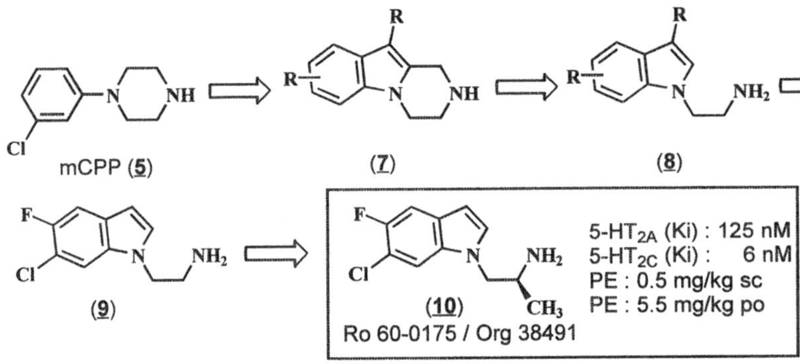


By searching the databases on the phenyl and piperazine rings as two isolated moieties, also the tertracyclic compounds **11** were found as hits. Again, ring-opening towards **12** increased the *in vitro* potency, whereas in this series the methoxy-substituent in compound **13** turned out to be optimal for selectivity. Also in this series of compounds, the introduction of an α-methyl substituent improved the *in vivo* activity. Unfortunately, the further development of Ro 60-0213 / Org 35032 (**14**), an agonist with a 100 to 150 fold selectivity for the 5-$HT_{2C}$ receptor, has been stopped due to toxicity.

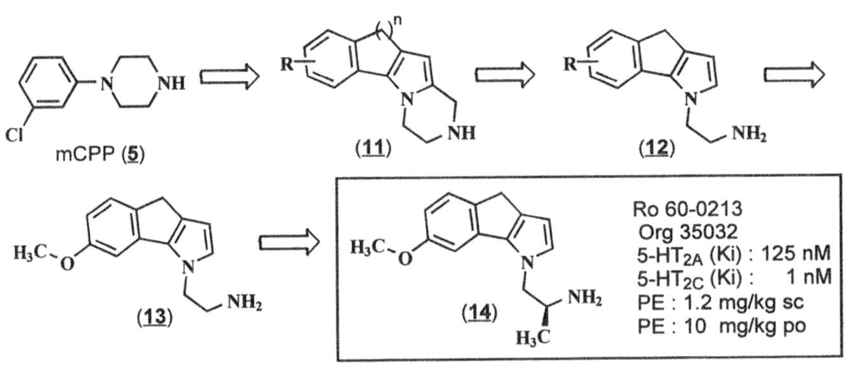

**3.4. Bio-isosterism**

At that time we had no new leads available at Organon, therefore we tried different techniques. The concept of bio-isosterism was applied. From the SAR obtained with the phenylpiperazines, it became clear that a protonated amine, an aromatic ring and a hetero-atom, with specific spatial orientations, are essential functional groups. These can be incorporated into amino-azetidines **15**.

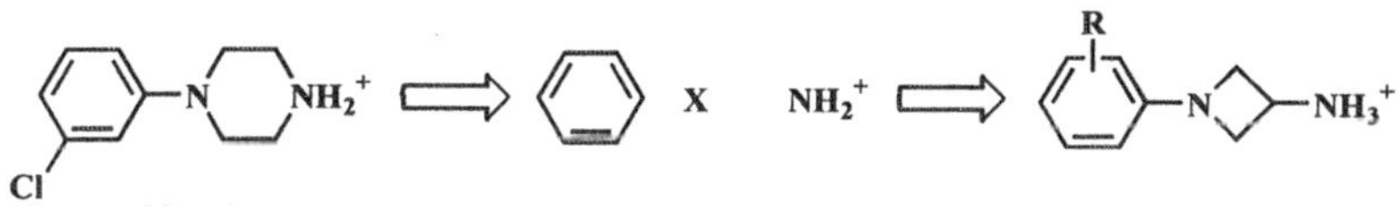

mCPP (**5**) (**15**)

From modelling studies with $N^1$-aryl-piperazines and *N*-aryl-3-amino-azetidines, it is clear that a nice fit can be obtained between the essential functional groups described above. However, by comparing the data obtained with compounds **16** and **17**, it can be concluded that the *N*-substituted-3-amino-azetidines are no bio-isosteres for the $N^1$-substituted-piperazines regarding 5-$HT_{2C}$ activity.

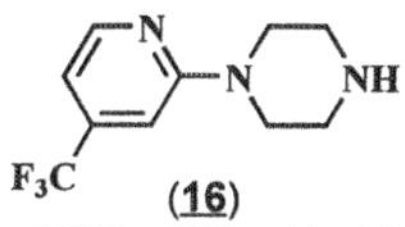

(**16**)
5-$HT_{2C}$ (Ki) : 40 nM

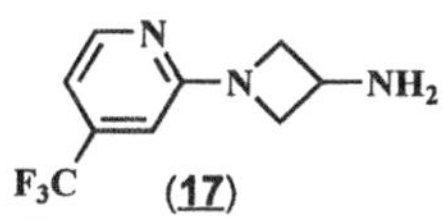

(**17**)
5-$HT_{2C}$ (Ki) : > 20.000 nM

As will be described later, not only the correct orientation of functional groups is essential, also the orientation of the hydrogen atoms of the protonated amine is of prime importance. Modelling studies reveal that the hydrogen atoms of azetidines **15** cannot adopt the orientation needed for activity. Therefore we examined the reversed azetidines **18**. Although the fit of the aromatic rings of **5** and **18** is less optimal, at least the hydrogen atoms can adopt the same orientation

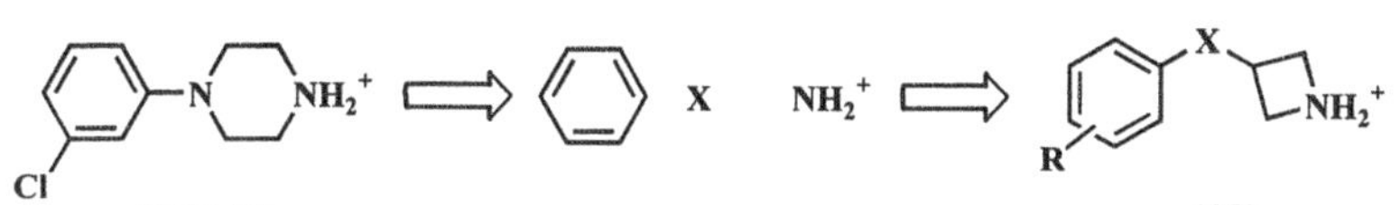

mCPP (**5**) (**18**)

A series of compounds were prepared.

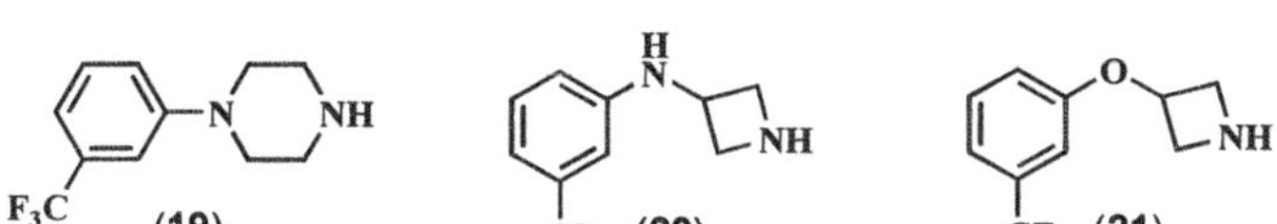

(**19**) 5-$HT_{2C}$ (Ki) : 60 nM

(**20**) 5-$HT_{2C}$ (Ki) : 200 nM

(**21**) 5-$HT_{2C}$ (Ki) : 300 nM

Although being somewhat less active as their piperazine counterparts, e.g. **19**, the azetidines **20** and **21** were active 5-HT$_{2C}$ ligands. As the oxyazetidines, such as **21**, had high selectivity 5-HT$_{2C}$/5-HT$_{2A}$, these were further optimised as 5-HT$_{2C}$ agonists. Introducing a second aromatic ring in **22** towards **23**, yielded compounds with drastically increased potencies. These derivatives **23** are also active as serotonin reuptake inhibitors, an activity which could be eliminated by saturating the annelated ring, e.g. in compounds **24**. Optimal selectivity could be obtained with the 2-methoxy-indanol part, affording Org 36262.

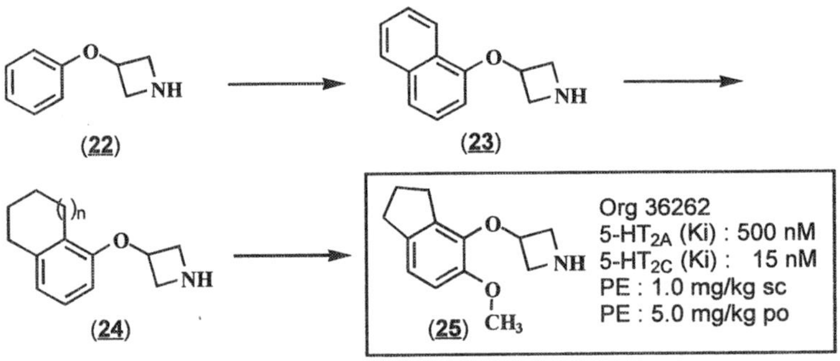


The synthesis of Org 36262 is depicted in the following scheme.

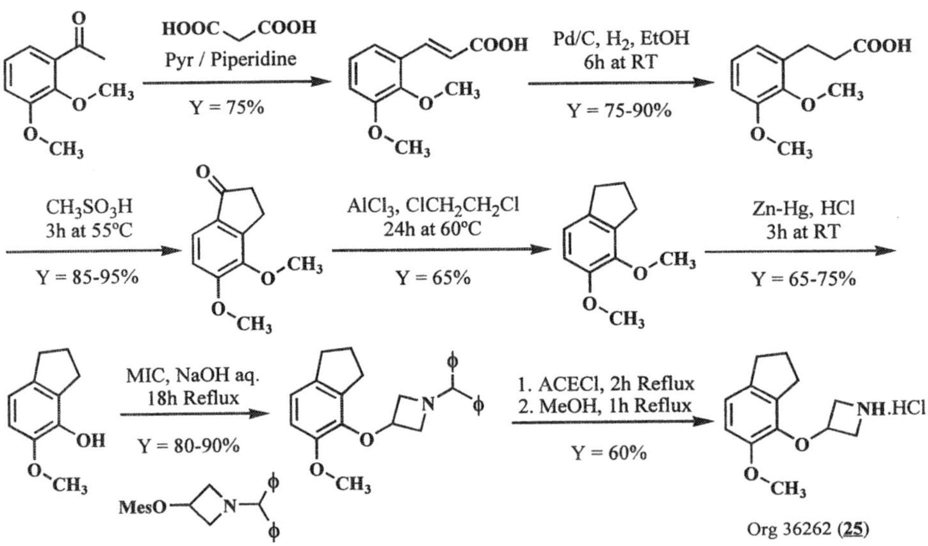


In a later stage of development it was observed that this compound induced skin reddening in test animals. When we examined compounds from different series with similar activities, we found that there might be some relation between this *in vivo* effect and the affinity for the 5-HT$_{1D\alpha}$ receptor. This receptor is also present in the coronary arteries. When tested at high concentrations, Org 36262 induced constriction of these arteries, an activity we obviously wanted to avoid. It was finally found that by replacing the azetidine ring by a pyrrolidine moiety, this unwanted effect could be eliminated. The preclinical development of Org 37684, e.g. the (S)-enantiomer **26**, will soon be completed.

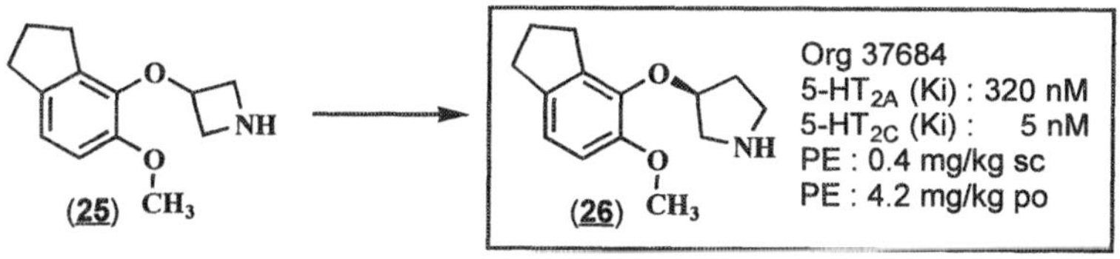


## 4. 3-D DATABASE SEARCHES

### 4.1. Approaches used

In order to find new 5-HT$_{2C}$ agonists we have identified the active conformation of the natural ligand 5-HT, which was used to perform 3-D database searches. This active conformation was determined by pharmacophore modelling and receptor docking.

Active conformation 5-HT ⇒ 3-D database search ⇒ lead finding

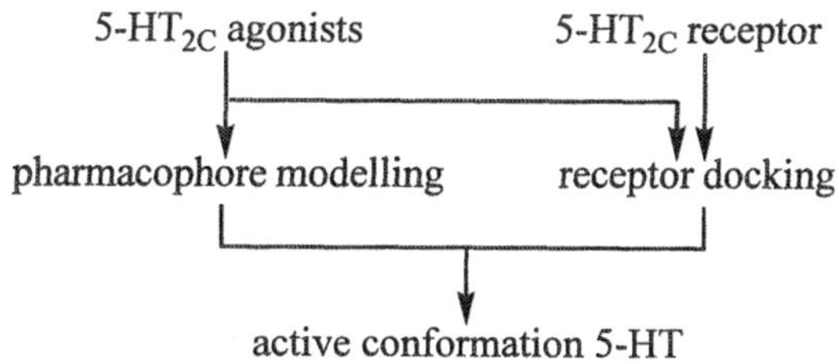


### 4.2. Pharmacophore modelling 5-HT$_{2C}$ agonists

For the construction of a pharmacophore model, we used the chiral reference compounds given in Table 3, as well as non-chiral ligands such as 5-HT and several phenylpiperazines.

The minimum energy conformations were determined by using X-ray analysis and MM287- and ToBaD calculations. These conformations are described by X- and Y-coordinates, which represent the distances between the centre of the aromatic nuclei and the counter-ion A$^-$ of the protonated amine. By doing so, we take into account the relative position of the amine function, as well as the orientation of the nitrogen-hydrogen bond.

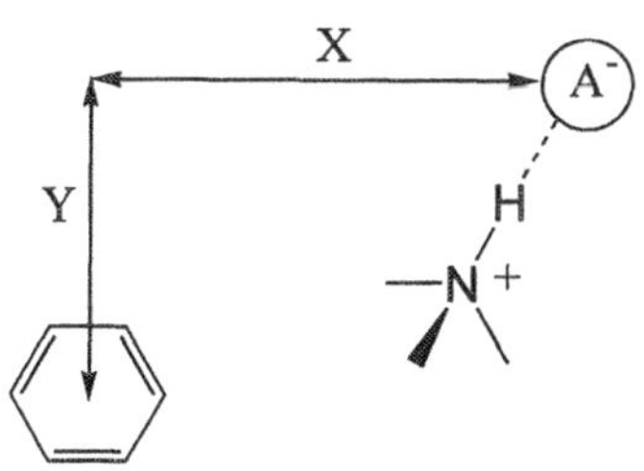

Table 3

Activities of chiral reference 5-$HT_{2C}$ agonists at 5-$HT_{2A}$ and 5-$HT_{2C}$ receptors

| Entry | Compound | | 5-$HT_{2A}$ Ki (nM) | 5-$HT_{2C}$ Ki (nM) | 5-$HT_{2C}$ selectivity |
|---|---|---|---|---|---|
| 1 | DOI 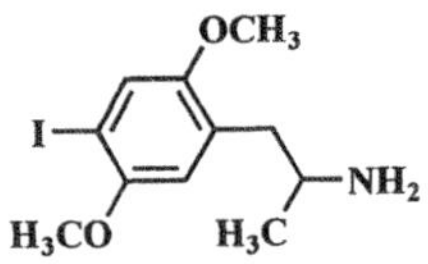 | S | 14 | 16 | 0.8 |
| | | R | 5 | 7 | 0.7 |
| | | ER | 2.8 | 2.3 | 0.8 |
| 2 | Pfizer [11] 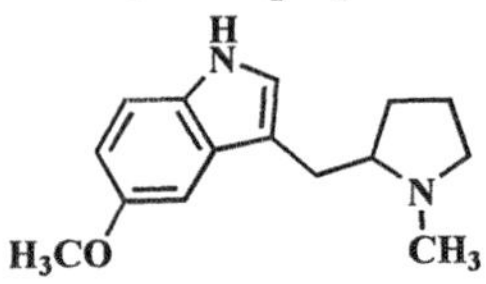 | S | 680 | 1,200 | 0.6 |
| | | R | 17 | 33 | 0.5 |
| | | ER | 40 | 36 | 0.9 |
| 3 | Normethyl SCH 23390/23388 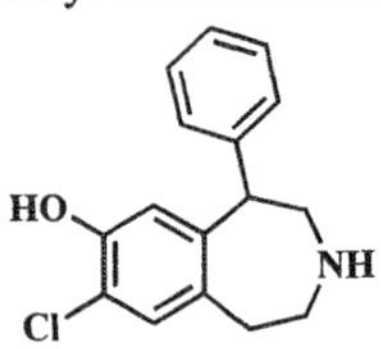 | S | 4,000 | 2,520 | 1.6 |
| | | R | 40 | 32 | 1.2 |
| | | ER | 100 | 80 | 0.8 |
| 4 | Flumexodol 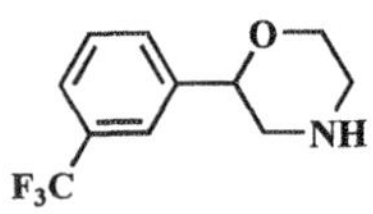 | (-) | 4,000 | 251 | 16 |
| | | (+) | 1,000 | 25 | 40 |
| | | ER | 4 | 10 | 2.5 |

A probability distribution was obtained, yielding two clusters of minimum energy conformations for all ligands used. The centres of these two clusters are : (7.2 Å, 0 Å) cluster I, (5.5 Å, 3.2 Å) cluster II. For the phenylpiperazines, such as mCPP (5), these two clusters indicate that the two hydrogens at the protonated amine are not equivalent at the chiral receptor site. Interaction can occur at the equatorial or at the axial hydrogen atom.

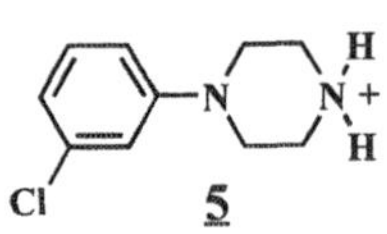

5

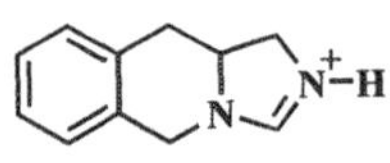

27

To find out which is the active cluster, we prepared the dihydro-imidazole **27**. This compound nicely fits upon the phenylpiperazines such as **5**. However, due to is $sp^2$ hybridisation, compound **27** can only interact via its pseudo-equatorial hydrogen. Compound **27** was totally inactive as a 5-$HT_{2C}$ agonist. Therefore we concluded that the 5-$HT_{2C}$ agonists must interact with the 5-$HT_{2C}$ receptor via their axial hydrogen atom, which represents cluster I. From this cluster the active conformation of 5-HT was deduced.

### 4.3. Receptor docking

The active conformation of 5-HT was also determined by receptor docking. In order to develop a model of the ligand binding site at the 5-$HT_{2C}$ receptor, a panel of single point mutant receptors was made. In the human 5-$HT_{2C}$ receptor we mutated D134, S182, S186, S219, F327 and F328 into alanines. A panel of various 5-$HT_{2C}$ ligands was tested on these mutant receptors. The data were analysed and translated into a model of the binding site for 5-$HT_{2C}$ agonists [12].

To check the validity of this theoretical model, several ligands were docked into it. The calculated interaction energies for these ligands correlates well with the experimentally determined 5-$HT_{2C}$ affinities. From the receptor docking experiments, the extended-away conformation of 5-HT (**1**) within the human 5-$HT_{2C}$ receptor was found. This conformation is essentially the same as the one obtained by pharmacophore modelling.

### 4.4. Results of 3-D database searches

The active conformation of 5-HT, as determined by the methods given in chapters **4.2.** and **4.3.**, was translated into 3-D parameters. A 3-D database search of our chemical database was performed and the hits were tested. The most interesting hit was the bridged benzocyclo-octeenamine **28**. As given in Table 4, the (-)-enantiomer, e.g. Org 8484, is active at 6 nM, having a 5-$HT_{2C}$/5-$HT_{2A}$ selectivity of 270. The absolute configuration of this eutomer has been determined as 5S,8**R**,9**S**.

Table 4
Activities of chiral 5-$HT_{2C}$ agonists at 5-$HT_{2A}$ and 5-$HT_{2C}$ receptors

| Compound | | 5-$HT_{2A}$ Ki (nM) | 5-$HT_{2C}$ Ki (nM) | 5-$HT_{2C}$ selectivity |
|---|---|---|---|---|
| 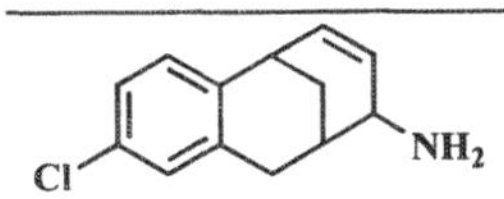 | (+) : Org 8483 | 6,300 | 630 | 10 |
| **28** | (-) : Org 8484 | 1,600 | 6 | 270 |
| | ER | 4 | 105 | 27 |

The eudismic ratios given in Table4 clearly illustrate that chirality has a profound effect upon the potency and selectivity of this type of 5-$HT_{2C}$ agonists.

## 5. *IN VITRO* ACTIVITIES

### 5.1. SAR aryloxyazetidines

The introduction of a 2-methoxy substituent in phenylpiperazine **29** and indanoxyazetidine **30** increased the affinity +/- 5-fold, e.g. compounds **31** and **25**. The introduction of a 3-chloro substituent increased the affinity in the phenylpiperazine series by a factor of 25, e.g. compound **32**, whereas in the indanoxyazetidines the affinity was reduced by a factor of 4, e.g. compound **33**. This apparent discrepancy can be explained by using the hydrofilic and lipofilic pockets, which according to the theoretical model of the human 5-$HT_{2C}$ receptor, are within the binding pocket of the agonists, e.g. S182/186/219 and F327/328 respectively.

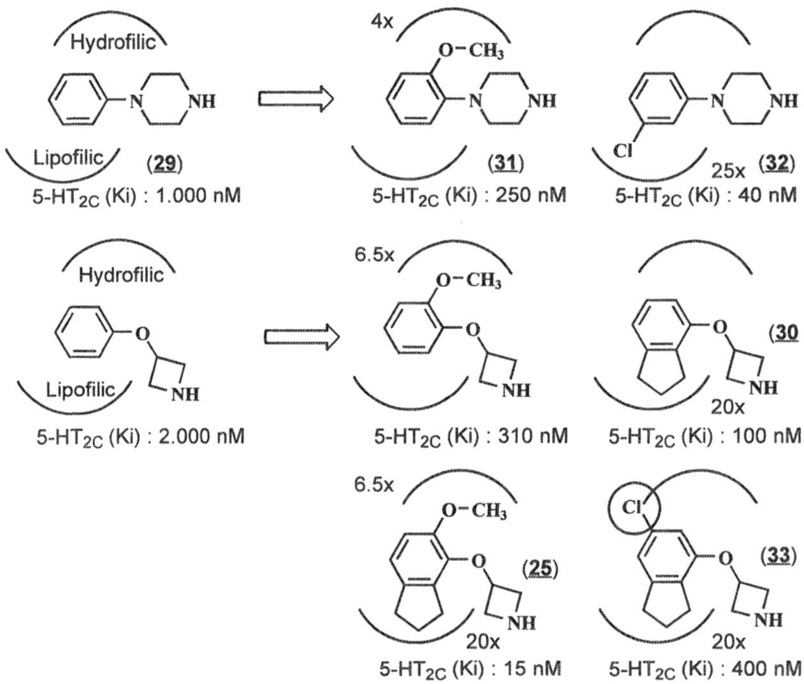


The data obtained with Org 36262 (**25**), clearly indicate that the impact of the appropriate hydrofilic and lipofilic substituents is additive. The decreased affinity of **33** is also explained.

### 5.2. Differences between binding- and activation sites, according to chemical class

Theoretically, when a compound uses the same binding sites for binding as for activating the receptor, there should be a 1:1 correlation between the pKi and $pEC_{50}$ values. By means of a regression analysis, the following relation was obtained for a set of 5-$HT_{2C}$ agonists :

$$pEC_{50} = 0.88 \times pKi + 0.66 \ (n = 28, R = 0.745)$$

By clustering the data points according to the chemical classes, two regression lines were obtained. Compounds belonging to class 1 cluster around a regression line, nearly equal to the theoretical 1:1 line. The regression line belonging to compounds of class 2 is shifted in a parallel fashion. The differences between these lines, at $pEC_{50}$ = 7.0, corresponds to 1.1 kcal/mole.

Class 1 : $pEC_{50} = 0.87 \times pKi + 0.97$ (n = 20, R = 0.82).
Class 2 : $pEC_{50} = 0.98 \times pKi - 0.67$ (n = 8, R = 0.991),

This could mean that compounds belonging to class 1 have equal pKi and $pEC_{50}$ values, e.g. the same sites are used for binding as for activation, whereas compounds belonging to class 2, for which the affinity is approximately 6-7 fold higher as the $EC_{50}$ for activation, use one extra site for binding, which has to be given up in order to activate the receptor. Further support for this hypothesis stems from the data obtained with 5-HT (**1**) and tryptamine (**34**). From the data given in Table 5, it is clear that 5-HT belongs to the class 2 compounds, whereas tryptamine is a member of class 1. Is there any molecular basis which can explain that 5-HT, in contrast to tryptamine, uses one site extra for binding? In our theoretical receptor model, the 5-OH function of 5-HT forms a hydrogen-bridge with S219 in helix V. We examined the effect of a point mutation where the serine residue at position 219 has been converted into an alanine residue, which cannot form a hydrogen bond.

Table 5
Activities of 5-HT and tryptamine at the wild type and A219 mutant $5\text{-}HT_{2C}$ receptor

| Ligand | Structure | $EC_{50}$ Wild type | Ki Wild type | Ki A219 mutant |
|---|---|---|---|---|
| 5-HT (**1**) | 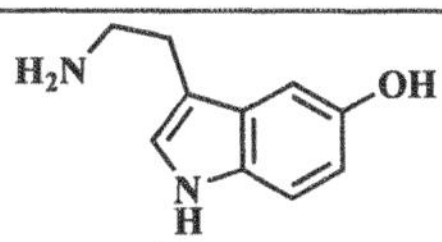 | 40 nM | 4 nM | 75 nM |
| Tryptamine (**34**) | 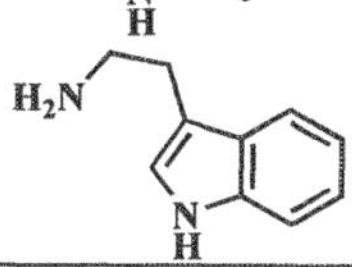 | 30 nM | 32 nM | 63 nM |

From the data given in Table 5 for the wild-type and the A219 mutant, it can be concluded that 5-HT, in contrast to tryptamine, indeed forms a hydrogen bridge with S219. The affinity of 5-HT for the A219 mutant is comparable to the $EC_{50}$ of 5-HT for the wild type and the affinity and $EC_{50}$ of tryptamine for both the wild type and the A219 mutant. Therefore the hydrogen bond of 5-HT with S219 in the wild type becomes broken in order to activate the receptor. For activation of the receptor we have indications that the hydrogen of S219 interacts with the $\pi$-cloud of the aromatic nuclei of the $5\text{-}HT_{2C}$ agonists, leading to the generally accepted conformational changes in the receptor.

## 6. *IN VIVO* ACTIVITY OF 5-$HT_{2C}$ AGONISTS IN ANIMAL MODELS

### 6.1. Animal models for depression

The compounds tested were active in the following models for depression :

- pharmaco-EEG,
- restoration of bulbectomy induced passive avoidance acquisition deficit,
- reversal of chronic mild stress induced self-stimulation reduction [13],
- differential reinforcement of low rates schedule, e.g. DRL 72.

### 6.2. Animal models for anxiety

6.2.1. General anxiety

The 5-$HT_{2C}$ agonist are inactive in general conflict tests. In sharp contrast to the SSRIs, the 5-$HT_{2C}$ agonists are not anxiogenic in the plus-maze model. The 5-$HT_{2C}$ agonists selectively inhibit the defensive burying behaviour.

6.2.2. Panic

Currently there is an increased clinical interest in the anti-panic effect of clomipramine. The 5-$HT_{2C}$ agonists are active in inhibiting the aversive brain stimulation escape.

6.2.3. Obsessive compulsive disorder (OCD)

Several antidepressants, such as the SSRIs, are active in OCD. Compared with the SSRIs, a greater potency and effect size was observed with the 5-$HT_{2C}$ agonists in following models :

- schedule-induced polydipsia paradigm,
- reduction of excessive scratching in monkeys.

## 7. CONCLUSIONS

By using different approaches, potent and selective 5-$HT_{2C}$ agonists were identified from different chemical series. Org 12962 and Ro 60-0175 (Org 38491) will soon enter phase 2 clinic, whereas the preclinical development of Org 37684 is almost completed. A theoretical model of the 5-$HT_{2C}$ receptor has proven to be useful for the identification of the active conformation of 5-HT, used to identify new leads. This model can describe the binding, as well as activation, of divergent agonists. As compared with fluoxetine, the 5-$HT_{2C}$ agonists have a better tolerability (not anxiogenic in rats and less nausea in monkeys) and a greater potency and effect size. There is no evidence of tolerance development. Future clinical trials need to clarify whether these compounds are indeed safe and effective antidepressants, with potential use in disorders such as panic and OCD.

## REFERENCES


1. D. Leysen and R. Pinder, Annu. Rep. Med. Chem., 29 (1994) 1.
2. C.L.E. Broekkamp, D. Leysen, B.W.M.M. Peetres and R.M. Pinder, J. Med. Chem., 38 (1995) 4615.

3. P. Blier, C. de Montigny and Y. Chaput, J. Clin. Psychiat., 51, Suppl. 4 (1990) 14.
4. T. de Boer, Int. Clin. Psychopharmacol., 10, Suppl. 4 (1995) 19.
5. J.H. Gaddumand and Z.P. Picarelli, Brit. J. Pharmacol., 12 (1957) 323.
6. H.H. Berendsen and C.L. Broekkamp, Eur. J. Pharmacol., 253 (1994) 83.
7. H.H. Berendsen, Pharmacol. Ther., 66 (1995) 17.
8. J.L. Moreau, F. Jenck, J.R. Martin, S. Perrin and W.E. Haefely, Psychopharmacol., 110 (1993) 140.
9. J. Leysen in Serotonin Receptor Subtypes : Pharmacological significance and Clinical Implications (Editors : S.Z. Langer, N. Brunello, C. Racagni and J. Mendlewics), Karger, Basel, 1992, 34-35.
10. A. Pazos, D. Hoyer and J.N. Palacios, Eur. J. Pharmacol., 106 (1984) 539.
11. J.E. Macor, J. Blake, C.B. Fox, C. Johnson, B.K. Koe, L.A. Lebel, J.M. Morrone, K. Ryan, A.W. Schmidt, D.W. Schulz and S.H. Zorn, J. Med. Chem., 35 (1992) 4503.
12. N.J. Stam, P. Vanderheyden, J. Kelder, T. de Boer, C. van Alebeek and D. Leysen, Soc. Neuroscience, 1159 (1994) 476.
13. J.L. Moreau, M. Bös, F. Jenck, J.R. Martin, P. Mortas and J. Wichmann, Eur. Neuropsychopharmacol., 6 (1996) 169.

H. van der Goot, (Editor)
TRENDS IN DRUG RESEARCH II



# 5-$HT_{1A}$- affinity, activity and selectivity *versus* $D_2$-receptors of flesinoxan and analogous *N*-Arylpiperazines†

Wilma Kuipers, Department of Medicinal Chemistry, Solvay Pharmaceuticals Research Laboratories, P.O. Box 900, 1380 DA Weesp, The Netherlands.

## 1. SUMMARY

5-$HT_{1A}$-receptor affinity of the agonist flesinoxan and its selectivity with respect to $D_2$ receptors were investigated. Effects of $N^4$-substitution are quite similar for 5-$HT_{1A}$- and $D_2$-receptor affinity, and are dominated by lipophilicity at a distance of four carbon atoms from the piperazine $N^4$-atom. The amide group of flesinoxan is unlikely to interact with the 5-$HT_{1A}$ receptor, and probably acts as a spacer. Selectivity for 5-$HT_{1A}$ *versus* $D_2$ receptors can be gained from the arylpiperazine substitution.

The bioactive conformation of flesinoxan at 5-$HT_{1A}$ receptors was studied by conformational analysis of a rigidized analog. The $N^4$-ethyl-(*p*-fluorobenzamide) substituent is probably directed *anti*-periplanar relative to the H(N4)-atom. Flesinoxan and two of its congeners were docked into a model of the 5-$HT_{1A}$ receptor in the putative bioactive conformation. Amino acid residues surrounding the $N^4$-ethyl-(*p*-fluorobenzamide) substituent are also present in $D_2$ receptors. In contrast, several residues that are in contact with the benzodioxane moiety, differ from those in $D_2$ receptors. These model-based observations agree with the 5-$HT_{1A}$ SAR data, and probably account for the selectivity of flesinoxan *versus* $D_2$ receptors.

## 2. INTRODUCTION

Flesinoxan (**1** Figure 1) is a selective agonist for 5-$HT_{1A}$ receptors.[1] In contrast, considerable affinity for dopamine $D_2$ receptors was reported for many other high-affinity $N^4$-substituted 5-$HT_{1A}$ arylpiperazines.[2] Flesinoxan, and a series of congeners were investigated to gain insight in requirements for 5-$HT_{1A}$ and $D_2$ receptor affinity. At first, structure-affinity relationships (SARs) were studied, and the influence of log P was investigated. Secondly, a conformational analysis of a rigidized flesinoxan analogue was used to study the bioactive conformation. This experiment was supported by the determination of its 3D crystal structure. Finally, SARs were rationalized in a modeling study, by docking flesinoxan and two of its congeners into a previously published 5-$HT_{1A}$-receptor model.[3]

---

†A detailed description of this study has been previously published in *J. Med. Chem.* **1997**, *40*, 300-312. Co-authors: CG Kruse, I v Wijngaarden, PJ Standaar, MThM Tulp, N Veldman, AL Spek, AP IJzerman.

**Figure 1.** *5-HT$_{1A}$ receptor ligands and model compounds 5 and 6.*

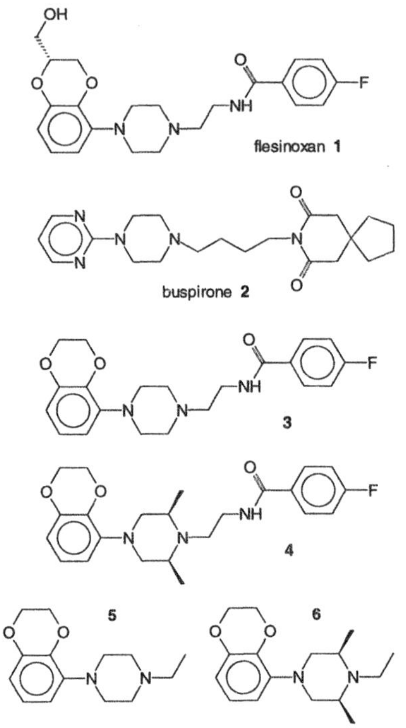


## 3. RESULTS AND DISCUSSION

### 3.1. 5-HT$_{1A}$ and D$_2$ receptors SARs

From Table 1 it appears that the lipophilic character of the phenyl ring of compound **8** is important for high affinities at 5-HT$_{1A}$ and dopamine D$_2$ receptors. Its removal in compound **7**, or replacement with more polar aromatic rings (as in **10-14**) reduces both 5-HT$_{1A}$ and D$_2$ receptor affinities. However, its replacement by 2-thiophene (as in **9**) or saturation to cyclohexane in **15**, has little effect on the affinity for both receptors. Lipophilicity may also play a role in the slight increase in affinities by replacement of the C=O of **15** in C=S in **16**. Indeed, replacement by the more polar C=NH in **17** also decreases both affinities.

Addition of a *p*-fluoro substituent (**18**) has no influence on 5-HT$_{1A}$ receptor affinity, and slightly increases D$_2$-receptor affinity. The amide moiety appears not to be important to either 5-HT$_{1A}$ or D$_2$ receptor affinities. The replacement of the NH of **18** with a methylene bridge in **19** has no effect. Further modification of the keto functionality of **19** to hydroxyl or ether in **20** and **24**, or even its complete removal in **21** and **22** also hardly affects either of the two affinities. Apparently, neither the electrostatic effect of the C=O group, nor the conformational effect of the amide function in **18** seem to influence the affinity for 5-HT$_{1A}$ or D$_2$ receptors. The presence of the ether oxygen in **23** gives a decrease in the affinities for both receptors. Apparently, an electronegative atom is unfavorable in this position.

None of the variations of the $N^4$ substituent in this series results in selective ligands.

**Table 1** *5-$HT_{1A}$- and $D_2$-receptor affinities*

| 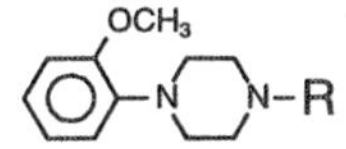 | | $K_i$ ± SEM (nM)[a] | |
|---|---|---|---|
| R | no. | 5-$HT_{1A}$ | $D_2$ |
| 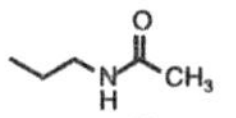 | **7** | 800 ± 70 | 1100 ± 40 |
| 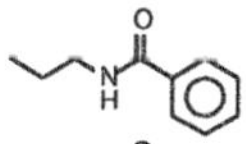 | **8** | 1.3 ± 0.3 | 13 ± 4 |
| 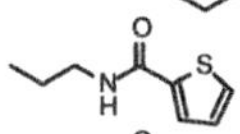 | **9** | 1.6 ± 0.3 | 17 ± 2 |
| 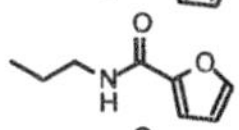 | **10** | 18 ± 4 | 73 ± 21 |
| 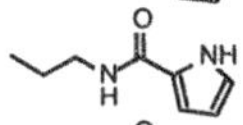 | **11** | 16 ± 3 | 64 ± 4 |
| 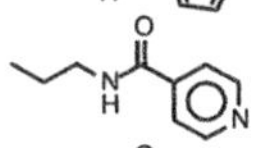 | **12** | 18 ± 1 | 230 ± 40 |
| 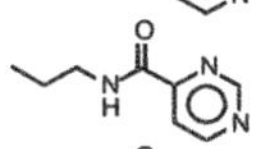 | **13** | 15 ± 4 | 250 ± 50 |
| 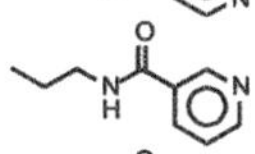 | **14** | 140 ± 40 | 570 ± 190 |
| 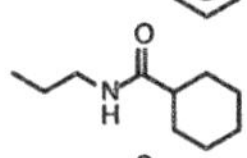 | **15** | 0.8 ± 0.3 | 23 ± 6 |
| 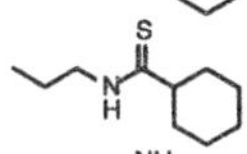 | **16** | 0.32 ± 0.04 | 6.4 ± 1.7 |
| 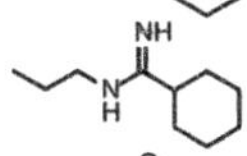 | **17** | 4.5 ± 0.9 | 130 ± 10 |
| 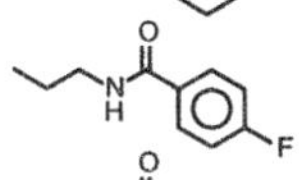 | **18** | 1.0 ± 0.3 | 5.0 ± 0.8 |
| 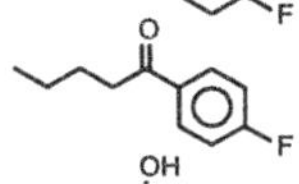 | **19** | 1.0 ± 0.2 | 3.1 ± 0.5 |
| 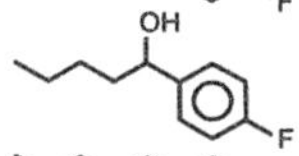 | **20** | 2.5 ± 0.2 | 3.4 ± 0.6 |
| 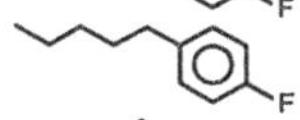 | **21** | 1.1 ± 0.3 | 3.3 ± 0.4 |
| 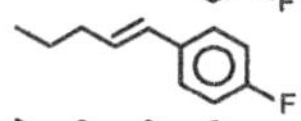 | **22** | 1.8 ± 0.2 | 4.1 ± 1.1 |
| 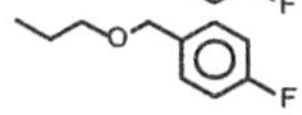 | **23** | 22 ± 5 | 10 ± 1 |

([a]) *$K_i$ values for the displacement of [$^3H$]-8-OH-DPAT from central 5-$HT_{1A}$-recognition sites in rat frontal cortex homogenates and of [$^3H$]-spiperone from $D_2$-binding sites in rat striatum. Values are based on n=3 determinations.*

Table 2 shows that differences between 5-HT$_{1A}$- and $D_2$-receptor affinities may arise from the aryl substitution pattern. Omitting the 2-methoxy group of compound **18**, giving **25**, enhances the selectivity for 5-HT$_{1A}$ receptors by a factor 4. Incorporation of the 2-methoxy group into a furan ring in **26** is highly favorable for 5-HT$_{1A}$-receptor affinity, whereas the affinity for $D_2$ receptors is not affected. The result is a significant increase in selectivity (ratio $D_2$/5-HT$_{1A}$ is 35). The benzodioxane **3** is slightly less potent than the benzofuran **26**, but is even more selective (ratio $D_2$/5-HT$_{1A}$ is 47). This selectivity is further increased to a factor 82 by the $CH_2OH$ group of flesinoxan. Thus, its selectivity seems to be largely due to the arylpiperazine aromatic ring system substitution rather than the $N^4$-substituent.

**Table 2** *5-HT$_{1A}$- and $D_2$-receptor affinities*

| 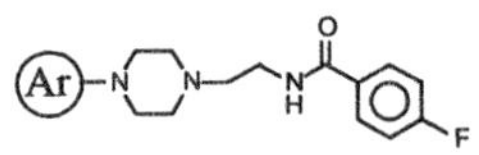 | | | | 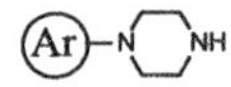 | | | |
|---|---|---|---|---|---|---|---|
| | | $K_i$ ± SEM (nM)[a] | | | | $K_i$ ± SEM (nM)[a] | |
| Ar | no. | 5-HT$_{1A}$ | $D_2$ | Ar | no. | 5-HT$_{1A}$ | $D_2$ |
| 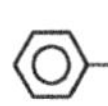 | **25** | 4.9 ± 1.5 | 92 ± 17 | 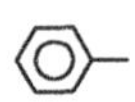 | **27** | 500 ± 20 | 6800 ± 600 |
| 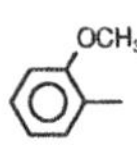 | **18** | 1.0 ± 0.3 | 5.0 ± 0.8 | 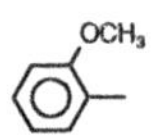 | **28** | 170 ± 20 | 1200 ± 40 |
| 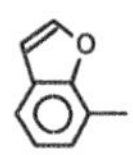 | **26** | 0.15 ± 0.01 | 5.3 ± 0.9 | 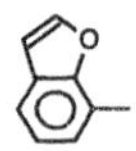 | **29** | 13 ± 2 | 780 ± 80 |
| 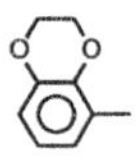 | **3** | 0.30 ± 0.04 | 14 ± 1 | 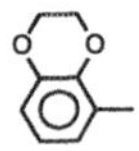 | **30** | 40 ± 5 | 1100 ± 90 |
| 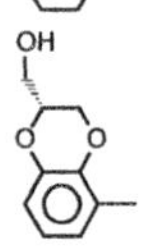 | **1** | 1.7 ± 0.2 | 140 ± 30 | 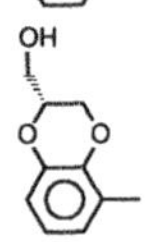 | **31** | 59 ± 11 | > 10,000 |

($^a$) $K_i$ values are based on $n$=3 determinations

## 3.2. Binding site and conformation analysis: 5-HT$_{1A}$ SARs

*$N^4$-substituted phenylpiperazines compared to nor-compounds.* Table 2 shows that the effects of substitution in the aromatic part of the arylpiperazine moiety are rather similar for the $N^4$-substituted arylpiperazines and the corresponding nor-compounds. Therefore, it seems likely that the arylpiperazine parts of these compounds address similar binding sites at 5-HT$_{1A}$ receptors. This hypothesis is corroborated by the fact that both nor-compounds **29** and eltoprazine **30**, as well as $N^4$-substituted arylpiperazines from this series, **3** and flesinoxan **1**, display agonist properties at 5-HT$_{1A}$ receptors.[1]

*Functional data of flesinoxan analogs.* Compound **3** and the *cis*-dimethyl substituted analog **4** are similarly potent 5-$HT_{1A}$ receptor agonists ($K_i$ = 0.3 and 1.0, and $EC_{50}$ = 0.90 and 0.78, respectively). Compound **3** is a full agonist like flesinoxan **1**; the intrinsic activity of **4** $\alpha$ = 0.86 ± 0.14. Apparently, the methyl groups do not affect the binding mode of **4** significantly, although these groups dramatically reduce the conformational freedom of its $N^4$-substituent. Therefore, **4** was used as a model compound for the study of the bioactive conformation of **1** and **3** as well.

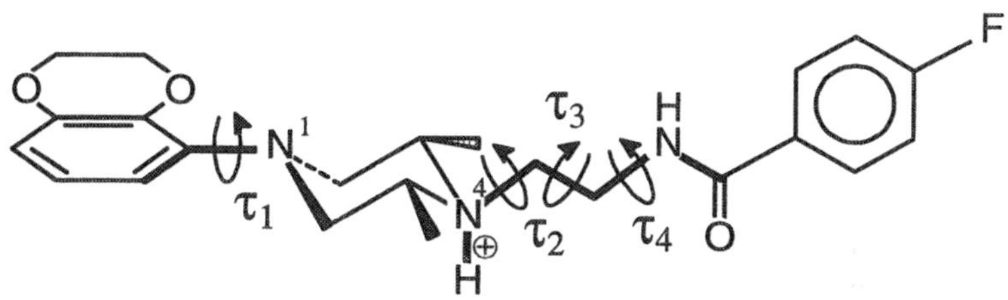


**Figure 2** *Definition of the torsion angles $\tau_1$, which determines the angle between the benzene and the piperazine ring, $\tau_2$, which primarily determines the direction of the $N^4$-substituent, $\tau_3$, and $\tau_4$. Defined angles are printed in bold.*

In order to rationalize SARs from the previous sections, flesinoxan, and the two potent agonist congeners **3** and **4** were docked into a 5-$HT_{1A}$ receptor model. For this purpose, the crystal structure of the most rigid analog **4** was determined at first. Then possibly bioactive conformations were generated, and a putative binding site of the three compounds in the model was identified. Conformational aspects that were investigated are depicted in Figure 2.

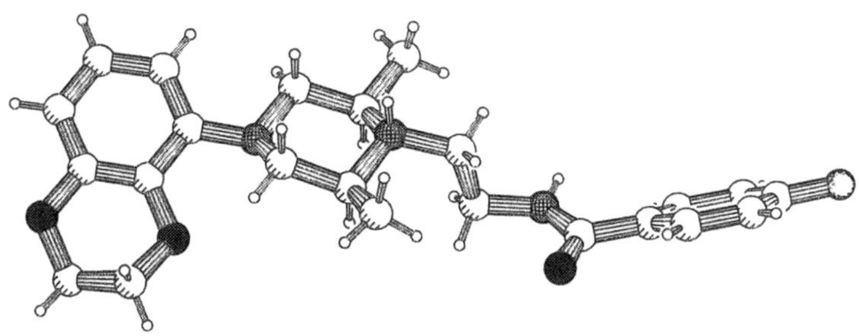

**Figure 3** *PLUTON[4] representation of the conformation of compound **4** in the crystal structure. The torsion angles $\tau_1$, $\tau_2$, $\tau_3$, and $\tau_4$ are −11.3° (3), 176.9° (3), 169.7° (2), and 80.0° (3), respectively (for definitions see Figure 2).*

### 3.4. Crystal structure verification of compound 4

The X-ray structure of the potent 5-$HT_{1A}$ agonist **4** is shown in Figure 3. The observed torsion angle $\tau_1$ is −11.3°(3). This is in agreement with previous results that show that singly *ortho*-substituted agonists phenylpiperazines bind at 5-$HT_{1A}$ receptors in a relatively coplanar conformation with torsion angle $\tau_1 \approx 0°$, with the plane angle between the piperazine and the benzene ring being approximately 30°.[3b]

### 3.5. Modeling

*Bioactive conformation of the $N^4$-substituent.* The conformation of the $N^4$-substituent is primarily determined by the torsion angles $\tau_2$, $\tau_3$, and $\tau_4$ (Figure 2). The benzamide part is fairly rigid; the plane angle between the phenyl and the amide groups being approximately 30°. The torsion angle $\tau_2$ was studied by semiempirical MOPAC/AM1 calculations[5] of the rotational barrier of the $N^4$-ethyl chain in the hypothetical compounds **5** and **6** (see Figure 1). In these compounds, the benzamide moiety of **3** and **4** is replaced by a hydrogen atom, allowing the study of the torsion angle $\tau_2$ independently from $\tau_3$ and $\tau_4$. Thus **5** and **6** may serve as model compounds for the investigation of $\tau_2$ of **3** and **4**, respectively. The results are shown in Figure 4.

Three staggered low-energy conformations are found for compound **5**. The *anti*-periplanar conformation at $\tau_2$ = 180° is also a minimum for compound **6**. The latter cannot adopt the other two minima of **5** (*gauche* conformers) because of severe steric hindrance between the C2″ methyl group and the two equatorial *cis*-methyl substituents. This is in agreement with the *anti*-periplanar conformation of the crystal structure of compound **4** (Figure 3). Therefore, the crystal structure of **4** ($\tau_2$ = 176.9° (3)) was used for further study.

Possible conformations of the complete $N^4$-ethyl-(*p*-fluorobenzamide) chain were generated by varying $\tau_3$ and $\tau_4$ in the RANDOMSEARCH option, using the crystal structure of compound **4** as the starting geometry. This option in the SYBYL program locates energy minima by randomly adjusting the chosen torsion angles (here $\tau_3$ and $\tau_4$) and optimizing the resulting geometry.[6] The conformation of the arylpiperazine moiety was constrained during this procedure because the Molecular Mechanics (MM) methods, as used in the RANDOMSEARCH option, are less reliable for calculation of this fragment.[3,7] The RANDOMSEARCH analysis of $\tau_3$ and $\tau_4$ yielded 20 conformations with energies (Tripos force field)[6] ranging from 116.7–119.2 kcal $mol^{-1}$. Subsequent full-energy minimization with MOPAC/AM1[5] further reduced the number of favorable conformations to 8 (A–H: see Table 3). The conformations A and B have $\tau_3$ values of approximately 180°; in conformation H, $\tau_3$ is −63.8°. Torsion angle $\tau_3$ for the conformations C–G is in the range 64.0–76.1°. Conformations C–G can be divided into two clusters, separated by their $\tau_4$ values. The torsion angle $\tau_4$ determines the orientation of the amide moiety (see Figure 2). Thus, conformations F and G are practically identical. Structure A closely resembles the crystal structure of **4**. This indicates that the approach we followed indeed yields accessible low-energy conformations. The 8 potentially bioactive conformations A–H were further investigated and docked into a 5-$HT_{1A}$ receptor model.

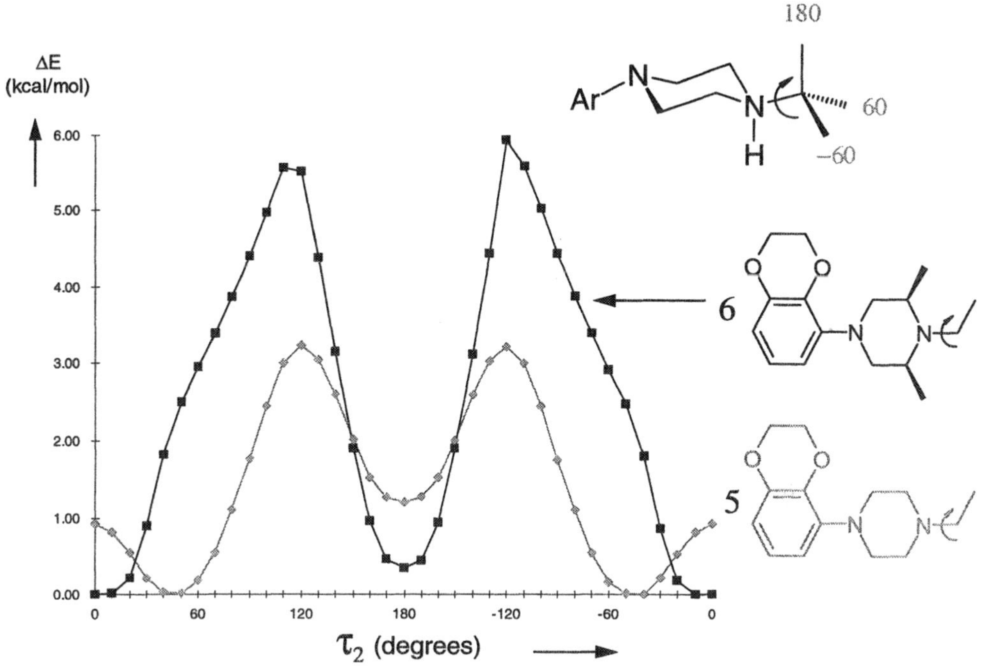


**Figure 4** *Results of MOPAC/AM1 calculations*[6] *of torsion angle* $\tau_2$ *of compounds* **5** *and* **6**, *which serve as model compounds for* **3** *and* **4**, *respectively. Energy is relative to the calculated absolute minimum. For compound* **5** *three staggered minimum energy conformations are calculated with* $\tau_2$ *values of –45°, 45°, and 180°, respectively. In conformations –45° and 45°, the C2″-atom is positioned gauche relative to the H(N4) atom: in the 180° minimum this orientation is anti-periplanar. Only the latter conformation is also energetically favored by compound* **6**, *because in the gauche conformations the C2″ atom is severely sterically hindered by the two methyl substituents at the piperazine ring. The second minimum of* **6** *is the eclipsed conformation at* $\tau_2 = 0°$, *in which the C2″-atom points in the same direction as the H(N4)-atom. This conformation is not an energy minimum for compound* **5**.

*5-HT$_{1A}$ receptor docking study.* It was assumed that the agonists flesinoxan **1**, **3** and **4** occupy a similar binding site as other 5-HT$_{1A}$ agonists. Binding sites of these other agonists, such as 5-HT and 8-OH-DPAT as well as several arylpiperazine compounds, have been hypothesized in previous studies.[3] The agonists flesinoxan **1**, **3**, and **4** were docked into the receptor model with the arylpiperazine moieties overlapping that of the nor-compounds **29** and **30**,[3b] in the same conformation (*i.e.*, also similar to the arylpiperazine conformation in the crystal structure of **4**). The 5-HT$_{1A}$-SAR data of Table 2 support the validity of this choice, since the effect of aryl substitution is highly similar for $N^4$-substituted arylpiperazines and the corresponding nor-compounds.

**Table 3** *Crystal structure and conformations (A–H) resulting from RANDOMSEARCH analysis*[6] *of $\tau_3$ and $\tau_4$ of compound **4***

| Conformation | $\tau_1$ (°) | $\tau_2$ (°) | $\tau_3$ (°) | $\tau_4$ (°) |
|---|---|---|---|---|
| crystal structure [a] | −11.3 | 176.9 | 169.7 | 80.0 |
| A | −12.8 | −178.9 | −178.4 | 66.2 |
| B | −16.1 | −177.5 | 174.0 | −64.3 |
| C | −16.7 | 167.9 | 76.1 | −104.2 |
| D | −14.6 | 168.2 | 75.2 | −105.9 |
| E | −14.1 | 166.1 | 67.2 | 65.6 |
| F | −13.1 | 170.2 | 68.1 | 75.5 |
| G | −18.3 | 163.8 | 64.0 | 66.9 |
| H | −16.4 | −163.5 | −63.8 | −63.9 |

([a]) Standard deviations are 3°, 3°, 2°, and 3°, for $\tau_1$, $\tau_2$, $\tau_3$, and $\tau_4$, respectively. ND=not determined.

Conformations A, B, and H of compound **4** and the close analogs flesinoxan **1** and **3** (Table 3) could not be accommodated in the 5-HT$_{1A}$-receptor model, since the benzamide part of their $N^4$-substituents caused severe steric hindrance with the backbones of helix VI (conformation H) or helix VII (conformations A and B). The complexes of the resulting conformations C–G with the 5-HT$_{1A}$-receptor model were further minimized. The conformations C–G all had $\tau_3$ values of approximately 70°, and consequently, occupied similar regions in the 5-HT$_{1A}$-receptor model. The conformations C and D of flesinoxan **1**, **3**, and **4** converged to stable receptor-ligand complexes. Figure 5 shows flesinoxan in the receptor model in conformation C. Conformations E–G fitted the model slightly worse.

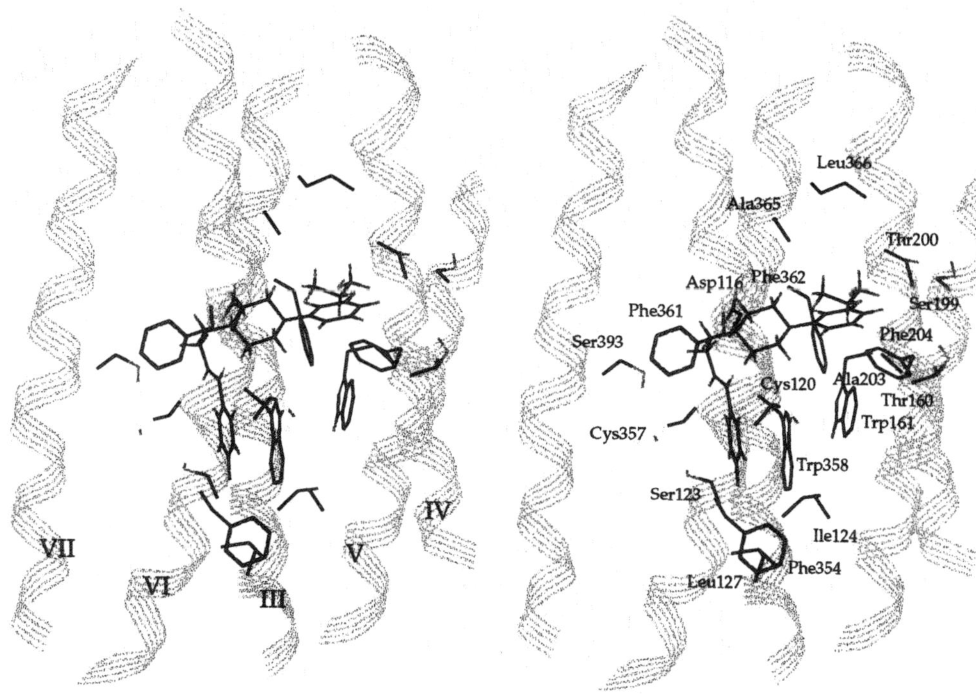


**Figure 5** *Stereo representation of flesinoxan docked in the receptor model in conformation C, looked upon from the side. The extracellular side is at the top. The hydrophobic residues Cys120, Ile124, Leu127, and Cys357 and the aromatic residues Phe354, Trp358, and Phe361 form a lipophilic pocket at one side of the* $N^4$*-substituent. At the other side of the pocket two hydrophilic serines are observed at positions 123 and 393.*

The conformations E–G occupy a similar binding region at the 5-$HT_{1A}$-receptor model as conformations C and D. Therefore, the residues surrounding the *p*-fluorobenzene moiety of compounds **1**, **3**, and **4** in conformations E–G are similar to those in conformations C and D. However, two possible orientations of the amide moiety in the model are found, in which the C=O group of the amide moiety points in the direction of lipophilic (conformations C and D), or hydrophilic residues (conformations E–G). The $N^4$-substituent's benzene rings are located in a lipophilic pocket, formed by Cys120, Ile124, Leu127, Phe354, Cys357, Trp358, and Phe361. This may account for the considerable contribution of lipophilicity to both 5-$HT_{1A}$- and $D_2$-receptor affinity.

| | | | |
|---|---|---|---|
| 5-$HT_{1A}$ | 38 | I T S L L L G T L I F C A V L G N A C V V A A | I |
| $D_2$ | 36 | Y Y A M L L T L L I F I I V F G N V L V C M A | |
| 5-$HT_{1A}$ | 74 | L I G S L A V T D L M V S V L V L P M A A L Y | II |
| $D_2$ | 72 | L I V S L A V A D L L V A T L V M P W V V Y L | |
| 5-$HT_{1A}$ | 109 | C D L F I A L **D** V L C **C** T S **S** **I** L H **L** C A I | III |
| $D_2$ | 107 | C D I F V T L **D** V M M **C** T A **S** **I** L N **L** C A I | |
| 5-$HT_{1A}$ | 153 | A A A L I S L **T** **W** L I **G** F L I S I P P M L | IV |
| $D_2$ | 152 | V T V M I A I ***V*** **W** V L ***S*** F T I S C P L L F | |
| 5-$HT_{1A}$ | 195 | Y T I Y **S** **T** F G **A** **F** Y I P L L I M L V L Y | V |
| $D_2$ | 189 | F V V Y **S** ***S*** I V ***S*** **F** Y V P F I V T L L V Y | |
| 5-$HT_{1A}$ | 347 | L G I I M G T **F** I L **C** **W** L P **F** **F** I V **A** **L** V L P F | VI |
| $D_2$ | 376 | L A I V L G V **F** I I **C** **W** L P **F** **F** I T ***H*** ***I*** L N I H | |
| 5-$HT_{1A}$ | 381 | L G A I I N W L G Y S N **S** L L N P V I Y A Y F N | |
| $D_2$ | 408 | L Y S A F T W L G Y V N **S** A V N P I I Y T T F N | VII |

**Figure 6** *Sequences of the putative transmembrane domains of the rat 5-$HT_{1A}$ receptor, that were used for construction of the receptor model, and the corresponding alignment with the rat $D_2$-receptor sequence.[8] Residues marked boldface were found to be part of the binding site (residues within a sphere of 4 Å) of flesinoxan 1, 3, and 4 in the 5-$HT_{1A}$-receptor model. Residues in the $D_2$-receptor sequence that differ from those in the modeled binding site are printed in italics. Residues in the $D_2$ sequence that correspond with those in the modeled 5-$HT_{1A}$-receptor binding site are underlined.*

The residues that surround the $N^4$-substituent in the 5-$HT_{1A}$-receptor model (*i.e.*, Asp116, Cys120, Ser123, Ile124, Leu127: helix III; Phe354, Cys357, Trp358, Phe361: helix VI and Ser393: helix VII) are also present in $D_2$ receptors (see alignment in Figure 6). This might explain the similar SARs of the $N^4$-substituent at 5-$HT_{1A}$ and $D_2$ receptors. In contrast, a number of residues in the model that are contacting the hetero-aryl moiety (*i.e.*, Thr160, Trp161, Gly164: helix IV; Ser199, Thr200, Ala203, Phe204: helix V; Phe362, Ala365, and Leu366: helix VI) differ from those in $D_2$ receptors. Observed differences between 5-$HT_{1A}$ and $D_2$ receptors in this region are Thr160↔Val, Gly164↔Ser, Thr200↔Ser, Ala203↔Ser, Ala365↔His, and Leu366↔Ile, respectively. This is in good agreement with SAR data presented in this study, which show that the arylpiperazine part itself is not very well accommodated by the $D_2$ receptor. In contrast, the 5-$HT_{1A}$ receptor can favor rather bulky arylpiperazine substituents like the dioxane ring in **3** and **4**, and even the additional $CH_2OH$ group of flesinoxan **1**. The $CH_2OH$ group has no additional interaction with residues in the receptor model. This is in agreement with SAR data that show no contribution of this group to 5-$HT_{1A}$-receptor affinity. The dioxane ring of the compounds makes a close contact with the small and lipophilic Ala365 in the 5-$HT_{1A}$-receptor model. This residue is replaced with a bulky and polar histidine in $D_2$ receptors (see Figure 6), which may account for the low tolerance for bulky arylpiperazine substituents by $D_2$ receptors. Thus, selectivity for 5-$HT_{1A}$ *versus* $D_2$ receptors is mainly caused by the aryl-substitution pattern of the arylpiperazine moiety.

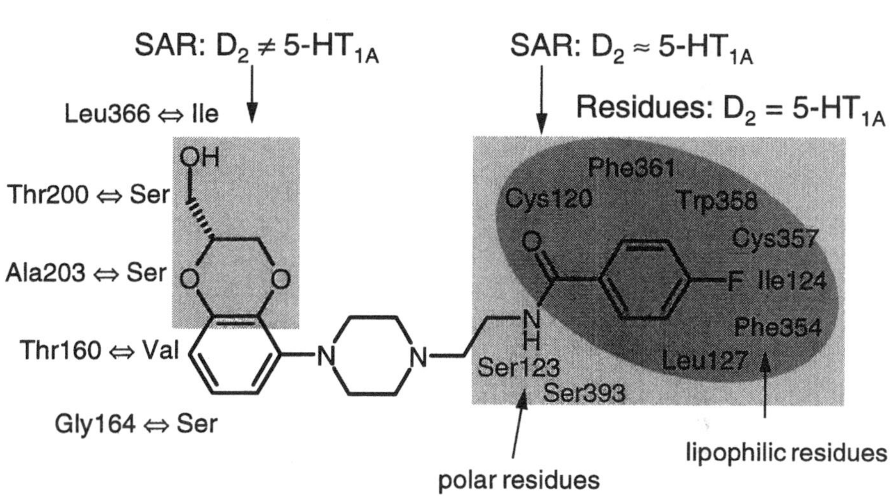


**Figure 7** *Schematic representation of the environment of flesinoxan 1 in the 5-$HT_{1A}$-receptor model (in conformations C-G). Hydrophobic residues form a lipophilic pocket at one side of the $N^4$-substituent. At the other side of the pocket two hydrophilic serines are observed. All residues that surround the $N^4$-substituent are identical to those at identical positions in the $D_2$ rat receptor according to the alignment of Figure 6 (i.e., Asp116, Cys120, Ser123, Ile124, Leu127, Phe354, Cys 357, Trp358, Phe361, and Ser393). Several of the residues that surround the $N^1$-aryl part are different from those in $D_2$ receptors (i.e., Thr160, Gly164, Thr200, Ala203, Ala365, and Leu366).*

## 4. CONCLUSIONS

5-$HT_{1A}$-receptor affinity and selectivity *versus* $D_2$ receptors of the potent 5-$HT_{1A}$ receptor agonist flesinoxan **1**, and a series of analogues were studied. The selectivity of flesinoxan **1** was shown to emerge mainly from the $N^1$-aryl substitution pattern. For both 5-$HT_{1A}$ and $D_2$ receptors, SARs at the $N^4$-substituent appeared to be dominated by lipophilicity. The $N^4$-substitutions did not yield selective compounds.

The bioactive conformation was investigated, and the results were used to study the interactions of the ligands in a 5-$HT_{1A}$-receptor model. The coplanar conformation of the arylpiperazine moiety, which was shown to be preferred by $N^4$-unsubstituted 5-$HT_{1A}$ receptor agonists,[3] can also be adopted by the $N^4$-substituted agonists flesinoxan **1**, **3**, and **4**. In this conformation, the plane angle between the benzene and piperazine ring is approximately 30°. In the bioactive conformation of the agonists **1**, **3**, and **4**, the $N^4$-substituent is probably directed *anti*-periplanar with respect to the H(N4)-atom. Eight of such *anti*-periplanar conformations (A–H), among which the crystal structure of **4**, could be identified as potentially bioactive at 5-$HT_{1A}$ receptors. Five of these conformations fitted the 5-$HT_{1A}$ receptor model well. The predominantly hydrophobic residues of helices III, VI, and VII, that surround the $N^4$-substituent, are identical for 5-$HT_{1A}$ and $D_2$ receptors. Six of the residues that contact the benzodioxin ring of flesinoxan in the 5-$HT_{1A}$ receptor model, are different from those in $D_2$ receptors. Thus, the positions of the 5-$HT_{1A}$ agonists flesinoxan **1**, **3**, and **4** in the receptor model, in the potentially bioactive conformations, agree with 5-$HT_{1A}$ SARs, and account for the selectivity of flesinoxan **1** *versus* $D_2$ receptors.

## REFERENCES


1 Van Wijngaarden, I.; Tulp, M.Th.M.; Soudijn, W. *Eur. J. Pharmacol. - Mol. Pharmacol. Sect.* **1990**, *188*, 301-312.

2 *e.g.*, (a) Scott, M.K.; Martin, G.E.; DiStefano, D.L.; Fedde, C.L.; Kubkla, M.J.; Barrett, D.L.; Baldy, W.J.; Elgin Jr., R.J.; Kesslick, J.M.; Mathiasen, J.R.; Shank, R.P.; Vaught, J.R. *J. Med. Chem.* **1992**, *35*, 552-558. (b) Peglion, J.; Canton, H.; Bervoets, K.; Audinot, V.; Brocco, M.; Gobert, A.; Le Marouille-Girardon, S.; Millan, M.J. *J. Med. Chem.* **1995**, *38*, 4044-4055.

3 a) Kuipers, W.; Wijngaarden, I. van; IJzerman, A.P. *Drug Design & Disc.* **1994**, *11*, 231-249. b) Kuipers, W.; Wijngaarden, I. van; Kruse, C.G.; ter Horst-van Amstel, M.; Tulp, M. Th.M.; IJzerman, A.P. *J. Med. Chem.* **1995**, *38*, 1942-1954.

4 Spek, A.L. PLUTON Molecular graphics program, Utrecht University, The Netherlands, **1995**.

5 Stewart, J.J.P. *J. Comp.-aided Mol. Design* **1990**, *4*, 1-105.

6 Sybyl package, Tripos Associates, Inc., St. Louis, Missouri, USA.

7 Dijkstra, G.D.H. *Recl. Trav. Chim. Pays-Bas* **1993**, *112*, 151-160.

8 Bairoch, A. Dept. Biochimie Medicinale, Centre Medicinal Universitaire, 1211 Geneva 4, Switzerland, and SWISS-PROT Protein sequence database, EMBL Data Library, D-69117 Heidelberg Germany.

H. van der Goot, (Editor)
TRENDS IN DRUG RESEARCH II



# Serotonin transmission in depression and anxiety disorders - new insights and potential new drugs

M. Briley[a] and C. Moret[b]

[a]Institut de Recherche Pierre Fabre, Parc Industriel de la Chartreuse, 81100 Castres, France

[b]Centre de Recherche Pierre Fabre, 17 Ave Jean Moulin, 8100 Castres, France

The selective serotonin reuptake inhibitors (SSRI) are effective antidepressants (Boyer and Feighner, 1991) with an efficacy generally similar to that of tricyclic antidepressants (TCA) except in more severe depression where they appear to be somewhat less efficacious. In vivo and certainly at clinical doses these compounds have a single acute pharmacological action, the inhibition of serotonin reuptake.

In spite of the rapid inhibition of serotonin uptake in man (within a few hours at most) the earliest signs of therapeutic improvement in depression appear only after about two weeks. SSRIs, when administered acutely to animals, surprisingly cause only modest increases in the extracellular levels of serotonin in the cortex or other brain areas innervated by the dorsal raphe nucleus, as measured by in vivo microdialysis in freely moving rats (Adell and Artigas, 1991; Bel and Artigas, 1992; Invernizzi et al., 1992).

## 1. SEROTONERGIC FEEDBACK SYSTEMS

This limited acute effect of the SSRIs appears to be the result of several of feedback systems. The firing of dorsal raphe neurones is under the control of 5-$HT_{1A}$ receptors located on somatodendrites in the dorsal raphe nucleus (Sprouse and Aghajanian, 1987), the stimulation of these receptors reducing the firing of these neurones. Increased levels of extracellular serotonin resulting from the inhibition of reuptake activate these autoreceptors in the dorsal raphe nucleus leading to a feedback inhibition of release in the terminal regions through decreased firing of the dorsal raphe serotonergic neurones. This feedback control on the firing rate of the dorsal raphe nucleus is not, however, the only negative feedback modulation of serotonergic neurotransmission.

### 1.1. 5-HT1B receptor feedback control of serotonin release

The release of serotonin from nerve terminals is under the control of inhibitory 5-HT1B autoreceptors (for reviews see Moret (1985) and Middlemiss (1988)). Local infusion of the relatively non-selective 5-HT1B receptor agonist, 5-carboxamidotryptamine (5-CT), reduces the extracellular levels of serotonin by 40-50% (Lawrence and Marsden, 1992) as measured by microdialysis. On the other hand infusion of the non-selective 5-HT1B receptor antagonist, methiothepin, into the guinea pig substantia nigra results in a large (as much as 10 fold) increase in the level of extracellular serotonin (Briley and Moret, 1993).

Similarly Hjorth and Tao (1991) have shown that CP-93,129 (3-(1,2,5,6-tetrahydropyrid-4-yl)pyrolol[3,2,6]pyrid-5-one), a selective 5-HT1B receptor agonist, when administered via the dialysis perfusion medium reduces serotonin output in the hippocampus of anaesthetised rats, an effect significantly antagonised by co-infusion of methiothepin. In the rat hypothalamus, methiothepin, when applied locally via a microdialysis probe, increased the extracellular levels of serotonin both in the absence and in the presence of the serotonin uptake inhibitor, citalopram, suggesting that in awake animals 5-HT1B autoreceptors in the terminal projection areas are tonically activated and exert a potent inhibitory tone on the release of serotonin (Moret and Briley, 1996).

### 1.2. 5-HT1B receptor feedback control of serotonin synthesis

The activity of the rate limiting enzyme of serotonin synthesis, tryptophane hydroxylase, is also under feedback control. Recently Hjorth et al. (1995) showed, in the rat, that the 5-HT1B/2C receptor agonist, trifluoro-methylphenylpiperazine (TFMPP), suppresses serotonin synthesis in vivo. This suppression, which was evident in terminal projection areas such as the limbic forebrain and striatum, was also observed in axotomised animals, indicating that it was independent of neuronal firing. Furthermore, a similar inhibitory effect of TFMPP on serotonin synthesis was found in vitro in slice preparations in the presence of depolarising concentrations of potassium. In vitro the effect of TFMPP was attenuated by the non-selective 5-HT1B receptor antagonist, methiothepin, as well as the 5-HT1B receptor antagonists, propranolol or cyanopindolol. These data thus suggest that the reduction of rat brain serotonin synthesis by TFMPP is mediated by 5-HT1B autoreceptors located on the serotonergic axon terminals, and that this is a direct effect and independent of the firing of serotonergic neurones.

## 2. THE EFFECT OF CHRONIC ADMINSTRATION OF SSRIs ON THE FEEDBACK SYSTEMS

Thus acutely the potential stimulation of serotonergic neurotransmission by a SSRI is severely attenuated by several presynaptic feedback systems. These findings are particularly interesting in the context of the latency of the therapeutic effects in depressive illness. This latency has been attributed to the need for adaptive changes to be brought about by long term treatment (for review, see Briley and Moret, 1993). One of these adaptive changes may be the desensitisation of the terminal 5-$HT_{1B}$ autoreceptor, with the subsequent rise of synaptic levels of 5-HT and the stimulation of one or more postsynaptic receptors which is thought to be an essential long-term action of these antidepressants.

### 2.1. The effects of chronic SSRI administration on the release of serotonin.

Administration of citalopram (50 mg/kg p.o.) to rats for 21 days followed by a washout of 24 h resulted in an increased in vitro stimulation-induced release of serotonin from hypothalamic slices preloaded with $[^3H]$5-HT (Moret and Briley, 1990). In addition, the concentration-effect curve of the agonist, LSD, was significantly shifted to the right compared with control animals, indicating a desensitisation of the autoreceptor for the agonist. It was suggested that repeated administration of the SSRI resulted in a decreased efficacy of the terminal autoreceptor, allowing an increased release of 5-HT (Moret and Briley, 1990).

In guinea pigs, Blier and Bouchard (1994) found that the electrically induced release of $[^3H]$5-HT was increased by a chronic treatment with the SSRI, paroxetine (14 days with 48 h withdrawal), in slices of hypothalamus, hippocampus and frontal cortex, and that the inhibitory effect of the non-selective 5-HT receptor agonist, 5-methoxytryptamine, was attenuated. Thus the terminal 5-$HT_{1B}$ autoreceptor also appears to be desensitised in the guinea pig after a long-term blockade of serotonin uptake.

The desensitisation of the terminal release controlling 5-$HT_{1B}$ autoreceptor has also been deduced from indirect electrophysiological measurements (Chaput et al., 1986; Blier et al., 1988; Chaput et al., 1991). Neurones in $CA_3$ region of the hippocampus possess postsynaptic 5-$HT_{1A}$ receptors which when stimulated by the serotonergic innervation from the raphe induce a hyperpolarisation of these cells. Thus by electrically stimulating the raphe neurones and measuring the hyperpolarisation of cells in the hippocampus it is possible to measure the overall efficiency of this serotonergic pathway. The iv administration of the autoreceptor antagonist, methiothepin, produces an increased hyperpolarisation through an increased release resulting from the blockade of the terminal autoreceptors mediating the feedback inhibition of release. The extent of the effect of methiothepin can thus be used to deduce the sensitivity of the terminal autoreceptor.

Three weeks administration of citalopram (20 mg/kg/day ip for 14 days; (Chaput et al., 1986)), fluoxetine (10 mg/kg/day ip for 14 days; (Blier et al., 1988)) or paroxetine (5 mg/kg/day ip for 21 days; (Chaput et al., 1991)) increases the efficiency of serotonergic neurotransmission by attenuation of the effect of the terminal autoreceptor. The increased efficacy of the stimulation of the raphe on the hyperpolarisation of the $CA_3$ hippocampal cells induced by an iv injection of the autoreceptor antagonist, methiothepin, was abolished in SSRI-treated rats (Chaput et al., 1991).

Recently, we (Moret and Briley, 1996) attempted to demonstrate the increase in synaptic 5-HT using in vivo microdialysis on freely moving rats after a chronic administration of citalopram under exactly the same conditions as in the previous in vitro study (Moret and Briley, 1990). Somewhat unexpectedly no change was seen in the basal extracellular levels of endogenous 5-HT in chronic drug-treated animals after washout. In addition, the enhancing effect of methiothepin, administered through the microdialysis probe, was similar in both control and chronically treated animals. These results suggest that under the conditions of this study, repeated administration of citalopram followed by a washout of 24 h does not lead to a desensitisation of the terminal 5-HT autoreceptor of sufficient magnitude for it to be measured in vivo, in contrast to the effects shown in vitro. At present no clear explanation exists for the discrepancy between in vitro and in vivo findings.

The effects of chronic administration of an SSRI were also studied without washout which is a closer approximation to the clinical situation. In this case extracellular levels of 5-HT were increased by both acute and repeated citalopram administration (Moret and Briley, 1996). In rats treated chronically, methiothepin (administered locally via the probe) had a greater maximal effect on outflow of serotonin than in rats receiving acute citalopram treatment. This study shows that a SSRI and an autoreceptor antagonist are both capable of increasing extracellular levels of serotonin. Furthermore these two effects are additive or possibly synergistic, suggesting that a terminal $5\text{-HT}_{1B}$ autoreceptor antagonist or a combination of such a drug with a SSRI would produce a greater increase of extracellular serotonin levels and thus be potentially useful in the treatment of depressive disorders resistant to therapy by a single drug.

## 2.2. The effects of chronic SSRI administration on the synthesis of serotonin.

Repeated administration of citalopram (50 mg/kg/day ip) for 21 days also modifies the synthesis of serotonin. This treatment results in an increased basal activity of tryptophane hydroxylase. Interestingly, however, acute administration of citalopram still inhibits the synthesis with a dose-response curve which is approximately parallel to that in control animals (Moret and Briley, 1992). Thus the activity of tryptophane hydroxylase would appear to be under a complex control. The chronic inhibition of serotonin synthesis produced by the repeated administration of a SSRI appears to result in an increased basal enzyme activity probably as a result of increased enzyme concentration due to enzyme induction. Serotonin synthesis is, however, still responsive, with apparently little or no change in its sensitivity, to temporarily increased levels of serotonin produced by the acute administration of an SSRI.

Thus, through down-regulation of the $5\text{-HT}_{1B}$ mediated feedback mechanisms antidepressant therapy leads, after a few weeks, to increased serotonergic neurotransmission. Total or partial inactivation of one or more of the various serotonergic feedback systems should therefore lead to a more rapid increase in serotonergic neurotransmission and consequently an alleviation of the symptoms of depression. Thus selective $5\text{-HT}_{1B}$ autoreceptor antagonists may represent an interesting new therapeutic class for the treatment of depression. Indeed there is considerable interest and activity within the pharmaceutical industry as recently reviewed by Halazy et al. (1997).

$5\text{-HT}_{1B}$ receptors appear to be localised mainly presynaptically, either as autoreceptors on serotonergic terminals (Engel et al., 1986; Moret, 1985) or as heteroreceptors on the terminals of other transmitters such as GABA (Hen, 1992) and acetylcholine (Maura et al., 1989). Thus it is unlikely that a $5\text{-HT}_{1B}$ receptor antagonist would block postsynaptic receptors that are the target of synaptic serotonin during antidepressant therapy.

# 3. ARE $5\text{-HT}_{1B}$ RECEPTORS MODIFIED IN DEPRESSION AND OTHER SEROTONIN RELATED DISORDERS?

5-HT autoreceptors are capable of modulating 5-HT neurotransmission via the control of the release and synthesis of 5-HT. Acutely they attenuate the action of antidepressants such as the SSRIs and as such play an important role in their therapeutic action. To what extent are $5\text{-HT}_{1B}$ autoreceptors involved in the pathophysiology of depression? There is, as yet, no data from depressed patients but evidence from animal studies and in patients with other serotonin-related disorders is suggestive of such a role.

The induction of the depression-like state of learned helplessness results in an increase in $5\text{-HT}_{1B}$ receptor density (Edwards et al., 1991) in various brain regions and an increase in $5\text{-HT}_{1B}$ receptor mRNA in the raphe (Neumaier et al., 1997). An increase in $5\text{-HT}_{1B}$ autoreceptor sensitivity is consistent with the decreased release of serotonin from these rats observed by microdialysis in the cortex (Petty et al., 1992). Interestingly, learned helpless rats not only show a number of "depressive" signs (Sherman et al., 1979), but also exhibit behaviour associated with high levels of "anxiety" (Vandijken et al., 1992a,b). Increased serotonergic neurotransmission is associated with an increased level of anxiety (Chopin and Briley, 1987; Briley and Chopin, 1994). The frequent co-existence of high levels of anxiety with depression, a supposedly hyposerotonergic state is, nevertheless, difficult to explain. In the case, however, of supersensitivity of $5\text{-HT}_{1B}$ auto- and heteroreceptors, a decreased release of serotonin resulting from an increased autoinhibition and an increased level of anxiety resulting from activation of supersensitive $5\text{-HT}_{1B}$ heteroreceptors decreasing GABA release for example, could be expected. This would correspond to the situation found in the learned helplessness model of depression where both $5\text{-HT}_{1B}$ auto- and heteroreceptors may be supersensitive. Repeated administration of 5-HT uptake blocking antidepressants would be expected to desensitise both auto- and hetero- $5\text{-HT}_{1B}$ receptors thus alleviating the symptoms of both depression and anxiety as seen with SSRIs.

A totally independent line of reasoning has led Zohar and co-workers (Dolberg et al., 1995) to the conclusion that $5\text{-HT}_{1B}$ receptors may be supersensitive in OCD and their desensitisation through long-term administration of SSRIs may be responsible for their therapeutic effect. The administration of the non-selective 5-HT receptor agonist, m-chlorophenylpiperazine (mCPP), to untreated patients suffering from OCD causes a marked and transient exacerbation of their symptoms (Zohar et al., 1987; Hollander et al., 1992) whereas the administration to healthy volunteers does not, in general, induce any symptomatology. This effect, which can be prevented by pretreatment with the non-selective 5-HT receptor antagonist, metergoline, (Pigott et al., 1991) has been suggested to result from stimulation of 5-HT receptors that are supersensitive in OCD patients. This idea has found support in the observation that the effects of mCPP are blunted in patients whose OCD has been successfully treated with clomipramine (Zohar et al., 1988) which presumably normalise these supersensitive receptors. mCPP has high affinity for $5\text{-HT}_{1A}$, $5\text{-HT}_{1B}$ and $5\text{-HT}_{2C}$ receptors. Since more selective serotonergic agonists, such as ipsapirone ($5\text{-HT}_{1A}$) and MK-212 (2-chloro-6-(1-piperazinyl) pyrazine) ($5\text{-HT}_{2C}$), do not produce the exacerbation of OCD symptoms (Bastani et al., 1990), it would appear that the receptors involved are probably of the $5\text{-HT}_{1B}$ subtype. In addition, since OCD symptoms are unaltered by acute modification of synaptic serotonin levels, through tryptophane depletion (Pigott et al., 1993; Barr et al., 1994), the receptors involved are probably not $5\text{-HT}_{1B}$ autoreceptors but more likely $5\text{-HT}_{1B}$ heteroreceptors. The putative implication of $5\text{-HT}_{1B}$ receptors has been recently tested by administering sumatriptan to OCD patients who reacted with a marked and transient aggravation of their OCD symptomatology (Dolberg et al., 1995). These studies suggest that supersensitive $5\text{-HT}_{1B}$ receptors may indeed be involved in the pathophysiology of OCD and may represent a potential target for its treatment. As in depression a significant gain in the delay of onset of action can be envisaged by the use of specific $5\text{-HT}_{1B}$ receptor antagonists to treat OCD.

## 4. CONCLUSIONS

The potential for a SSRI to increase the synaptic levels of serotonin are acutely attenuated to a considerable extent by a variety of feedback systems. Two of these, the control of the release of serotonin and the control of its synthesis are mediated through presynaptically located $5\text{-HT}_{1B}$ receptor. The chronic administration of SSRIs results in the desensitisation or down-regulation of these receptors and the feedback system they mediate. This has the effect of allowing the SSRI to fully increase synaptic serotonin levels. The time required to down-regulate these receptors may be related to the latency of onset of the antidepressant effects of SSRIs and other antidepressants.

The co-morbidity in patients and the co-existence in animal models such as the learned helpless rats, of symptoms of anxiety and depression is, at first sight, hard to reconcile with the hypotheses that anxiety arises from a serotonergic hyperstimulation whereas depression is related to decreased serotonergic neurotransmission. The potential existence of hypersensitive auto- and hetero $5\text{-HT}_{1B}$ receptors does however offer a potential mechanistic explanation. Hypersensitive autoreceptors, through an increased feedback control on serotonin release and synthesis would result in decreased serotonergic neurotransmission leading to depressive symptomatology. Hypersensitive heteroreceptors on GABA terminals for example could lead to reduced GABA neurotransmission resulting in increased anxiety. There is no data at this time to indicate whether $5\text{-HT}_{1B}$ receptors are supersensitive in depressed patients. Several lines of evidence from animal models and other serotonin related disorders such as OCD suggest, however, that changes in the sensitivity of 5-$HT_{1B}$ auto- and heteroreceptors may be fundamental to a variety of serotonin related psychiatric disorders. Further investigation into the role of 5-$HT_{1B}$ receptor function in psychiatric disorders would appear to be potentially rewarding. In addition the development of selective 5-$HT_{1B}$ receptor antagonists may lead to important therapeutic advances in depression, OCD and anxiety by providing more rapid and more potent therapy.

## REFERENCES

Adell A, Artigas F (1991) Differential effects of clomipramine given locally or systemically on extracellular 5-hydroxytryptamine in raphe nuclei and frontal cortex - An in vivo brain microdialysis study. Naunyn-Schmiedeberg's Arch of Pharmacol

Barr LC, Goodman WK, McDougle CJ, Delgado PL, Heninger GR, Charney DS, Price LH (1994) Tryptophan depletion in patients with obsessive-compulsive disorder who respond to serotonin reuptake inhibitors. Arch Gen Psychiatr 51: 309-317

Bastani B, Nash F, Meltzer H (1990) Prolactin and cortisol responses to MK-212, a serotonin agonist, in obsessive-compulsive disorder. Arch Gen Psychiatry 47: 946-951

Bel N, Artigas F. (1992) Fluvoxamine preferentially increases extracellular 5-hydroxytryptamine in the raphe nuclei: An in vivo microdialysis study. Eur J Pharmacol 229: 101-103.

Blier P, Chaput Y, De Montigny C. (1988) Long-term 5-HT reuptake blockade, but not monoamine oxidase inhibition, reduces the function of the terminal 5-HT autoreceptor: An electrophysiological study in the rat brain. Naunyn-Schmiedeberg's Arch Pharmacol; 337: 246-254

Blier P, Bouchard C (1994) Modulation of 5-HT release in the guinea pig brain following long-term administration of antidepressant drugs. Br J Pharmacol 113: 485-495

Boyer WF, Feighner JP. (1991) The efficacy of selective serotonin reuptake inhibitors in depression. In: Feighner JP, Boyer WF, eds Selective Serotonin Reuptake Inhibitors. Chichester: Wiley,pp 89-108.

Briley M, Moret C (1993) Neurobiological mechanisms involved in antidepressant therapies. Clin Neuropharmacol 16: 387-400

Briley M, Chopin P (1994) Is anxiety is associated with a hyper- or hypo-serotonergic state? In: Palomo T, Archer T (eds) Strategies for studying brain disorders, Vol 1: "Depression, anxiety and drug abuse disorders". Editorial Complutence, Donoso Cortés, Madrid, pp 197-209

Chaput Y, De Montigny C, Blier P. (1986) Effects of a selective 5-HT reuptake blocker, citalopram, on the sensitivity of 5-HT autoreceptors: electrophysiological studies in rat brain. Naunyn-Schmiedeberg's Arch Pharmacol 333: 342-348.

Chaput Y, De Montigny C, Blier P. (1991) Presynaptic and postsynaptic modifications of the serotonin system by long-term administration of antidepressant treatments -an in vivo electrophysiologic study in the rat. Neuropsychopharmacology 5: 219-229.

Chopin P, Briley M (1987) Animal models of anxiety: the effect of compounds that modify 5-HT neurotransmission. Trends Pharmacol Sci 8: 383-388

Dolberg OT, Sasson Y, Cohen R, Zohar J (1995) The relevance of behavioural probes in obsessive-compulsive disorder. Eur Neuropsychopharmacol 5: 161-162

Edwards E, Harkins K, Wright G, Henn FA (1991) 5-$HT_{1B}$ receptors in an animal model of depression. Neuropharmacology 30: 101-105

Engel G, Göthert M, Hoyer D, Schlicker E, Hillenbrand K (1986) Identity of inhibitory presynaptic 5-Hydroxytryptamine (5-HT) autoreceptors in the rat brain cortex with 5-$HT_{1B}$ binding sites. Naunyn-Schmiedeberg's Arch Pharmacol 332: 1-7

Halazy S, Lamothe M, Jorand-Lebrun, C (1997) 5-$HT_{1B/D}$ antagonists and depression. Exp. Opin. Ther. Patents 7: 339-352

Hen R. (1992) Of Mice and Flies - Commonalities among 5-HT receptors. Trends Pharmacol Sci 13: 160-165.

Hjorth S, Tao R (1991) The putative 5-$HT_{1B}$ receptor agonist CP-93,129 suppresses rat hippocampal 5-HT release in vivo - Comparison with RU 24969. Eur J Pharmacol 209: 249-252

Hjorth S, Suchowski CS, Galloway MP (1995) Evidence for 5-HT autoreceptor-mediated, nerve impulse-independent, control of 5-HT synthesis in the rat brain. Synapse 19: 170-176

Hollander E, DeCaria CM, Nitescu A (1992) Serotonergic function in obsessive-compulsive disorder. Arch Gen Psychiatry 49: 21-28

Invernizzi R, Belli S, Samanin R. (1992) Citalopram's ability to increase the extracellular concentrations of serotonin in the dorsal raphe prevents the drug's effect in the frontal cortex. Brain Res; 584: 322-324.

Lawrence AJ, Marsden CA (1992) Terminal autoreceptor control of 5-Hydroxytryptamine release as measured by in vivo microdialysis in the conscious guinea pig. J Neurochem 58: 142-146

Maura G, Fedele E, Raiteri M. (1989) Acetylcholine release from rat hippocampal slices is modulated by 5-hydroxytryptamine. Eur J Pharmacol 165: 173-179.

Middlemiss DN (1988) Autoreceptors regulating serotonin release. In: Sanders-Bush E (ed) The serotonin receptors. Humana Press, Clifton, NJ, pp 210-224

Moret C (1985) Pharmacology of the serotonin autoreceptor. In: Green AR (ed) Neuropharmacology of Serotonin. Oxford University Press, Oxford, pp 21-49

Moret C, Briley M (1990) Serotonin autoreceptor subsensitivity and antidepressant activity. Eur J Pharmacol 180: 351-356

Moret C, Briley M. (1992) Effect of antidepressant drugs on monoamine synthesis in brain in vivo. Neuropharmacology 31: 679-684

Moret C, Briley M (1996) Effects of acute and repeated administration of citalopram on extracellular levels of serotonin in rat brain. Eur J Pharmacol 295: 189-197

Neumaier JF, Petty F, Kramer GL, Szot P, Hamblin MW (1997) Learned helplessness increases 5-hydroxytryptamine$_{1B}$ receptor mRNA in the rat dorsal raphe nucleus. Biol. Psychiatry 41: 668-674

Petty F, Kramer G, Wilson L (1992) Prevention of learned helplessness - In vivo correlation with cortical serotonin. Pharmacol Biochem Behav 43: 361-367

Pigott TA, Zohar J, Hill JL (1991) Metergoline blocks the behavioral and neuroendocrine effects of orally administered m-chlorophenylpiperazine in patients with obsessive-compulsive disorder. Biol Psychiatry 29: 418-426

Pigott TA, Murphy DL, Brooks A (1993) Pharmacological probes in OCD: support for selective 5-HT dysregulation. Abstracts of the First International Obsessive Compulsive Disorder Congress ,Capri, Italy

Sherman AD, Allers GL, Petty F, Henn FA (1979) A neuropharmacologically relevant animal model of depression. Neuropharmacology 18: 891-894

Sprouse JS, Aghajanian GK. (1987) Electrophysiological responses of serotonergic dorsal raphe neurones to 5-HT1A and 5-HT1B agonists. Synapse; 1: 3-9.

Vandijken HH, Mos J, van der Heyden JAM, Tilders FJH (1992a) Characterization of stress-induced long-term behavioural changes in rats - Evidence in favor of anxiety. Physiol Behav 52: 945-951

Vandijken HH, van der Heyden JAM, Mos J, Tilders FJH (1992b) Inescapable footshocks induce progressive and long-lasting behavioural changes in male rats. Physiol Behav 51: 787-794

Zohar J, Mueller EA, Insel TR (1987) Serotonin responsivity in obsessive compulsive disorder. Arch Gen Psychiatry 44: 946-951

Zohar J, Insel TR, Zohar-Kadoush RC, Hill JL, Murphy DL (1988) Serotonergic responsivity in obsessive-compulsive disorder: Effects of clomipramine treatment. Arch Gen Psychiatry 45: 167-172

H. van der Goot, (Editor)
TRENDS IN DRUG RESEARCH II



# Pharmacokinetics and metabolism in drug development: current and future strategies

D.D. Breimer

Leiden/Amsterdam Center for Drug Research, Division of Pharmacology, Leiden University, P.O.Box 9502, 2300 RA Leiden, The Netherlands

## 1. INTRODUCTION

Innovation in drug research is often associated with drug discovery and drug design, i.e. with the creation of new biologically active molecules. In essence however the core business of the innovative pharmaceutical industry is not new molecules *per se*, but the knowledge on how to effectively and safely use them in individual patients. The enormous knowledge base required to achieve this takes place during the development stage and innovation is equally needed in this phase, i.e. new concepts and technologies in the study of drug disposition (pharmacokinetics), drug delivery, drug safety and (clinical) effectiveness. There is currently a lot of pressure to make the drug development process more informative, as well as more efficient. In addition the progress in combinatorial chemistry and high-throughput screening for receptor binding activity implies that numerous biologically active compounds are identified, but that optimization of pharmacokinetic properties and safety profile represent major new challenges.

Scientific progress in recent years particularly in the fields of pharmacokinetics (and its relationship with pharmacodynamics) and in drug metabolism, will be contributing very much to make the drug development process more efficient and more informative. This will be based on the application of new concepts and new technologies. An example is the use of multiple component analysis by sophisticated analytical techniques (LC-MS/MS) which makes it possible to screen for in vivo pharmacokinetic properties of a number of drug substances given simultaneously. In drug absorption the use of intestinal epithelial cell lines of human origin can be employed to assess absorption properties. There are numerous examples where highly active and promising compounds exhibit poor absorption characteristics or suffer from extensive "first-pass" elimination on oral administration. Consequently there is poor and highly variable bioavailability, which is undesirable for safe and effective clinical application. Optimization of absorption properties at a very early stage of development is of crucial importance for the further development process. Strategies of absorption enhancement can also be used, for example specific formulation techniques for compounds which are very poorly soluble in water, or the association with absorption enhancers for highly soluble drugs (peptide and protein drugs). Alternatively, non-oral routes of administration may be investigated and decided upon at an early phase of development. Much research in the drug delivery area in recent years has clearly indicated that there may be excellent opportunities for drug candidates to be given by the transdermal, pulmonary, nasal, buccal or other route of administration. The reader is referred to the Proceedings of the annual meetings of the Controlled Release Society to take detailed note of what progress has been made.

In this brief review some emphasis will placed on the potential contribution that studies on the relationship between pharmacokinetics and pharmacodynamics (PK/PD), preclinically and clinically, may make to a greater degree of informativeness and efficiency of the drug development process. In addition some important recent progress made in the field of drug metabolism will be highlighted.

## 2. PHARMACOKINETIC/PHARMACODYNAMIC MODELLING

Pharmacokinetics, i.e. the assessment of the time course of drug concentration in body fluids, is not an objective in itself but its relevance can only be judged if the relationship between drug concentration in vivo and pharmacological (toxicological) effects is established [1]. Modelling techniques have been developed to achieve this and the objective of PK/PD modelling is to identify key properties of a drug in vivo, which allow the characterization and prediction of the time course of drug effects under physiological and pathophysiological conditions. Hence it represents the basis of a dosage regimen design and allows individualization of dosing, if the factors contributing to intersubject variability have sufficiently been studied. In Figure 1 the relationship between pharmacokinetics and pharmacodynamics is schematically presented and a distinction is made between pharmacological effects in general (which may often be of a surrogate nature, e.g. in animal studies) and those relevant to the improvement of the disease process.

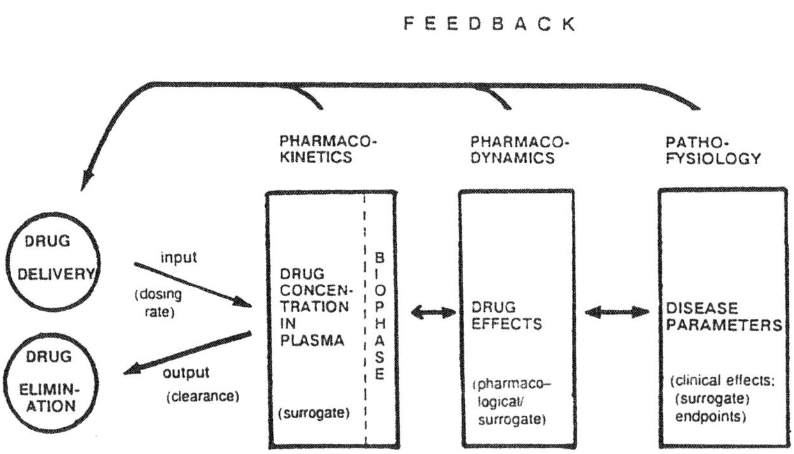


*Figure 1. Schematic representation of the relationship between pharmacokinetics and pharmacodynamics (PK/PD)*

In terms of application, PK/PD modelling has been proposed to be particularly relevant for incorporation throughout all sstages of drug development and to strongly influence decision-making at key transition steps in the development process and thereby rationalize and enhance the efficiency of this process [2]. In particular the lack of being able to identify effective and safe dosing regimens at early phases of drug development and prior to large clinical trials, has proved to be a major drawback of the current strategies. At the preclinical stage already relevant information can be obtained, because for various drugs it has been shown that the (unbound) plasma concentration required to produce a certain pharmacological effect is quite similar in experimental animals and man, whereas the dosages involved may be very different due to major inter-species differences in pharmacokinetics [3]. Therefore preclinical PK/PD information could be used to predict effective and toxic drug concentrations in man, which should be useful in the initial dose selection and escalation in Phase I studies. Preclinical studies can be undertaken toidentify the most appropriate PK/PD model(s), including equilibrium delays with the site of action, effect delays, dose and time dependencies, influence of route and rate of drug administration, the development of tolerance, etc. Importantly, such studies can also allow the detection of active metabolite formation (still a major problem in drug development) and elucidate potential differences in in vivo activity between enantiomers which differ kinetically. PK/PD information can also be used on a comparative basis to identify in vivo whether a compound is a full or partial agonist or antagonist or inverse agonist [4]. Furthermore, if various effects are being measured, an assessment can be made on the potential of selectivity of drug action in vivo.

Phase I clinical studies (healthy subjects or patients) should be designed with the aid of the provisional PK/PD data base obtained preclinically, i.e. the initial dose selection and escalation studies to determine any drug effects as a function of dose and plasma concentration. This will provide relevant information on dose and concentration limits of drug tolerability, as well as PK/PD relationships for pharmacological effects in man. Here the original PK/PD model may have to be adjusted or refined and, in combination with initial studies in patients (Phase II), the key pharmacokinetic and pharmacodynamic properties of the drug in man may be estimated under physiological and pathophysiological conditions. Furthermore, relevant information on inter-subject and inter-patient variability is to be obtained, leading to the identification of special populations or clinical conditions that exhibit altered PK/PD. In principle the Phase II clinical studies should be concentration-response studies with designs that expose a substantial fraction of the study group to more than one dose level, all in all exploring the fully tolerated dose range. The PK/PD data base thus obtained in Phase I and II should become the backbone of designing appropriate and relevant Phase III trials to substantiate efficacy and safety. It is anticipated that this will lead to considerable rationalization of clinical trials (fewer and better) leading to much improved efficiency and cost-effectiveness of the drug development process.

## 3. DRUG METABOLISM

Major advances have been made during the past 10-15 years on a better understanding of the mechanisms and regulation of drug metabolism at the molecular level, in particular with respect to the important oxidative enzyme system, the Cytochrome P450 (CYP) family. Around 30 human CYPs have been identified, of which at least 7 play a clinically relevant role in drug (xenobiotic) metabolism. These are CYP1A2, CYP2C8, CYP2C9, CYP2C19, CYP2D6, CYP3A4 and CYP2E1. These enzymes catalyse different oxidative reactions and have different (but partially overlapping) catalytic specificity with respect to substrate binding and to product formation. In Figure 2 the circles schematically represent the different human CYPs with their "probe" substrates and (selective) inducers and inhibitors [5].

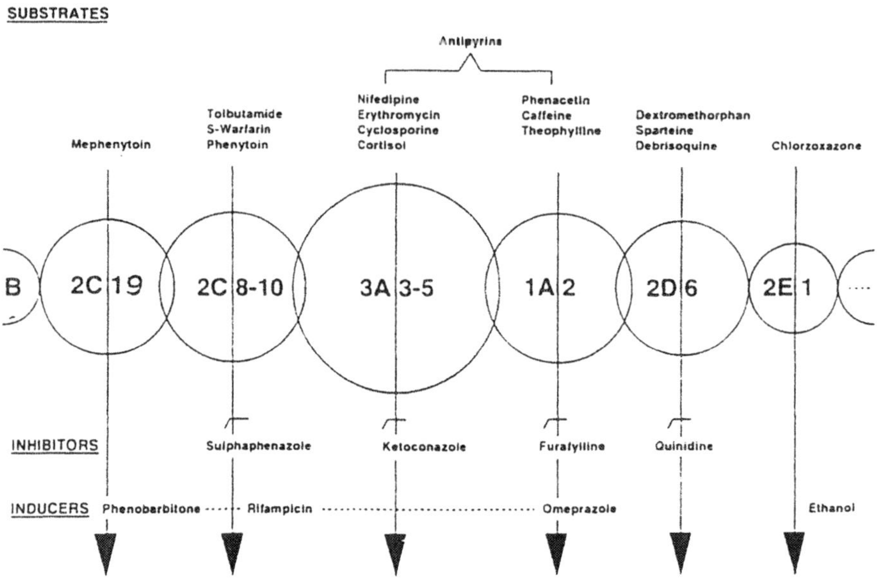


*Figure 2. Schematic representation of human Cytochrome P450 enzymes (CYPs) with model substrates, inhibitors and inducers*

The size of the circles does not represent the relative amount of that CYP present in human liver under healthy conditions. In fact, CYP3A4 is the most abundant with about 25% of total enzyme content and CYP1A2, CYP2C8 and CYP2C9 with each around 10% and CYP2D6 and CYP2C19 around 5%. However, all enzymes exhibit major variability in their activity (10- to more than 100-fold) between different subjects, which is predominantly caused by genetic factors (in particular for those enzymes exhibiting genetic polymorphism: CYP2D6 and CYP2C19) and by environmental factors (CYP3A4).

For many important drugs it is currently known by which enzyme(s) they are predominantly metabolized And for the newly developed drugs it is mandatory that this knowledge becomes available at a very early stage of development. This can be achieved by the application of new methodologies using human material, e.g. liver microsomes, hepatocytes, cell lines, purified enzymes, specific antibodies and by enzyme specific inhibitors. These are the tools to identify which enzyme is catalyzing which metabolic pathway and also indicate the relative importance of the pathway in the overall metabolism of a compound. In vitro inhibition studies with known substrates of high binding affinity of specific enzymes have proved to be very helpful in this respect. For CYP2D6, for example, quinidine may be used and strong inhibition is a clear indication that the inhibited substance may be subject to genetic polymorphism in vivo.

The identification of a specific enzyme catalysing an important metabolic pathway of a new drug may also be very helpful in predicting which drug-drug interactions are likely to occur and thereby help to plan the appropriate (selective) studies in vivo. The in vitro strategies also allow for an early recognition of the fact that a new drug will be subject to genetic polymorphism in (part of) its metabolism. If the drug is judged to be of great therapeutic potential this should not necessarily lead to stopping further development. Alternatively a "back-up" compound not exhibiting polymorphism may be preferred for further investigation and development.

## REFERENCES


1. D.D. Breimer and M. Danhof, Relevance of the application of PK/PD modelling concepts in drug development: the wooden shoe paradigm. Clin. Phamacokin. 32 (1997) 259.
2. C. Peck et al., Opportunities for integration of pharmacokinetics, pharmacodynamics and toxicokinetics in rational drug development. Clin. Pharmacol. Therap. 51 (1992) 465.
3. G. Levy, The case for preclinical pharmacodynamics. In: Integration of pharmacokinetics, pharmacodynamics and tocicokinetics in rational drug development (A. Yacobi, J.P. Skelly, V.P. Shah & L.Z. Benet, editors). Plenum Press, New York (1993) p. 7.
4. J.W. Mandema, M.T. Kuck and M. Danhof, Differences in intrinsic efficacy of benzodiazepines are reflected in their concentration-EEG-effect relationship. Br. J. Pharmacol. 105 (1992) 164.
5. D.D. Breimer, An integrated molecular and kinetic/dynamic approach to metabolism in drug development. In: Proceedings European Conference on Specificity and Variability in Drug Metabolism (G. Alvan et al., editors). European Commission, Luxemburg (1995) p. 3.
6. D.D. Breimer, Genetic polymorphisms in drug metabolism: clinical implications and consequences in ADME studies. In: The Relevance of Ethnic Factors in the Clinical Evaluation of Medicines (S. Walker, C. Lumley & N. McAuslane, editors). Kluwer Academic Publishers, Dordrecht/London (1994) p.13.

H. van der Goot, (Editor)
TRENDS IN DRUG RESEARCH II



# Novel approaches towards anti-HIV chemotherapy

Erik De Clercq

Rega Institute for Medical Research, Katholieke Universiteit Leuven, Minderbroedersstraat 10, B-3000 Leuven, Belgium

The compounds that are currently used for the treatment of human immunodeficiency virus (HIV) infections are targeted at the virus-associated reverse transcriptase or protease. Here, I will describe several new therapeutic approaches that are targeted at the following specific viral events: (i) virus adsorption to the cells; (ii) virus-cell fusion; (iii) virion uncoating (disassembly); (iv) integration of proviral DNA into the cellular genome; (v) viral RNA transcription and translation; and (vi) virion packaging (assembly).

## 1. INTRODUCTION

The replicative cycle of HIV encompasses a number of virus-specific events that could readily serve as targets for chemotherapeutic intervention. Among these targets, particularly the reverse transcription and viral maturation step, catalyzed by the reverse transcriptase (RT) and viral protease, respectively, have been most intensively pursued. This has led to the discovery of an impressive array of both nucleoside/nucleotide RT inhibitors (NRTIs) and non-nucleoside RT inhibitors (NNRTIs) as well as protease inhibitors (PIs) which have either been formally approved or may be approved in the near future for the treatment of HIV infections [1-5].

Of the NNRTIs, AZT, ddI, ddC, d4T and 3TC have been approved, and PMEA (adefovir), its oral prodrug form bis(POM)-PMEA (adefovir dipivoxil), PMPA and its oral prodrug form bis(POC)-PMPA, and 1592U89 (which can be considered as the prodrug of carbovir monophosphate) are under clinical development. Of the NNRTIs, nevirapine and delavirdine have been approved, whereas others, such as loviride, I-EBU (MKC-442), DMP-266 and quinoxaline HBY 097, are under clinical development. Of the PIs, saquinavir, ritonavir, indinavir and nelfinavir have been approved, and others such as VX-478 may soon follow. The NRTIs, NNRTIs and PIs that are currently used or pursued for the treatment of HIV infections are listed in the "Current antiviral agents factfile" [6]. Their structures are depicted in Figs. 1-3 of reference [5].

This review will be focussed on new classes of anti-HIV agents that are targeted at various steps in the HIV replicative cycle other than the reverse transcriptase step and viral protease step.

## 2. NOVEL TARGETS FOR HIV INHIBITORS

### 2.1. Virus adsorption/fusion to the cells

Virus adsorption depends on the interaction of the viral envelope glycoprotein gp120 with the CD4 receptor at the cell surface. This process can be blocked by various polyanionic substances such as polysulfates, polysulfonates, polycarboxylates, polyphosphates and polyoxometalates. Typical examples of this class of polyanionic substances are dextran sulfate, polyvinylalcohol sulfate, polyvinyl sulfonate and naphthalene sulfonate polymers (Figure 1). Clinical testing with the poly(naphthalene sulfonate) PRO 2000 is now under way [7]. Dextran sulfate and its congeners have been generally regarded as non-specific in their antiviral action. Yet, it has proved feasible to make HIV resistant to dextran sulfate due to the accumulation of a number of amino acid substitutions in the viral envelope gp120 glycoprotein (including the mutations K269E, Q278H, N293D in the V3 loop) [8]. This resistance was obtained following repeated passage of HIV in the presence of dextran sulfate (DS), and the DS-resistant virus also appeared to be cross-resistant to polyanionic compounds that are structurally related to dextran sulfate.

Polyanionic substances may not only interfere with the interaction of the viral envelope gp120 with the CD4 receptor, but also with the subsequent interaction of gp120 with the second cellular receptor, i.e. the chemokine receptor CCR-5, thus blocking virus-cell fusion. This has been clearly shown for the 17-mer oligodeoxynucleotide AR 177 (zintevir) (Figure 2) [9]. Upon repeated passage of HIV in the presence of zintevir, resistance arose to zintevir, and the resistant phenotype was associated with the emergence of different mutations (i.e. K148E, Q278H, K290Q and F391I) in the viral envelope gp120 molecule [10]. The latter observations suggest that zintevir is targeted at the viral gp120 rather than the viral integrase (see *infra*).

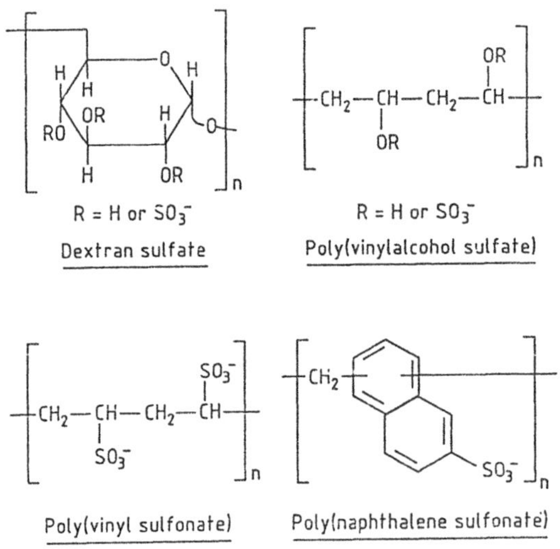


Figure 1. Dextran sulfate, poly(vinylalcohol sulfate), poly(vinyl sulfonate) and poly-(naphthalene sulfonate).

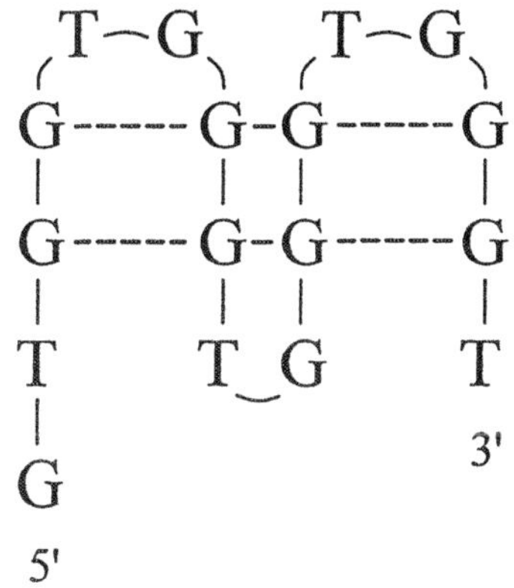

Figure 2. Zintevir, AR 177 (previously designated T30177): 17-mer oligodeoxynucleotide, composed of only deoxyguanosine and deoxythymidine with single phosphorothioate internucleoside linkages at its 5' and 3' ends.

Another class of compounds that have been found to interfere with the virus adsorption/fusion process are the negatively charged albumins (NCAs) [11]. Typical examples of this class of compounds are succinylated and aconitylated human serum albumins (Suc-HSA, Aco-HSA) (Figure 3), lactalbumins and lactoglobulins [12-14]. These negatively charged proteins interact with the V3 loop of the viral gp120 glycoprotein and inhibit virus-cell binding and, especially, virus-cell fusion. Whether the succinylated and aconitylated proteins, like dextran sulfate and zintevir, are able to elicit resistance mutations in gp120 and/or prevent the binding of gp120 to its co-receptor (i.e. CCR-5), remain issues for further investigation.

**2.2. Virus-cell fusion/uncoating**

The bicyclams represent a unique class of compounds that specifically interfere with the virus fusion process, while not affecting virus adsorption [15-17]. This interaction must be highly specific and highly potent, as AMD3100 (previously referred to as JM3100 or SID791), the prototype of the bicyclam derivatives (Figure 4) is able to inhibit HIV replication at nanomolar concentrations that are more than 100,000-fold lower than the concentrations required to impair host cell viability [16]. Furthermore, AMD3100 has been shown to cause a dose-dependent decrease in viremia *in vivo*, in SCID-hu Thy/Liv mice infected with HIV-1 [18]. It has taken more than 60 passages (300 days) *in vitro*, in MT-4 cells, for the virus to become only partially resistant to JM3100 [19]. Several mutations, most of them located in the V3 loop of the viral gp120 glycoprotein (i.e. K269N, N270S, R272T, S274R, Q278H, I288V, N293H, and A297T) were detected, which together must have contributed to the resistant phenotype [20]. Recent evidence [21] suggests that the bicyclams specifically interact with the co-receptor CXCR-4 (fusin) for the HIV envelope glycoprotein gp120.

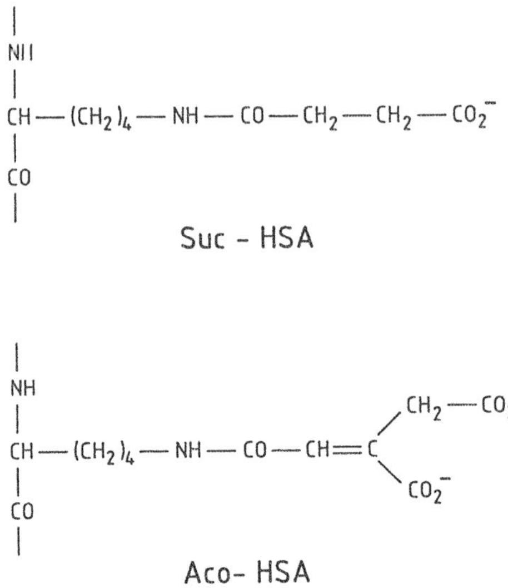


Figure 3. Succinylated (Suc) and aconitylated (Aco) human serum albumin (HSA), obtained through treatment of HSA with succinic anhydride or cis-aconitic anhydride, respectively.

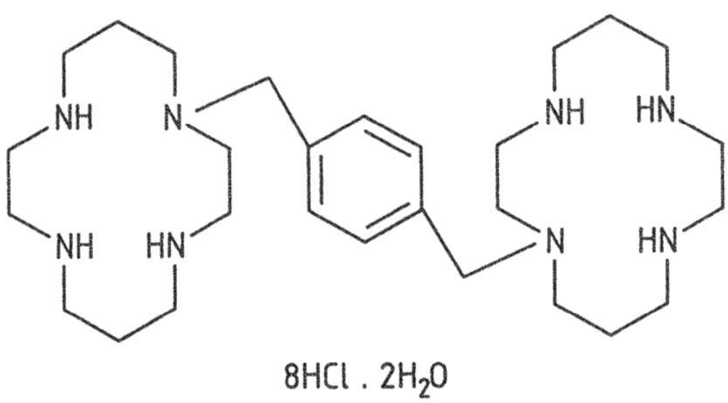


Figure 4. Bicyclam AMD3100 (JM3100): 1,1'-[1,4-phenylenebis-(methylene)]-bis-1,4,8,11-tetraazacyclotetradecane octahydrochloride dihydrate.

Various compounds have been postulated to interact with the virus-cell fusion process: i.e. triterpene derivatives such as betulinic acid RPR 103611 (Figure 5) [22], polyphemusin derivatives such as T22 [Tyr-5,12,Lys-7] polyphemusin II [23] and a 21-mer tricyclic peptide from *Streptomyces* termed siamycin I (or NP-06) (Figure 6) [24,25]. Resistance to siamycin I could be obtained by *in vitro* passage of virus in the presence of increasing concentrations of the compound [25]. No attempts were made to elicit resistance to RPR 103611, but HIV-2 and some HIV-1 strains (i.e. NDK and ELI) were found to be resistant *ab initio* to the betulinic acid derivative [22].

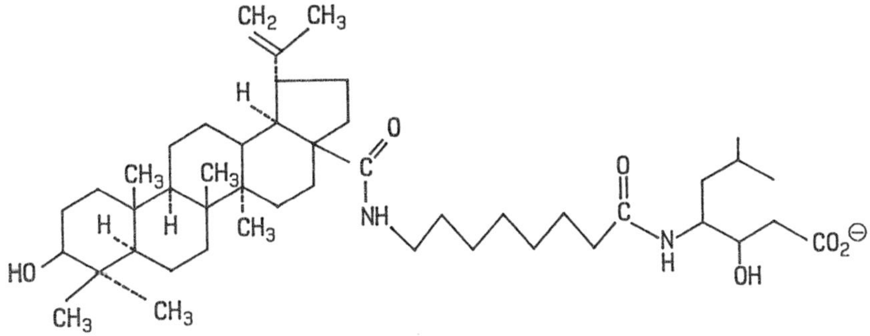


Figure 5. Betulinic acid derivative RPR 103611: N'-{N-[3β-hydroxyl-20(29)-ene-28-oyl]-8-aminooctanoyl}-L-statine.

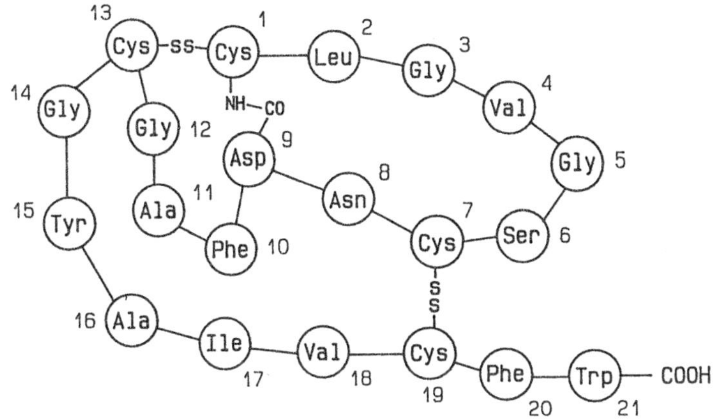


Figure 6. Siamycin I (NP-06).

Another set of fusion inhibitors consists of the synthetic peptides DP-107 and DP-178, which represent the leucine zipper domain (DP-107) and a second putative helical domain (DP-178) of the transmembrane glycoprotein gp41 of HIV-1 (Figure 7). These gp41 derivatives may interfere with the HIV fusogenic process through a molecular clasp mechanism [26]. From DP-178, synthetic 35-residue peptides were derived that inhibited the infectivity of other fusogenic viruses such as respiratory syncytial virus, parainfluenza virus type 3 and measles virus [27].

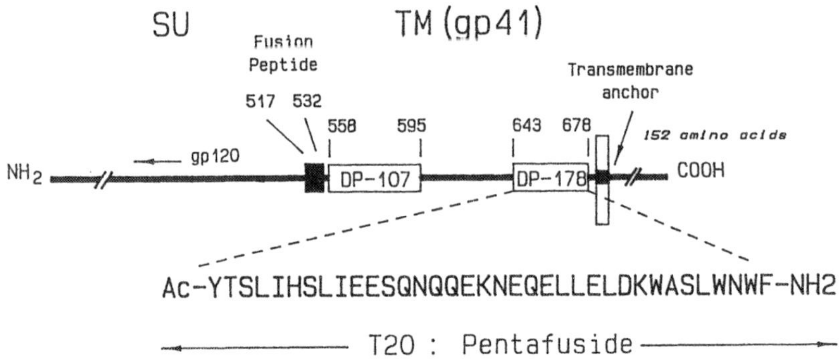


Figure 7. Pentafuside (T20, DP-178).

**2.3. Virus uncoating (disassembly) and packaging (assembly)**

The two zinc [Cys-$X_2$-Cys-$X_4$-His-$X_4$-Cys (CCHC; X, any amino acid)] fingers in the nucleocapsid (NCp7) protein has been considered as the molecular target for Zn-ejecting compounds such as disulfide benzamides (DIBAs) [28], i.e. DIBA-1 (also termed PD022551) (Figure 8). The DIBAs have been found to inhibit the binding of NCp7 to ψ RNA [29]. They are able to interfere with both early and late phases of retroviral replication. Their effect at the late phase (assembly) results in an abnormal processing of the gag precursors (due to the formation of intermolecular cross-linkages among the zinc fingers of adjacent Gag molecules) and the release of noninfectious virus particles [30]. Their effect at the early phase (disassembly) may be ascribed to extensive cross-linkage among adjacent zinc fingers of adjacent NCp7 molecules. DIBAs are able to enter intact virions, and the cross-linkage of NCp7 in virions correlates with loss of infectivity and decreased priviral DNA synthesis during acute infection [31].

The relative rates of HIV inactivation by various disulfide benzamides correlate with their relative kinetic rates of NCp7 zinc ejection, which is consistent with the nucleocapsid protein being the target of action of these compounds [32]. However, disulfide benzamides can be easily reduced yielding two monomeric (R-SH) structures that no longer react with the zinc figures. Therefore, non-dissociable tethered dithiane compounds have been conceived [33]. These compounds, like the DIBA's, specifically attack the retroviral zinc fingers. They directly inactivate HIV-1 virions and block production of infectious virus from chronically infected cells [33]. Similarly, macrocyclic diamides possessing the disulfide linkage [i.e. SRR-SB3 (Figure 8)] have been found to inhibit a late stage of the HIV replicative cycle, which most probably corresponds to the viral RNA packaging (assembly). The fact that SRR-SB3 inhibits virus assembly rather than disassembly (uncoating) may be related to the accessibility and/or affinity of NCp7 for SRR-SB3 during the process of assembly versus disassembly [34]. As a rule, Zn-finger-reactive compounds, such as DIBAs and SRR-SB3, that are targeted at NCp7, may be assumed to interfere with viral infectivity at the level of either virus disassembly or assembly, or both.

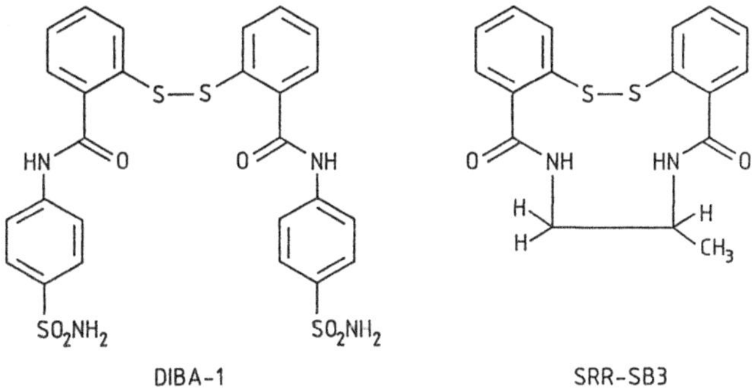


Figure 8. 2,2'-dithiobisbenzamides (i.e. DIBA-1 or PD022551) and macrocyclic 2,2'-dithiobisbenzamides (i.e. SRR-SB3).

The nonimmunosuppressive cyclosporin A analogue [MeIle[4]]cyclosporin (SDZ NIM 811), which differs from cyclosporin A by substitution of N-methyl-L-isoleucine for N-methyl-L-leucine at position 4 (Figure 9), represents another class of molecules that interfere at two stages of the HIV replicative cycle: (i) translocation of the preintegration complex from the cytoplasm to the nucleus and (ii) production of infectious virus particles [35]. The target for the anti-HIV action of [MeIle[4]]cyclosporin is cyclophilin A which, through its peptidyl-prolyl *cis-trans* isomerase activity, assists in the folding of proteins [36]. Cyclophilin A interacts with the HIV-1 Gag polyprotein and is incorporated in the HIV-1 virions. [MeIle[4]]cyclosporin disrupts the Gag-cyclophilin A interaction [37] and inhibits cyclophilin A incorporation into virions [38]. Consequently, [MeIle[4]]cyclosporin causes a dose-dependent inhibition of the formation of infectious virus particles [39,40], and this dose-dependent reduction in infectivity is directly correlated with a reduction in the amounts of virion-associated cyclophilin A [41].

## 2.4. Proviral DNA integration

The aforementioned oligonucleotide AR177 (also termed T30177 or zintevir) (Figure 2) [10] has been shown to inhibit HIV-1 integrase in the concentration range of 0.05-0.1 μM [42]. The overall integration reaction can be dissected in several steps, i.e. (i) formation of an initial stable complex, (ii) 3'-processing (cleavage of dinucleotide at 3'end), (iii) DNA strand transfer and (iv) (dis)integration. Zintevir would inhibit the HIV-1 integrase by preventing the formation of the initial stable complex between integrase and substrate DNA [43]. It is not obvious, however, that the integrase-inhibiting effect of zintevir would contribute to its anti-HIV activity in cell culture, as time-of-addition experiments [42] and resistance mutation studies [10] point to the viral envelope glycoprotein gp120, rather than the integrase, as the main target for the anti-HIV action of zintevir [10]. Likewise, dextran sulfate has been found to inhibit HIV-1 integrase in a cell-free system [43], although there is ample evidence, based on virus-cell binding studies, time-of-addition experiments [15], as well as the analysis of resistance mutations [8], that dextran sulfate owes its anti-HIV activity to inhibition of the virus adsorption/fusion process.

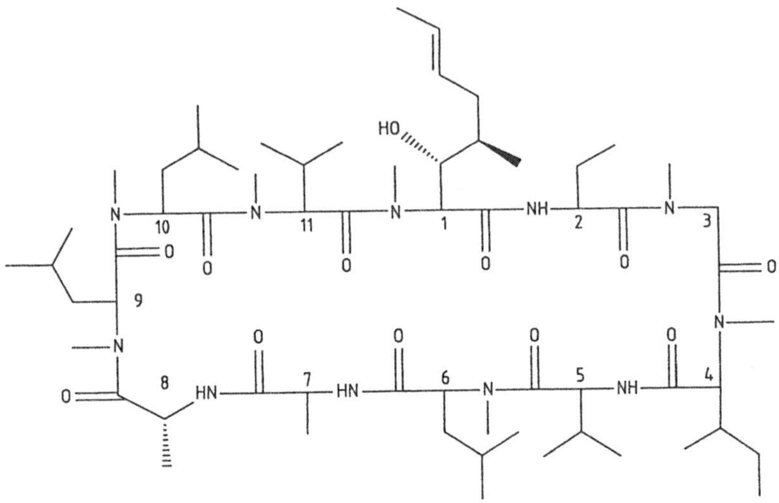

Figure 9. Nonimmunosuppressive cyclosporin A analogue [MeIle[4]]cyclosporin (SDZ NIM 811).

There are a number of compounds, i.e. 3,5-dicaffeoylquinic acid, 1-methoxyoxalyl-3,5-dicaffeoylquinic acid and L-chicoric acid (Figure 10) that have been reported to inhibit both HIV-1 integrase (at concentrations ranging from 0.06-0.66 μg/ml) and HIV-1 replication in cell culture (at 1-4 μg/ml) [44]. These compounds were found to inhibit the core catalytic domain of HIV-1 integrase and the calculated change in internal free energy of the ligand-integrase complex correlated with the ability of the compounds to inhibit HIV-1 integrase [45]. What remains to be shown, however, i.e. by time-of-addition experiments, whether the inhibitory effects of the dicaffeoylquinic acid derivatives on HIV-1 integrase and HIV replication are causally or only co-incidentally related.

The diarylsulfones represent a novel class of HIV-1 integrase inhibitors [46], although it is not clear to what, if any, extent, the inhibitory effects of the diarylsulfones on HIV-1 integrase contribute to their inhibitory effects on HIV replication. In fact, no correlation was found between the 50% inhibitory concentrations ($IC_{50}$) of these compounds for the viral integrase and their $IC_{50}$ values for virus replication. Other compounds, i.e. NSC 158393 (which contains four 4-hydroxycoumarin residues), have been reported to inhibit both HIV-1 integrase and protease [47]. While such dually targeted action could, in principle, be considered as an interesting approach to achieve synergistic anti-HIV activity and reduce the risk of resistance development, it is not evident whether the antiviral activity of these compounds in cell culture is due to inhibition of either, both or neither of these two enzymes.

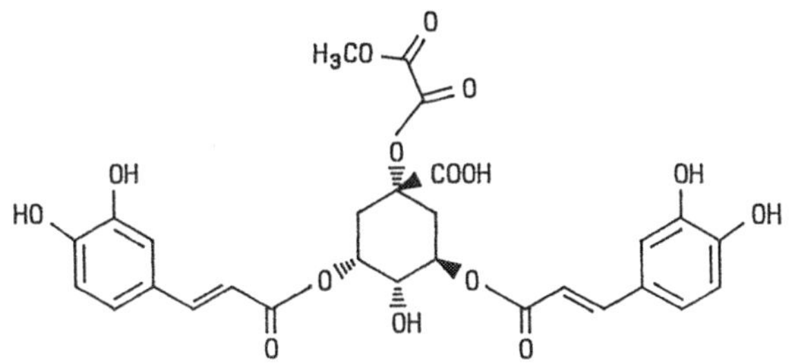

1-Methoxyoxalyl-3, 5-dicaffeoylquinic acid

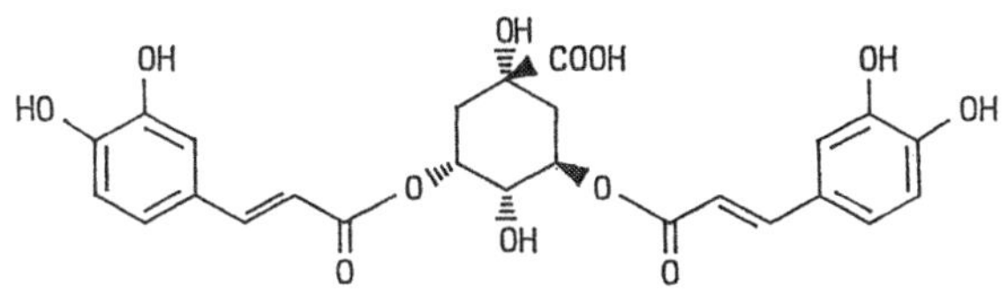

3, 5-Dicaffeoylquinic acid

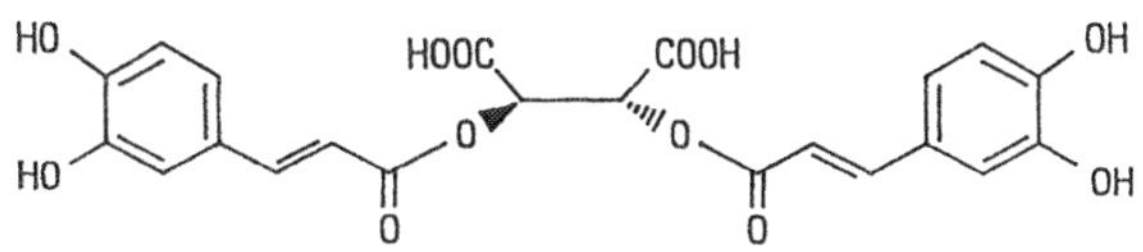

L-Chicoric acid

Figure 10. Dicaffeoylquinic acid derivatives.

**2.5. Viral RNA transcripton and translation**

At the level of viral RNA transcription, the Tat transactivation process that stimulates the DNA-dependent RNA transcription starting from the long terminal repeat (LTR) promotor has been envisaged as an appropriate target for chemotherapeutic intervention. Both benzodiazepin (i.e. Ro 5-3335) and non-benzodiazepin derivatives [i.e. 2-glycineamide-5-chlorophenyl 2-pyrryl ketone (GCPK)] (Figure 11) have been identified as Tat-dependent transcription inhibitors ("Tat antagonists") [48,49]. These compounds are effective against both acute and chronic HIV-1 infections and may be assumed to inhibit HIV-1 replication through their anti-Tat activity. The exact molecular target(s) of the Tat antagonists remain to be defined. This target may well be a cellular protein(s) involved in the Tat transactivation process, as the effectiveness of the Tat antagonists in blocking HIV-1 replication strongly depends on the host cell type [50].

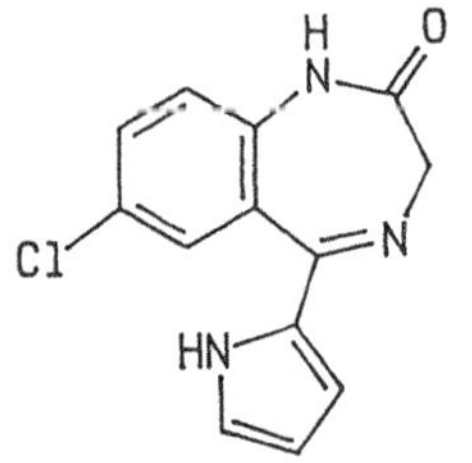

Ro 5-3335

7-Chloro-5-(2-pyrryl)-
3H-1,4-benzodiazepin-2(H)-one

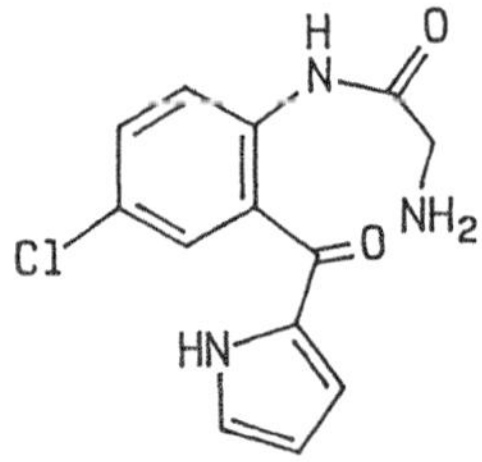

GCPK

2-Glycinamide-5-chlorophenyl
2-pyrrylketone

Figure 11. Benzodiazepin and non-benzodiazepin derivatives.

S-adenosylhomocysteine (AdoHcy) hydrolase inhibitors, such as neplanocin A, 3-deazaneplanocin A and carbocyclic 3-deazaadenosine (Figure 12), have since long been recognized as broad-spectrum antiviral agents [51]. Their antiviral activity spectrum also encompasses HIV-1 [52], and their anti-HIV activity may be mediated by inhibition of Tat transactivation [53]. Indeed, we have recently demonstrated a close correlation between the inhibitory effects of these adenosine analogues on: (i) HIV-1 Tat transactivation, (ii) HIV-1 replication and (iii) AdoHcy hydrolase activity [53]. We, therefore, presume that, through inhibition of S-adenosylmethionine-dependent methylation reactions, the AdoHcy hydrolase inhibitors may inhibit the transactivation-driven viral mRNA transcriptin from the LTR promotor and thus block HIV replication.

Antisense oligonucleotides can be designed to target specific sequences of the HIV genome and may interfere with several steps in the HIV replicative cycle, i.e. reverse transcription, viral RNA transcription, viral RNA translation, and even virus adsorption (the latter through a non sequence-specific effect). For example, oligonucleotides targeted to the $U_5$ region (within the LTR promotor) or to a site adjacent to the primer binding site inhibit the reverse transcription reaction and thus proviral DNA synthesis [54]. However, most of the antisense oligonucleotides have been designed with the aim of inhibiting the translation of viral mRNAs to viral proteins [55]. A representative example is the oligodeoxynucleotide phosphorothioate GEM 91 (GEM standing for "gene expression modulator") which is complementary to the AUG site of the *gag* region of the HIV-1 genome (Figure 13). GEM 91 may inhibit HIV replication by blocking translation of the *gag* mRNA. It has been found to inhibit the replication of various strains of HIV-1 in different cell systems within the concentration range of 0.1-1 μM [55,56]. GEM 91 serves as the prototype antisense oligonucleotide to be developed as a therapeutic agent for the treatment of HIV-1 infections in humans [57]. Antisense oligonucleotides offer plenty of opportunities for structural modifications in either the base or sugar moieties, or the internucleotide linkages, in order to optimize the molecules in terms of cellular permeability, tissue distribution, nucleolytic stability, and affinity and specificity for their target sequences.

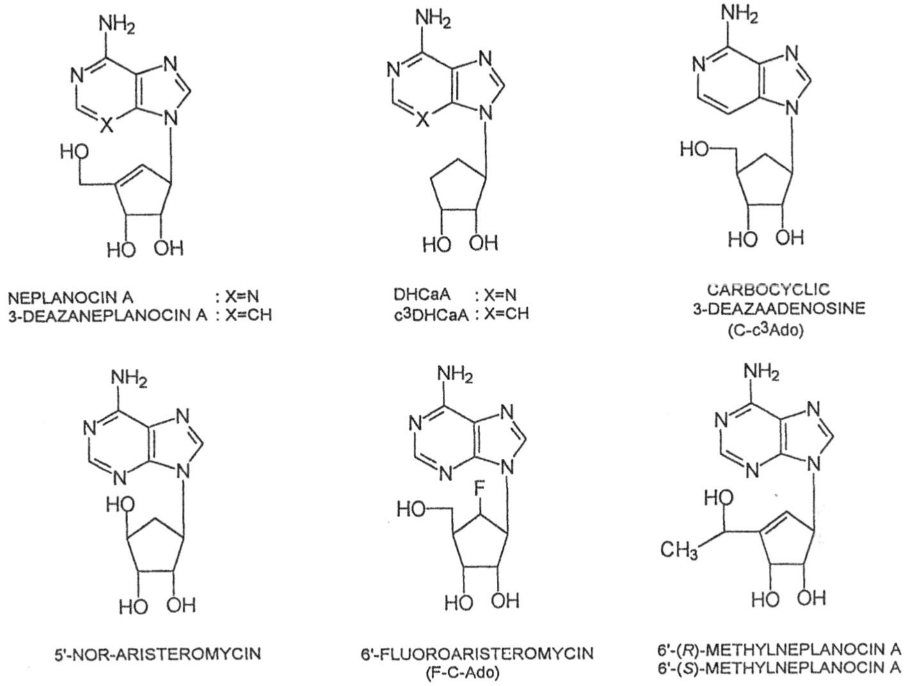


Figure 12. S-adenosylhomocysteine hydrolase inhibitors.

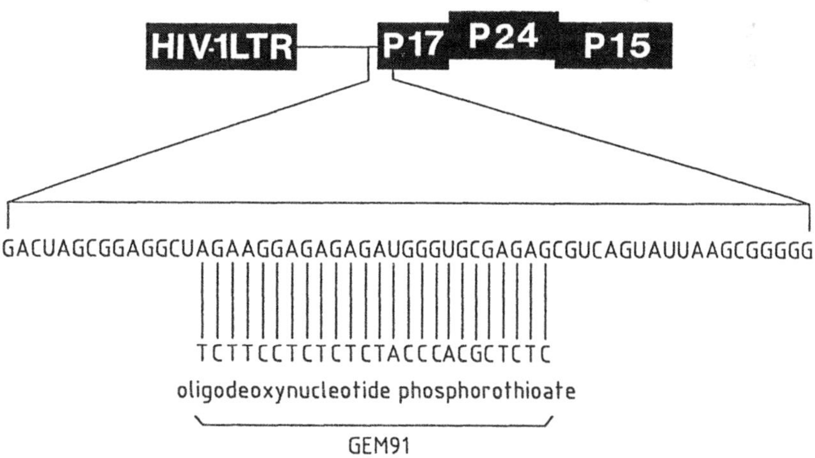


Figure 13. GEM91: 25-mer oligodeoxynucleotide phosphorothioate, complementary to the gag mRNA of HIV-1 at the initiator codon (AUG).

## 3. CONCLUSION

All the compounds that are currently used for the chemotherapy of HIV infections are targeted at either the reverse transcriptase (RT) or viral protease. The RT inhibitors fall into three categories: (i) nucleoside type of RT inhibitors (i.e. AZT, ddI, ddC, ...); (ii) nucleotide type of RT inhibitors (i.e. PMEA, PMPA, ...); (iii) and non-nucleoside type of RT inhibitors (i.e. nevirapine, delavirdine, loviride, ...). Whereas the non-nucleoside RT inhibitors and protease inhibitors can interact directly with their target site at the reverse transcriptase or protease, the nucleoside and nucleotide type of compounds need respectively three and two phosphorylation steps to be converted to their active metabolites, the triphosphate derivatives before they can interact (competitively) with the natural substrate binding site of RT. In addition to the RT and viral protease step, several other virus-specific events have proved to be adequate targets for chemotherapeutic agents, such as the virus adsorption/fusion process (i.e. dextran sulfate, polyvinylsulfate and -sulfonate, zintevir, Suc-HSA, Aco-HSA, ...), the fusion/uncoating process (i.e. pentafuside T20, bicyclam AMD3100, ...), viral assembly/disassembly (i.e. dithiobisbenzamides and cyclosporin A analogues), proviral DNA integration (zintevir, dicaffeoylquinic acid derivatives), Tat transactivation (AdoHcy hydrolase inhibitors) and viral mRNA translation (antisense oligonucleotides). The revelation of new targets and development of new compounds that specifically interact with these targets offer new therapeutic opportunities for the treatment of HIV infections.

## ACKNOWLEDGMENTS

The original investigations of the author have been supported by the Biomedical Research Programme of the European Commission, the Belgian Nationaal Fonds voor Wetenschappelijk Onderzoek, the Belgian Geconcerteerde Onderzoeksacties and the Janssen Research Foundation. I thank Christiane Callebaut for her dedicated editorial assistance.

## REFERENCES

1. E. De Clercq, J. Med. Chem., 38 (1995) 2491.
2. E. De Clercq, Clin. Microbiol. Rev., 8 (1995) 200.
3. E. De Clercq, Rev. Med. Virol., 6 (1996) 97.
4. E. De Clercq, *In*: "Insight in HIV Disease Management", Issue 6:1, C.S. Crumpacker (ed.). Meniscus Health Care Communications, Bala Cynwyd, Pennsylvania, USA, in press (1997).
5. E. De Clercq, Int. J. Antimicrob. Agents, submitted for publication (1997).
6. D. Kinchington, C. Minshull and C. Drummond, Int. Antiviral News, 4 (1996) 132.
7. S. Rusconi, M. Moonis, D.P. Merrill, P.V. Pallai, E.A. Neidhardt, S.K. Singh, K.J. Willis, M.S. Osburne, A.T. Profy, J.C. Jenson and M.S. Hirsch, Antimicrob. Agents Chemother., 40 (1996) 234.
8. J.A. Esté, D. Schols, K. De Vreese, K. Van Laethem, A.-M. Vandamme, J. Desmyter and E. De Clercq, Mol. Pharmacol., in press (1997).

9. J.A. Esté, P. Proost, S. Struyf, A. Wuyts, J. Van Damme, E. De Clercq and D. Schols, Abstracts of the Tenth International Conference on Antiviral Research, Atlanta, Georiga, April 6-11, 1997, Antiviral Res., 34 (1997) p. A41, no. 1.
10. J.A. Esté, D. Schols, K. De Vreese, P. Cherepanov, M. Witvrouw, C. Pannecouque, Z. Debyser, R.F. Rando, J. Desmyter and E. De Clercq, Mol. Pharmacol., submitted for publication (1997).
11. M.E. Kuipers, J.G. Huisman, P.J. Swart, M.-P. de Béthune, R. Pauwels, H. Schuitemaker, E. De Clercq and D.K.F. Meijer, J. Acquir. Immune Defic. Syndr. Hum. Retrovir., 11 (1996) 419.
12. R.W. Jansen, G. Molema, R. Pauwels, D. Schols, E. De Clercq and D.K.F. Meijer, Mol. Pharmacol., 39 (1991) 818.
13. R.W. Jansen, D. Schols, R. Pauwels, E. De Clercq, and D.K.F. Meijer, Mol. Pharmacol., 44 (1993) 1003.
14. P.J. Swart, M.E. Kuipers, C. Smit, R. Pauwels, M.P. de Béthune, E. De Clercq, D.K.F. Meijer and J.G. Huisman, AIDS Res. Hum. Retrovir., 12 (1996) 769.
15. E. De Clercq, N. Yamamoto, R. Pauwels, M. Baba, D. Schols, H. Nakashima, J. Balzarini, Z. Debyser, B.A. Murrer, D. Schwartz, D. Thornton, G. Bridger, S. Fricker, G. Henson, M. Abrams and D. Picker, Proc. Natl. Acad. Sci. USA, 89 (1992) 5286.
16. E. De Clercq, N. Yamamoto, R. Pauwels, J. Balzarini, M. Witvrouw, K. De Vreese, Z. Debyser, B. Rosenwirth, P. Peichl, R. Datema, D. Thornton, R. Skerlj, F. Gaul, S. Padmanabhan, G. Bridger, G. Henson and M. Abrams, Antimicrob. Agents Chemother., 38 (1994) 668.
17. K. De Vreese, D. Reymen, P. Griffin, A. Steinkasserer, G. Werner, G.J. Bridger, J. Esté, W. James, G.W. Henson, J. Desmyter, J. Anné and E. De Clercq, Antiviral Res., 29 (1996) 209.
18. R. Datema, L. Rabin, M. Hincenbergs, M.B. Moreno, S. Warren, V. Linquist, B. Rosenwirth, J. Seifert and J.M. McCune, Antimicrob. Agents Chemother., 40 (1996) 750.
19. J.A. Esté, K. De Vreese, M. Witvrouw, J.-C. Schmit, A.-M. Vandamme, J. Anné, J. Desmyter, G.W. Henson, G. Bridger and E. De Clercq, Antiviral Res., 29 (1996) 297.
20. K. De Vreese, V. Kofler-Mongold, C. Leutgeb, V. Weber, K. Vermeire, S. Schacht, J. Anné, E. De Clercq, R. Datema and G. Werner, J. Virol., 70 (1996) 689.
21. D. Schols, J.A. Esté, G. Henson and E. De Clercq, Antiviral Res., in press (1997).
22. J.-F. Mayaux, A. Bousseau, R. Pauwels, T. Huet, Y. Hénin, N. Dereu, M. Evers, F. Soler, C. Poujade, E. De Clercq and J.-B. Le Pecq, Proc. Natl. Acad. Sci. USA, 91 (1994) 3564.
23. H. Nakashima, M. Masuda, T. Murakami, Y. Koyanagi, A. Matsumoto, N. Fujii and N. Yamamoto, Antimicrob. Agents Chemother., 36 (1992) 1249.
24. S. Chokekijchai, E. Kojima, S. Anderson, M. Nomizu, M. Tanaka, M. Machida, T. Date, K. Toyota, S. Ishida, K. Watanabe, H. Yoshioka, P.P. Roller, K. Murakami and H. Mitsuya, Antimicrob. Agents Chemother, 39 (1995) 2345.
25. P.-F. Lin, H. Samanta, C.M. Bechtold, C.A. Deminie, A.K. Patick, M. Alam, K. Riccardi, R.E. Rose, R.J. White and R.J. Colonno, Antimicrob. Agents Chemother., 40 (1996) 133.
26. C.-H. Chen, T.J. Matthews, C.B. McDanal, D.P. Bolognesi and M.L. Greenberg, J. Virol., 69 (1995) 3771.
27. D.M. Lambert, S. Barney, A.L. Lambert, K. Guthrie, R. Medinas, D.E. Davis, T. Bucy, J. Erickson, G. Merutka and S.R. Petteway Jr., Proc. Natl. Acad. Sci. USA, 93 (1996) 2186.
28. W.G. Rice, J.G. Supko, L. Malspeis, R.W. Buckheit Jr., D. Clanton, M. Bu, L. Graham, C.A. Schaeffer, J.A. Turpin, J. Domagala, R. Gogliotti, J.P. Bader, S.M. Halliday, L. Coren, R.C. Sowder II, L.O. Arthur and L.E. Henderson, Science, 270 (1995) 1194.
29. P.J. Tummino, J.D. Scholten, P.J. Harvey, T.P. Holler, L. Maloney, R. Gogliotti, J. Domagala and D. Hupe, Proc. Natl. Acad. Sci. USA, 93 (1996) 969.
30. J.A. Turpin, S.J. Terpening, C.A. Schaeffer, G. Yu, C.J. Glover, R.L. Felsted, E.A. Sausville and W.G. Rice, J. Virol., 70 (1996) 6180.
31. J.A. Turpin, C.A. Schaeffer, S.J. Terpening, L. Graham, M. Bu and W.G. Rice, Antiviral Chem. Chemother., 8 (1997) 60.
32. P.J. Tummino, P.J. Harvey, T. McQuade, J. Domagala, R. Gogliotti, J. Sanchez, Y. Song and D. Hupe, Antimicrob. Agents Chemother., 41 (1997) 394.
33. W.G. Rice, D.C. Baker, C.A. Schaeffer, L. Graham, M. Bu, S. Terpening, D. Clanton, R. Schultz, J.P. Bader, R.W. Bucheit Jr., L. Field, P.K. Singh and J.A. Turpin, Antimicrob. Agents Chemother., 41 (1997) 419.
34. M. Witvrouw, J. Balzarini, C. Pannecouque, S. Jhaumeer-Laulloo, J.A. Esté, D. Schols, P. Cherepanov, J.-C. Schmit, Z. Debyser, A.-M. Vandamme, J. Desmyter, S.R. Ramadas and E. De Clercq, Antimicrob. Agents Chemother., 41 (1997) 262.

35. A. Billich, F. Hammerschmid, P. Peichl, R. Wenger, G. Zenke, V. Quesniaux and B. Rosenwirth, J. Virol., 69 (1995) 2451.
36. M. Thali, A. Bukovsky, E. Kondo, B. Rosenwirth, C.T. Walsh, J. Sodroski and H.G. Göttlinger, Nature, 372 (1994) 363.
37. E.K. Franke and J. Luban, Virology, 222 (1996) 279.
38. E. Mlynar, D. Bevec, A. Billich, B. Rosenwirth and A. Steinkasserer, J. Gen. Virol., 78 (1997) 825.
39. B. Rosenwirth, A. Billich, R. Datema, P. Donatsch, F. Hammerschmid, R. Harrison, P. Hiestand, H. Jaksche, P. Mayer, P. Peichl, V. Quesniaux, F. Schatz, H.-J. Schuurman, R. Traber, R. Wenger, B. Wolff, G. Zenke and M. Zurini, Antimicrob. Agents Chemother., 38 (1994) 1763.
40. A. Steinkasserer, R. Harrison, A. Billich, F. Hammerschmid, G. Werner, B. Wolff, P. Peichl, G. Palfi, W. Schnitzel, E. Mlynar and B. Rosenwirth, J. Virol., 69 (1995) 814.
41. S.R. Bartz, E. Hohenwalter, M.-K. Hu, D.H. Rich and M. Malkovsky, Proc. Natl. Acad. Sci. USA, 92 (1995) 5381.
42. J.O. Ojwang, R.W. Buckheit, Y. Pommier, A. Mazumder, K. De Vreese, J.A. Esté, D. Reymen, L.A. Pallansch, C. Lackman-Smith, T.L. Wallace, E. De Clercq, M.S. McGrath and R.F. Rando, Antimicrob. Agents Chemother., 39 (1995) 2426.
43. P. Cherepanov, J.A. Esté, R.F. Rando, J.O. Ojwang, G. Reekmans, R. Steinfeld, G. David, E. De Clercq, and Z. Debyser, Mol. Pharmacol., submitted for publication (1997).
44. W.E. Robinson, Jr., M.G. Reinecke, S. Abdel-Malek, Q. Jia and S.A. Chow, Proc. Natl. Acad. Sci. USA, 93 (1996) 6326.
45. W.E. Robinson, Jr., M. Cordeiro, S. Abdel-Malek, Q. Jia, S.A. Chow, M.G. Reinecke and W.M. Mitchell, Mol. Pharmacol., 50 (1996) 846.
46. N. Neamati, A. Mazumder, H. Zhao, S. Sunder, T.R. Burke, Jr., R.J. Schultz and Y. Pommier, Antimicrob. Agents Chemother., 41 (1997) 385.
47. A. Mazumder, S. Wang, N. Neamati, M. Nicklaus, S. Sunder, J. Chen, G.W.A. Milne, W.G. Rice, T.R. Burke, Jr. and Y. Pommier, J. Med. Chem., 39 (1996) 2472.
48. M.-C. Hsu, A.D. Schutt, M. Holly, L.W. Slice, M.I. Sherman, D.D. Richman, M.J. Potash and D.J. Volsky, Science, 254 (1991) 1799.
49. T. Kira, K.-I. Hashimoto, M. Baba, T. Okamoto and S. Shigeta, Antiviral Res., 32 (1996) 55.
50. M. Witvrouw, R. Pauwels, A.-M. Vandamme, D. Schols, D. Reymen, N. Yamamoto, J. Desmyter and E. De Clercq, Antimicrob. Agents Chemother., 36 (1992) 2628.
51. E. De Clercq, Biochem. Pharmacol., 36 (1987) 2567.
52. D.L. Mayers, J.A. Mikovits, B. Joshi, I.K. Hewlett, J.S. Estrada, A.D. Wolfe, G.E. Garcia, B.P. Doctor, D.S. Burke, R.K. Gordon, J.R. Lane and P.K. Chiang, Proc. Natl. Acad. Sci., 92 (1995) 215.
53. D. Daelemans, J.A. Esté, M. Witvrouw, C. Pannecouque, H. Jonckheere, C.-F. Perno, E. De Clercq and A.-M. Vandamme, Mol. Pharmacol., submitted for publication (1997).
54. B. Bordier, M. Perala-Heape, G. Degols, B. Lebleu, S. Litvak, L. Sarih-Cottin and C. Hélène, Proc. Natl. Acad. Sci. USA, 92 (1995) 9383.
55. J. Temsamani and S. Agrawal, *In*: Advances in Antviral Drug Design, vol. 2, E. De Clercq (ed.). JAI Press, Inc., Greenwich, Connecticut, USA (1996) p. 1
56. J. Lisziewicz, D. Sun, F.F. Weichold, A.R. Thierry, P. Lusso, J. Tang, R.C. Gallo and S. Agrawal, Proc. Natl. Acad. Sci. USA, 91 (1994) 7942.
57. R. Zhang, R.B. Diasio, Z. Lu, T. Liu, Z. Jiang, W.M. Galbraith, and S. Agrawal, Biochem. Pharmacol., 49 (1995) 929.

H. van der Goot, (Editor)
TRENDS IN DRUG RESEARCH II



# PMEA and PMPA: Acyclic Nucleoside Phosphonates with Potent anti-HIV Activity.

T. Cihlar and N. Bischofberger

Gilead Sciences, Inc., 333 Lakeside Drive, Foster City, California 94404, USA

## INTRODUCTION

Dramatic progress has been made in the last few years in the chemotherapy of HIV-infection and AIDS. The main contributors to this progress are the acceptance of combination therapy with nucleosides (which provides longer duration of response by reducing resistance development) and the introduction of protease inhibitors (which provide additional power in combination therapy to a reduction in viral burden). In such combinations nucleosides continue to play a crucial role which has led to a renewed interest in new agents with improved properties. The currently approved nucleosides for HIV-therapy are AZT, 3TC, d4T, ddC and ddI (Figure 1). The liabilities of these nucleosides include toxicities, development of viral resistance, limited efficacy and inconvenience of dosing for the patients. Thus, new agents with improved properties are urgently needed.

Figure 1. (T=thymine, C=cytosine, Hx=hypoxanthine)

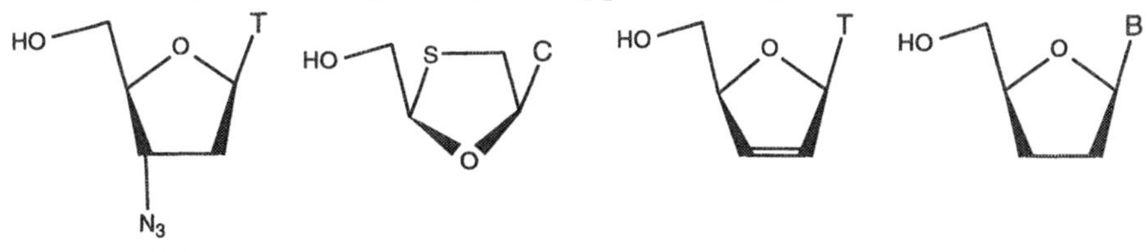


**AZT** **3TC** **d4T** **B=Hx ddI**
**B=C ddC**

This manuscript will describe two promising new agents, PMEA and PMPA. Both are acyclic nucleoside phosphonates and thus represent nucleotide analogs which entails them distinct properties that are different from nucleosides.

## 1. CHEMISTRY

AZT, 3TC, d4T, ddC and ddI exert their anti-HIV activity by inhibiting HIV reverse transcriptase at the level of the corresponding nucleoside triphosphate.

Consequently, these nucleosides have to undergo metabolic activation intracellularly to the triphosphate, and this activation is carried out in three distinct phosphorylation steps by kinases or transferases. E.g., AZT is phosphorylated to its monophosphate by the enzyme thymidine kinase and then to the di- and the triphosphate by other kinases. The conversion to the monophosphate by thymidine kinases can be limiting, especially in resting cells which have only low levels of thymidine kinase present. In contrast, nucleotides, which are phosphorylated nucleosides, bypass this first activation step and could potentially be active in a larger range of cells and tissues resulting in increased potency.

While the rationale for exploring nucleotides is straightforward, in practice, progress in the area of nucleotides has been limited by the number of structural modifications which allow antiviral activity.

Nucleoside phosphates *per se* (e.g., AZT-monophosphate) are metabolically unstable and become degraded rapidly to the parent nucleosides. Phosphonates are chemically and metabolically stable nucleotide analogs but can suffer from greatly decreased affinity to kinases and/or the target HIV-reverse transcriptase (HIV-RT).

Figure 2.

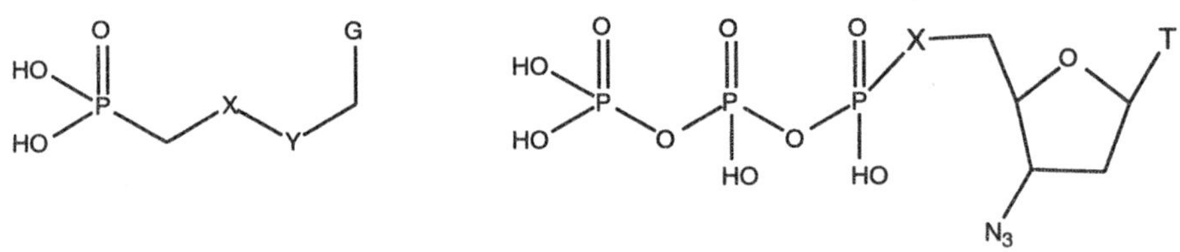


Kcat/Km ($\mu M^{-1} s^{-1}$)

| | | | | |
|---|---|---|---|---|
| 1 X=O | Y=$CH_2$ | | 3 X=O | 0.3 |
| 2 X=$CH_2$ | Y=O | | 4 X=$CH_2$ | 0.0001 |

E.g., compound 1 is phosphorylated by guanylate kinase to the corresponding diphosphate analog fairly efficiently, whereas the isomer 2 is not (Figure 2) (1). The change in affinity for RT by small modifications is illustrated by the fact that 3 (AZT-triphosphate) is a potent inhibitor of HIV-RT, whereas the corresponding methylenephosphonate 4, shows a 3000x decrease in inhibitory activity (Figure 2) (2).

One structural class of nucleotide analogs with unique properties are acyclic phosphonomethylether nucleosides which contain a phosphorous-carbon-oxygen linkage (Figure 3) (3). This class of compounds was pioneered by Holy and De Clercq (4) and contains compounds with a broad spectrum of biological activities, including antiviral, antiproliferative, antiparasitic and immunomodulatory activities. Examples of this class include HPMPA which has shown broad spectrum antiherpesvirus activity, HPMPC (Cidofovir) (5) which is approved for the treatment of CMV retinitis in AIDS patients, PMEG which has antiproliferative activity and compounds with potent antiretrovirus activity D4API, PMEA and PMPA (6) (Figure 3).

Figure 3. (A=adenine, G=guanine, C=cytosine)

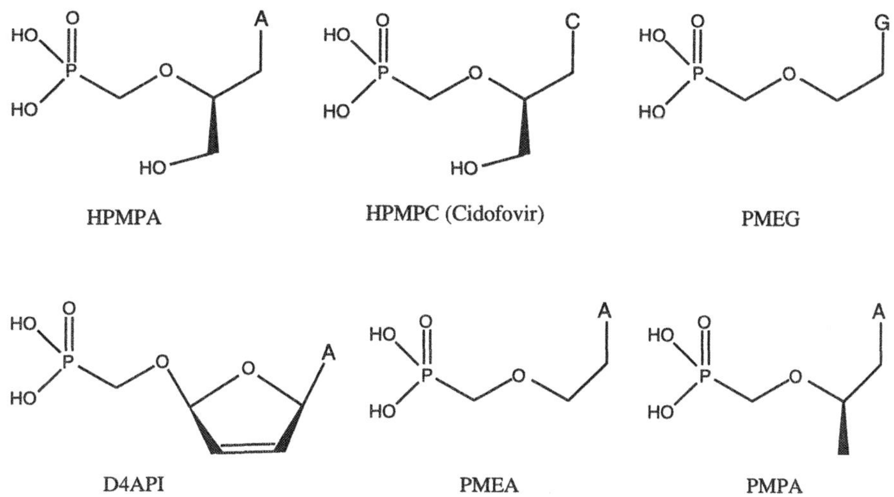


## 2. MECHANISM OF ACTION

The antiviral activity of acyclic nucleoside phosphonates is dependent on three distinct steps: cellular uptake, activation inside the host cell and interaction with the target viral enzyme (Figure 4). These steps determine the efficiency and spectrum of antiviral activity of each specific nucleoside phosphonate.

### 2.1. Cellular Uptake

Nucleoside analogs are transported across the plasma membrane either via facilitated diffusion (e.g. AZT) (7) or by membrane nucleoside transporters (e.g. ddC) (8). In contrast, cellular uptake of nucleoside phosphonates is slower and less efficient due to the negative charge of the phosphonate moiety. PMEA transport has been studied *in vitro* and its mechanism is dependent on the cell type. In H-9 (9) and CEM (10) human T-cells, nonspecific fluid-phase endocytosis was suggested as the primary mechanism of PMEA uptake based on its kinetics, temperature sensitivity, unsaturability and dependence on intracellular concentration of ATP. By contrast, HeLa S3 human carcinoma epithelial cells take up PMEA via receptor-mediated endocytosis. This process is highly specific, since PMEG and other PME- structures efficiently inhibited PMEA uptake, while nucleoside phosphonates with a substitution in their acyclic moiety (e.g. PMPA, HPMPA or HPMPC) were unable to interfere with this process. A 50 kDa protein in the plasma membrane of HeLa S3 cells was identified as a candidate receptor capable of PMEA binding (11). Interestingly, in Vero monkey kidney cells, PMEA uptake is mediated by specific $Na^+$-dependent active transport (12), but HPMPC is transported into these cells via fluid-phase endocytosis (13). These findings suggest that acyclic nucleoside phosphonates can enter cells by different mechanisms depending on the cell type and the specific phosphonate.

Figure 4. Mechanism of antiviral action of PMEA

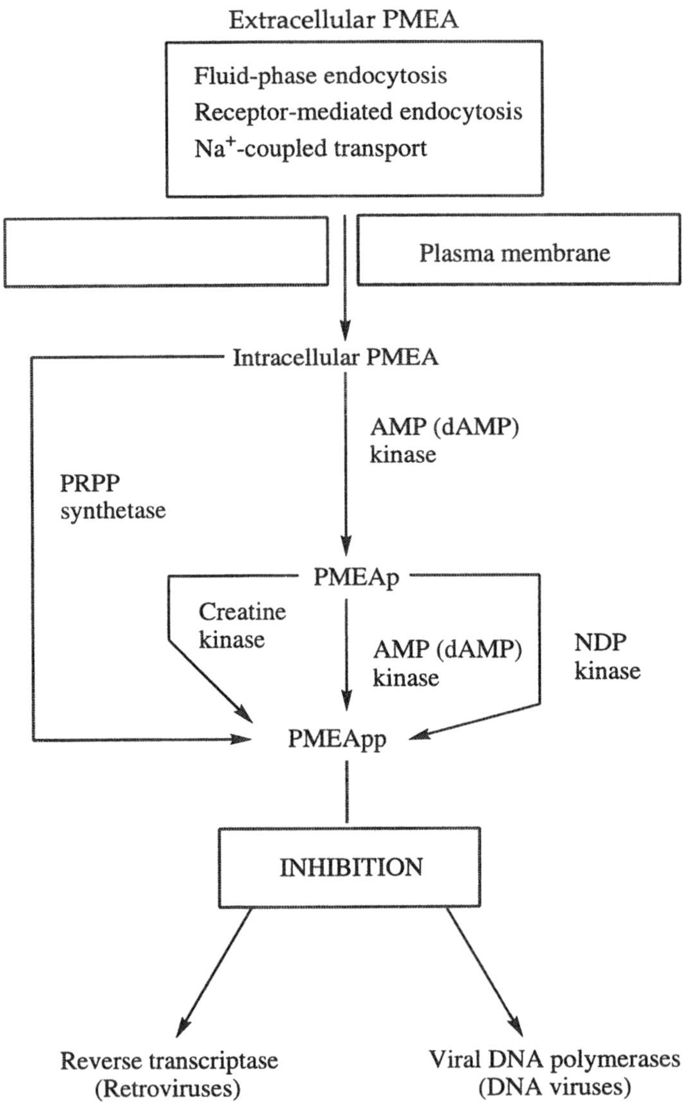

### 2.2. Intracellular Metabolism

After entering the cells, nucleoside phosphonates are activated by phosphorylation to their diphosphoryl derivatives. *In vitro* enzymatic studies suggested at least two different pathways for PMEA phosphorylation. Merta *et al.* showed that AMP(dAMP) kinase purified from murine T-leukemia cells is capable of phosphorylating PMEA and several other adenine nucleoside phosphonates to their corresponding diphosphates in two distinct steps. The monophosphate derivative is an intermediate and the overall phosphorylation efficiency of PMEA to PMEA-diphosphate (PMEApp) is lower by three orders of magnitude than the phosphorylation of AMP (14). As shown by cell fractionation, the mitochondrial form of AMP kinase phosphorylates PMEA 2-fold more efficiently than the cytosolic enzyme (15). In addition, creatine kinase is able to convert PMEA-monophosphate (PMEAp) to PMEA-diphosphate (PMEApp) (14). It is likely, that this phosphorylation step can also be catalyzed by nucleoside diphosphate kinase which has broad substrate specificity and can phosphorylate various nucleoside diphosphate analogs including acyclic structures (16). Independently, Balzarini *et al.* suggested a possible role of 5-phosphoribosyl-1-pyrophosphate (PRPP) synthetase in the activation of various adenine nucleoside phosphonates (17).

In metabolic studies with radiolabeled PMEA, only PMEA, PMEAp and PMEApp were detected in different cell types (18,19), suggesting that no other metabolism of PMEA occurs intracellularly. An important aspect of PMEA metabolism is a persistent intracellular level of its active metabolite PMEApp. Its half-life in T-cells was determined to be approximately 16 hrs (18) in comparison with 4 hrs for d4TTP and even shorter time for AZTTP (20). The long intracellular half-life of the active metabolites is a common feature of the acyclic nucleoside phosphonates and may explain their sustained antiviral activity.

### 2.3. Target Enzymes

Once phosphorylated, the diphosphates of acyclic nucleoside phosphonates function inside the cells as analogs of natural deoxynucleoside triphosphates. As such, they are substrates and/or potent competitive inhibitors of HIV-RT or other viral polymerases. Table 1 shows the inhibition constants ($K_i$) of PMEApp and PMPApp against HIV-1 RT and various viral and cellular DNA polymerases. PMEApp is a slightly more potent inhibitor of HIV RT than PMPApp with a $K_i$ value in the same concentration range as AZTTP (21). Similar to other nucleoside HIV RT inhibitors, nucleoside phosphonates inhibit the reverse transcription step more efficiently than the DNA polymerization step. Due to lack of the 3′-like hydroxyl in their structure, both PMEA and PMPA act as chain terminators after being incorporated by HIV RT into the nascent DNA chain. Both PMPApp and PMEApp efficiently target the HBV DNA polymerase also (22). In addition, PMEApp is a potent competitive inhibitor of herpesviral DNA polymerases, specifically of HSV-1 (23) and HCMV (24). In these cases, its inhibitory effect is comparable to acyclovir- and ganciclovir-triphosphate, respectively. Inhibition of HSV-1 ribonucleotide reductase by PMEApp was also detected (25).

Table 1
Inhibition of viral target enzymes and host DNA polymerases by PMEApp and PMPApp

| | $K_i$ [μM] | | |
|---|---|---|---|
| Enzyme | PMEApp | PMPApp | Reference |
| HIV-1 reverse transcriptase | | | |
| - RNA template | 0.012 (0.23) | 0.022 (0.93) | (21) |
| - DNA template | 0.98 (0.21) | 1.55 (0.33) | (21) |
| HSV-1 DNA polymerase | 0.105 (0.15) | | (23) |
| HCMV DNA polymerase | 0.45 (1.5) | 0.62 (2.0) | (24) |
| HBV DNA polymerase | 0.10* | | (22) |
| Human DNA polymerase α | 1.18 (0.45) | 5.2 (1.92) | (21) |
| Human DNA polymerase β | 70.4 (12.5) | 81.7 (14.3) | (21) |
| Human DNA polymerase γ | 0.97 (1.35) | 59.5 (100) | (21) |

Numbers in parentheses represent $K_i/K_m$ ratio.
* This represents $IC_{50}$ value.

Diphosphates of nucleoside phosphonates interact also with the host DNA polymerases. To some extent, PMEApp inhibits all the replicative DNA polymerases including mitochondrial DNA polymerase γ (21). PMPApp is a less potent inhibitor of each of the cellular polymerases than PMEApp with the most substantial difference in the inhibition of DNA polymerase γ. As shown recently, both diphosphates are also substrates for DNA polymerases α, β and γ, and are able to terminate DNA elongation after incorporation into the primer (26). The poor affinity of PMPApp toward these enzymes seems to, at least in part, explain the lower in vitro cytotoxicity of PMPA in comparison with PMEA.

## 3. *IN VITRO* ANTIVIRAL ACTIVITY

### 3.1. Antiviral Spectrum

Among nucleoside and nucleotide analogs, PMEA is a unique example of an agent exhibiting a dual antiviral effect against both retroviruses and DNA viruses (Table 2). Its antiretroviral spectrum covers HIV-1 and HIV-2, simian immunodeficiency virus (SIV), Rauscher murine leukemia virus, sheep visna virus and several others. PMEA inhibits efficiently *in vitro* HIV-1 replication both in T-lymphocytes and in monocyte/macrophages (27). While the anti-HIV activity of AZT and d4T is much reduced in resting peripheral blood mononuclear cells (PBMCs), PMEA exhibits similar efficiency regardless of the activation of PBMCs (28). This difference is due to the fact that activation of PMEA does not rely on cellular thymidine kinase, an enzyme whose expression is tightly cell cycle regulated.

In addition to its antiretroviral activity, PMEA has been shown to be active *in vitro* against at least six types of human herpesviruses with the most potent activity shown against Epstein-Barr virus (29).

Table 2
In vitro antiviral activity of PMEA and PMPA

| | $EC_{50}$ [µM] | | |
|---|---|---|---|
| Virus | PMEA | PMPA | Reference |
| Retroviruses | | | |
| HIV-1 | | | |
| - T-Cells | 2.5 - 10 | 4.2 - 17 | (30) |
| - Monocytes/Macrophages | 0.02 | 0.04 | (30) |
| - Resting PBMCs | 0.019 | | (28) |
| - Activated PBMCs | 0.05 | | (28) |
| Herpesviruses | | | |
| HSV-1 | 26 | 522 | (31, 32) |
| HCMV | 28 | >300 | (33, 32) |
| Hepadnaviruses | | | |
| Human HBV | 1.2 | 1.5 | (34) |

Unlike PMEA, PMPA and other PMP- structures exhibit potent activity against retroviruses without any significant antiherpesviral activity (32). PMPA exhibits a slightly higher selectivity index in cell culture which could be explained by the low affinity of PMPApp toward host DNA polymerases (see chapter 2.3).
Both PMEA and PMPA were identified also as equally potent inhibitors of HBV production in a human HepG2 hepatoma cell line (34) which is consistent with the similar $K_i$ values of PMEApp and PMPApp determined against HBV DNA polymerase. Among the other hepadnaviruses, replication of duck hepatitis B virus in duck primary hepatocytes is sensitive to inhibition by both of these nucleoside phosphonates.

**3.2. *In Vitro* Resistance**

Development of resistance to drugs by HIV is a major issue for the long term response to AIDS therapy. Searching for the most effective treatment strategies including combination therapies has spurred detailed investigations of the mechanisms of HIV resistance to different inhibitors as well as the characterization of cross-resistance profiles of drug resistant HIV strains.

HIV-1 with decreased susceptibility to PMEA was selected in cell culture in the presence of increasing concentrations of PMEA. Depending on the selection procedure, two independent mutations were identified in the RT gene of selected viruses. The novel K70E amino acid change confers about a 10-fold decreased susceptibility to PMEA and this mutation was also found to be associated with moderate resistance to 3TC (35). The K65R mutation causes a 16-fold decreased susceptibility to PMEA (36). This mutation confers cross-resistance to 3TC and was previously shown to be associated with decreased susceptibility to ddC and ddI (37). The K65R mutant shows only moderately decreased susceptibility to PMPA, while the K70E virus is fully sensitive to PMPA. Of the other known RT mutations, the virus with M184V amino acid change conferring significant (over 600-fold) resistance to 3TC was found to be fully susceptible to PMEA and PMPA (38). In addition, it was also demonstrated that the highly AZT-resistant HIV strains carrying multiple mutations in four different codons of RT gene are only 4- to 8-fold less sensitive to PMEA, while those strains with only a single amino acid change in RT and moderate resistance to AZT, remain fully susceptible to PMEA (39). In addition, SIV carrying the Q151M mutation in the RT gene was fully susceptible to PMEA and PMPA (40). The same Q151M substitution in HIV RT is associated with resistance to multiple nucleosides ("multidrug resistance phenotype"). To date, attempts to select for a HIV strain resistant to PMPA *in vitro* were unsuccessful.

## 4. ANIMAL MODELS FOR THE STUDY OF *IN VIVO* ANTIVIRAL ACTIVITY

The antiviral activity spectrum of PMEA and PMPA as defined *in vitro* in various cell culture systems was further demonstrated in many animal models in which PMEA and PMPA were found to be highly effective and selective agents for both treatment and prophylaxis of viral infections. Recently, a comprehensive review covering in detail all the *in vivo* models in which nucleoside phosphonates have been evaluated was published (41).

The antiherpesviral activity of PMEA was demonstrated in a significant number of animal models including genital herpes, herpetic keratitis or retinitis, and systemic cytomegalovirus infections in immunocompromised animals. To date, the antihepadnaviral activity of PMEA was characterized in a duck hepatitis B virus model, in which the drug caused a significant decline of serum viral DNA in treated animals.

One of the first models used for evaluation of *in vivo* antiretroviral activity of PMEA was Moloney murine sarcoma virus (MSV) infection in newborn mice. In this model, PMEA suppressed the MSV-induced tumor development with 25-fold higher efficiency and 5-fold better selectivity than AZT (42). The study of treatment schedule in this model revealed that infrequent dosing was optimal for anti-MSV activity and that the drug can be effective even when given prophylactically. The antiretroviral activity of PMEA was confirmed in two additional murine models using infection with Friend leukemia virus (43) or with LP-BM5 retrovirus complex ('murine AIDS'). In the latter model, efficient *in vivo* dual antiviral activity of PMEA was proven since the drug not only inhibited retroviral replication, but also suppressed various herpesviral infections in coinfection experiments (44). The efficacy of PMEA has also been demonstrated in feline retrovirus models (45).

Visna virus infection of sheep is an attractive *in vivo* model of neurotropic retrovirus. Infected lambs treated with PMEA showed reduction in brain lesions and suppression of viral replication in central nervous system (46). These data suggest that PMEA is able to cross efficiently the blood-brain barrier.

Most importantly, both PMEA and PMPA were tested against SIV infection in rhesus macaques. This model is considered to be the most relevant animal model for HIV infection in humans. Initial experiments in this model showed the ability of PMEA to suppress induction of anti-SIV gp120 antibodies in SIV-infected animals (47). Later studies indicated a significant preexposure prophylactic effect of PMEA treatment which could completely prevent the development of acute SIV infection in more than 80% of animals treated subcutaneously for 4 weeks beginning 48 hours before SIV inoculation (48). Comparative experiments with AZT under identical conditions showed prevention of SIV infection only in 6% of experimental animals. The higher potency of PMEA in this model may be explained by the long intracellular half-life of its active metabolite and/or by retaining its anti-HIV activity in resting lymphocytes. Since the phosphorylation of AZT in resting PBMCs is significantly decreased, the therapeutic effect of PMEA may occur in these cells whereas AZT is ineffective. More recently, experiments with PMPA in the SIV model revealed its postexposure prophylactic activity. Even when the treatment started 24 hours after infection, PMPA showed 100% efficacy in prevention of SIV acute infection at a dose that showed no toxic effects (49). These results suggest a potential role for PMPA postexposure prophylaxis against HIV infection in humans. Both PMEA and PMPA were found to be effective also against established SIV infection. In macaques chronically infected with SIV, either of the two compounds significantly or completely inhibited viral replication during the time of treatment (50). Potent antiviral activity of PMPA was demonstrated also in newborn macaques infected with SIV (51).

## 5. HUMAN STUDIES

Both PMEA and PMPA are not adequately bioavailable by the oral route, presumably due to the negatively charged phosphonate groups. Consequently, both have been converted into prodrugs which have good pharmaceutical properties, are orally absorbed and result in liberating the parent compounds in the systemic circulation. From a large number of prodrugs evaluated, bis(POM)PMEA (adefovir dipivoxil) and bis(POC)PMPA were chosen for clinical development (Figure 5) (52).

Adefovir dipivoxil was found to be 30-40% orally bioavailable in humans. In a Phase I/II clinical study, adefovir dipivoxil administered as a single 125 mg oral tablet once a day showed a statistically significant reduction in plasma HIV viral RNA with a concomitant increase in CD4 cells (53). Adefovir dipivoxil is currently under evaluation in Phase II/III studies in HIV infected individuals.

Bis(POC)PMPA is currently undergoing evaluation for pharmacokinetics, safety and efficacy in a Phase I/II study in HIV infected individuals (54). PMPA, given intravenously, was found to dramatically decrease plasma HIV RNA levels after only 8 doses.

Figure 5.

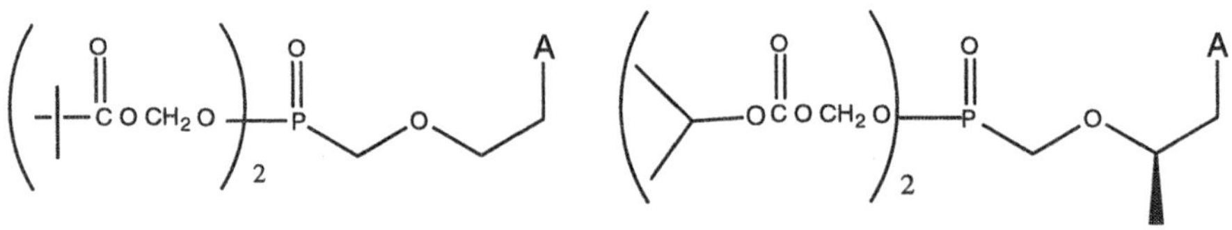


bis(POM)PMEA
adefovir dipivoxil

bis(POC)PMPA

**SUMMARY**

PMEA and PMPA are acyclic nucleoside phosphonates with interesting biological properties. Being nucleotides, their uptake into cells and their intracellular metabolism to the triphosphate analogs is distinct from nucleosides. PMEApp and PMPApp are selective inhibitors of viral polymerases and show less inhibition against human polymerases. Both compounds are efficacious against a number of viruses in a range of different animal models. Orally bioavailable prodrugs of PMEA and PMPA are currently being evaluated for the treatment of HIV infections in humans.

**REFERENCES**

1. C.U. Kim, B.Y. Luh, P.F. Misco, J.J. Bronson, M.J.M. Hitchcock, I. Ghazzouli, and J.C. Martin, J. Med. Chem., 33 (1990) 1207.
2. G.A. Freeman, J.L. Rideout, W. H. Miller, and J.E. Reardon, J. Med. Chem., 35 (1992) 3192.
3. A. Holy, Advances in Antiviral Drug Design, 1 (1993) 179, JAI Press Inc.
4. E. De Clercq, A. Holy, I. Rosenberg, T. Sakuma, J. Balzarini, and P.C. Maudgal, Nature, 323 (1986) 464.
5. M.J.M Hitchcock, H.S. Jaffe, J.C. Martin, and R. J. Stagg, Antiviral Chemistry Chemother., 7 (1996) 115.
6. J. Balzarini and E. De Clercq, in Antiviral Chemotherapy, D.J. Jeffries and E. De Clercq (eds.), John Wiley & Sons Ltd., Chapter 2 (1995).
7. T.P. Zimmerman, W.B. Mahony, and K.L. Prus, J. Biol. Chem., 262 (1987) 5748.
8. B. Ullman, Adv. Exp. Med. Biol., 253B (1989) 415.
9. G. Palú, S. Stefanelli, M. Rassu, C. Parolin, J. Balzarini, and E. De Clercq, Antiviral Res., 16 (1991) 115.
10. L. Olsanská, T. Cihlar, I. Votruba, and A. Holy, Collect. Czech. Chem. Commun., 62 (1997) 1.
11. T. Cihlar, I. Rosenberg, I., Votruba, and A. Holy, Antimicrobial Agents Chemother. 39 (1995) 117.

12. K.L. Prus, E.M. Hill, and M.N. Ellis, Abstract 188, 4th ICAR, New Orleans, Louisiana (1991).
13. M.C. Connelly, B.L. Robbins, and A. Fridland, Biochem. Pharmacol., 46 (1993) 1053.
14. A. Merta, I. Votruba, J. Jindrich, A. Holy, T. Cihlar, I. Rosenberg, M. Otmar, and T.Y. Herve, Biochem. Pharmacol., 44 (1992) 2067.
15. B.L. Robbins, J. Greenhaw, M.C. Connelly, and A. Fridland, Antimicrobial Agents Chemother., 39 (1995) 2304.
16. W.H. Miller and R.L. Miller, Biochem. Pharmacol., 31 (1982) 3879.
17. J. Balzarini, J.F. Nave, M.A. Becker, M. Tatibana, and E. De Clercq, Nucleosides & Nucleotides, 14 (1995) 1861.
18. P.P. Aduma, M.C. Connelly, R.V. Srinivas, and A. Fridland, Mol. Pharmacol., 47 (1995) 816.
19. J. Balzarini, Z. Hao, P. Herdewijn, D.G. Johns, and E. De Clercq, Proc. Natl. Acad. Sci., 88 (1991) 1499.
20. J. Balzarini, Pharmacy World & Sci, 16 (1994) 113.
21. J.M. Cherrington, S.J.W. Allen, N. Bischofberger, and M.S. Chen, Antiviral Chemistry Chemother., 6 (1995) 217.
22. T. Yokota, K. Konno, S. Shigeta, A. Holy, J. Balzarini, and E. De Clercq, Antiviral Chemistry Chemother., 5 (1994) 57.
23. A. Merta, I. Votruba, I. Rosenberg, M. Otmar, H. Hrebabecky, R. Bernaerts, and A. Holy, Antiviral Res., 13 (1990) 209.
24. X.F. Xiong and M.S. Chen, Abstract H29, 36th ICAAC, New Orleans, Louisiana, (1996).
25. J. Cerny, I. Votruba, V. Vonka, I. Rosenberg, M. Otmar, and A. Holy, Antiviral Res., 13 (1990) 253.
26. T. Cihlar and M.S. Chen, Antiviral Chemistry Chemother., 8 (1997) 187.
27. J. Balzarini, C.G. Perno, D. Schols, and E. De Clercq, Biochem. Biophys. Res. Commun., 178 (1991) 329.
28. T. Shirasaka, S. Chokekijchai, A. Yamada, G. Gosselin, J.L. Imbach, and H. Mitsuya, Antimicrobial. Agents Chemother., 39 (1995) 2555.
29. J.C. Lin, E. De Clercq. and J.S. Pagano, Antimicrobial. Agents Chemother., 31 (1987) 1431.
30. J. Balzarini, S. Aquaro, C.F. Perno, M. itrouw, A. Holy, and E. De Clercq, Biochem. Biophys. Res. Commun., 219 (1996) 337.
31. E. De Clercq, T. Sakuma, M. Baba, R. Pauwels, J. Balzarii, I. Rosenberg, A. Holy, Antiviral Res., 8 (1987) 261.
32. J. Balzarini, A. Holy, J. Jindrich, L. Naesens, R. Snoeck, D. Schols, and E. De Clercq, Animicrobial. Agents Chemother., 37 (1993) 332.
33. R. Snoeck, G. Andrei, and E. De Clercq, Eur. J. Clin. Microbiol. Infect. Dis., 15 (1996) 574.
34. R.A. Heijtink, J. Kruining, G.A. DeWilde, J. Balzarini, E. DeClercq, and S.W. Schalm, Antimicrobial. Agents Chemother., 38 (1994) 2180.
35. J.M. Cherrington, A.S. Mulato, M.D. Fuller, and M.S. Chen, Antimicrobial. Agents Chemother., 40 (1996) 2212.

36. A. Foli, M. Sopocio, B. Anderson, M. Kavlick, M.W. Saville, M.A. Wainberg, Z. Gu, J.M. Cherrington, H. Mitsuya, and R. Yarchoan, Antiviral Res., 32 (1996) 91.
37. Z. Gu, S. Horacio, J.M. Cherrington, A. S. Mulato, M.S. Chen, R. Yarchoan, A. Foli, K.M. Sogocio, and M.A. Wainberg. Antimicrobial Agents Chemother., 39 (1995) 1888.
38. J.M. Cherrington, A.S. Mulato, M.D. Fuller, T. Cihlar, and M.S. Chen, Abstract 59, 10th ICAR, Atlanta, Georgia (1997).
39. Y.F. Gong, D.R. Marshall, R.V. Srinivas, and A. Fridland, Antimicrobial. Agents Chemother., 38 (1994) 1683.
40. K.A. Van Rompay, J.L. Greenier, M.L. Marthas, M.G. Otsyula, R.P. Tarara, C.J. Miller, and N.C. Pedersen, Antimicrobial. Agents Chemother., 41 (1997) 278.
41. L. Naesens, R. Snoeck, G. Andrei, J. Balzarini, J. Neyts, and E. De Clercq, Antiviral Chemistry Chemother., 8 (1997) 1.
42. J. Balzarini, L. Naesens, and E. De Clercq, Proc. Natl. Acad. Sci., 86 (1989) 332.
43. L. Naesens, J. Neyts, J. Balzarini, A. Holy, I. Rosenberg, and E. De Clercq, J. of Med. Virol., 39 (1993) 167.
44. J.D. Gangemi, R.M. Cozens, E. De Clercq, J. Balazrini, and H.K. Hochkeppel, Antimicrobial Agents Chemother. 33 (1989) 1864.
45. E.A. Hoover, J.P. Ebner, N.S. Zeidner, and J.I. Mullins, Antiviral Res., 16 (1991) 77.
46. H. Thormar, G. Georgsson, P.A. Palsson, J. Balzarini, L. Naesens, S. Torsteinsdottir, and E. De Clercq, Proc. Natl. Acad. Sci., 92 (1995) 3283.
47. J. Balzarini, L. Naesens, J. Slachmuylders, H. Niphuis, I. Rosenberg, A. Holy, H. Schellekens, and E. De Clercq, AIDS, 5 (1991) 21.
48. C.C. Tsai, K.E. Follis, R. Grant, A. Sabo, R. Nolte, C. Bartz, N. Bischofberger, and R. Benveniste, J. of Med. Primatol., 23 (1994) 175.
49. C.C. Tsai, K.E. Follis, A. Sabo, T.W. Beck, R.F. Grant, N. Bischofberger, R.E. Benveniste, R. Black, Science, 270 (1995) 1197.
50. N. Bischofberger, C.C. Tsai, K.E. Follis, A. Sabo, R.F. Grant, T.W. Beck, P.J. Dailey, and R. Black, Antiviral Res., 30 (1996) A42.
51. K. Van Rompay, J.M. Cherrington, M.L. Marthas, C.J. Berardi, A.S. Mulato,A. Spinner, R.P. Tarara, D.R. Canfield, S. Telm, N. Bischofberger, and N.C. Pedersen, Antimicrobial Agents and Chemother., 40 (1996) 2586.
52. R.J. Jones and N. Bischofberger, Antiviral Res., 27 (1995) 1.
53. S. Deeks, J. Lalezari, A. Pavia, D. Rodrique, H.S. Jaffe, J. Toole, and J. Kahn, 3rd Conference on Retroviruses and Opportunistic Infections, 407, Washington D.C., (1996).
54. N. Bischofberger, L. Naesens, D. De Clercq, A. Fridland, R.V. Srinivas, B.L. Robbins, M. Arimilli, K. Cundy, C. Kim, S. Lacy, W. Lee, J.P. Shaw, 4th Conference on Retroviruses and Opportunistic Infections, 214, Washington, D.C. (1997).

H. van der Goot, (Editor)
TRENDS IN DRUG RESEARCH II



# HBY 097 - a second-generation nonnucleoside inhibitor of the HIV-1 reverse transcriptase

Jörg Peter Kleim

Central Pharma Research / HIV, Hoechst AG, D-65926 Frankfurt, Germany

Nonnucleoside reverse transcriptase inhibitors (NNRTIs) of HIV-1 were first described in 1990 [1]. In contrast to nucleoside analogs (NRTIs) such as 3′-azido-3′-deoxythymidine (AZT, zidovudine) or 2′,3′-dideoxy-3′-thiacytidine (3TC, lamivudine) the NNRTIs do only inhibit the HIV-1 reverse transcriptase (RT) and viral replication. There is virtually no activity detected against HIV-2 or other retroviruses. Strikingly, the whole group of NNRTIs comprises compounds derived from different chemical classes.

First generation NNRTIs helped much to understand the biochemical and structural properties of the target enzyme:

Efforts to resolve the crystal structure of the HIV-1 reverse transcriptase holoenzyme were first successful when the dipyridodiazepinone BI-RG-587 (nevirapine) was used to obtain crystals suitable for X-ray structure determination [2]. It was thereby shown that NNRTIs bind to a small hydrophobic pocket within the p66 subunit of the RT. This pocket is located in close proximity to the polymerase active site, however, the exact mode of enzyme inhibition caused by the binding of an NNRTI to this lipophilic pocket is still unknown.

Two of the first-generation compounds, namely BI-RG-587 (nevirapine) and the bis(heteroaryl)piperazine derivative U-90152 (delavirdine), have recently been approved for the treatment of HIV-1 infection in combination therapy regimens.

A series of quinoxaline derivatives was identified which share most of the features of first-generation NNRTIs outlined above. However, in a set of experiments the lead compound S-2720 (6-chloro-3,3-dimethyl-4-(isopropenyloxycarbonyl)-3,4-dihydroquinoxaline-2(1H)-thione, Fig. 1) was found to be approximately 10-fold more active against laboratory strains of HIV-1, when compared to i.e. BI-RG-587 or U-90152. In addition, S-2720 retained good inhibitory potency when it was tested against HIV mutants bearing resistance-conferring mutations which were selected by other NNRTIs [3, 4].

In vitro dose-escalation studies gave rise to viral variants which were more than $10^3$-fold less sensitive against the quinoxaline and which were not inhibited by the highest drug concentrations applied (20 $\mu$g/ml). These mutants were still sensitive to NRTIs but there was cross-resistance against all other NNRTIs tested. The genotypic basis of the resistant phenotype was shown to be a point mutation leading to a G190‡E amino acid substitution in the RT. In contrast to other drug-resistance mutations this change resulted in a clearly decreased polymerase activity of the mutant RT and in retarded viral growth properties [3, 4, 5].

HBY 097 ((S)-4-isopropoxycarbonyl-6-methoxy-3-(methylthiomethyl)-3,4-dihydroquinoxal-ine-2(1H)-thione, Fig. 1) was selected as a clinical candidate quinoxaline since this compound - in addition to its favourable antiviral properties - was shown to possess a good pharmacokinetic profile in two animal species (6).
Median $IC_{90}$ values measured against a set of 41 clinical isolates of HIV-1, including NRTI-resistant strains, were in the range of 0.9-21 nM. The drug was also shown to be highly active against different subtypes of HIV-1, such as viruses belonging to classes C, E, and O (6).
Drug-resistance was analysed in detail with HBY 097. The detrimental RT G190‡E mutant was found to develop with this compound in cell culture resistance selection experiments.
Other NNRTIs were previously reported to cause the appearance of different mutants affecting amino acid position 190. Therefore, site-specific mutagenesis was performed to introduce various residues instead of the natural glycine 190, which occurs in all published HIV-1 RT sequences, giving rise to a panel of recombinant enzymes. It was observed that the degree of resistance increased with the length of the side chain, whereas the RT polymerase activity decreased with more bulky residues at posisiton 190. Interestingly, BI-RG-587 was found to be inactivated already by conservative changes, such as G190‡A or G190‡S (5).
Dose-escalation studies with HBY 097 revealed the introduction of secondary changes, not affecting the NNRTI-binding site: The appearance of G190‡E was followed by changes at positions 74 and 75. These sites map in a different subdomain of the large RT subunit which is located more than 30 Å away from the location of the primary resistance substitutions and are not in contact with the drug binding region. On the other hand, the sites of these secondary mutations were previously known to be involved in NRTI resistance. In addition, for one of the secondary changes (V75→I) a compensatory effect was measured, restoring at least part the polymerase activity (7).
Therefore, a link was established between resistance against two types of compounds which act by different modes of inhibition, though the significance of changes at positions 74 and 75 is clearly different, comparing NRTI- vs. quinoxaline resistance.

More recently obtained data clearly demonstrate that the genotypic resistance pathways followed by HIV-1 to become insensitive to inhibition by HBY 097 can be manipulated in a defined way: Reducing the selective pressure exerted by the drug leads to an accumulation of up to five different NNRTI-resistance mutations, which together also caused high-level quinoxaline resistance. These results lead to the conclusion that HIV-1 adapts to low selective pressure conditions by avoiding the detrimental G190→E mutant (8).
Therefore, pharmacokinetic parameters will influence the type of mutations occurring with HBY 097 in vivo.

The fact that HBY 097 can exert extremely high pressure on HIV-1 replication *in vitro* makes it a valuable compound within drug cocktails put together to suppress HIV viral load *in vivo*.

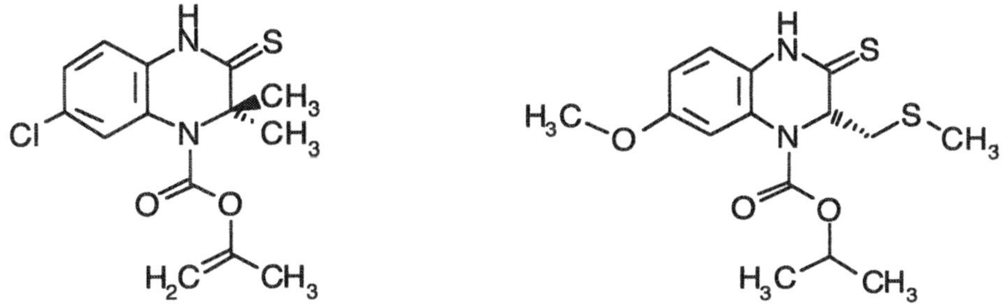


S-2720 HBY 097

Figure 1. Chemical structures of S-2720 and HBY 097

## REFERENCES


1. R. Pauwels, K. Andries, J. Desmyter, D. Schols, M.J. Kukla, H.J. Breslin, A. Raeymackers, J. van Gelder, R. Woestenborghs, J. Heykants, K. Schellekens, M.A.C. Janssen, E. de Clercq, and P.A.J. Janssen. Nature 343 (1990) 470.
2. L.A. Kohlstaedt, J. Wang, J.M. Friedman, P.A. Rice, and T.A. Steitz. Science 256 (1992) 1783.
3. J.-P. Kleim, R. Bender, U.-M. Billhardt, C. Meichsner, G. Rieß, M. Rösner, I. Winkler, and A. Paessens. Antimicrob. Agents Chemother. 37 (1993) 1659.
4. J. Balzarini, A. Karlsson, C. Meichsner, A. Paessens, G. Rieß, E. de Clercq, and J.-P. Kleim. J. Virol. 68 (1994) 7986.
5. J.-P. Kleim, R. Bender, R. Kirsch, C. Meichsner, A. Paessens, and G. Rieß. Virology 200 (1994) 696.
6. J.-P. Kleim, R. Bender, R. Kirsch, C. Meichsner, A. Paessens, M. Rösner, H. Rübsamen-Waigmann, R. Kaiser, M. Wichers, K. E. Schneweis, I. Winkler, and G. Rieß. Antimicrob. Agents Chemother. 39 (1995) 2253.
7. J.-P. Kleim, M. Rösner, I. Winkler, A. Paessens, R. Kirsch, Y. Hsiou, E. Arnold and G. Rieß. Proc. Natl. Acad. Sci. USA 93 (1996) 34.
8. J.-P. Kleim, I. Winkler, M. Rösner, R. Kirsch, H. Rübsamen-Waigmann, A. Paessens and G. Rieß. Virology (1997), in press.

H. van der Goot, (Editor)
TRENDS IN DRUG RESEARCH II



# The HIV Tat-TAR interaction, a novel target for drug discovery

Jonathan Karn[a], Nicholas J. Keen[a], Mark J. Churcher[a], Fareed Aboul-ela[a], Gabriele Varani[a], François Hamy[b], Eduard R. Felder[b], Gerhard Heizmann[b] and Thomas Klimkait[b]

[a]MRC Laboratory of Molecular Biology, Hills Road, Cambridge, CB2 2QH, UK.[b]Novartis, Ltd. Pharmaceuticals, Pharma Research, CH-4002 Basle, Switzerland

## 1. Introduction

The human immunodeficiency virus (HIV) pandemic represents a grave threat to public health and presents a formidable medical challenge. HIV infections are now endemic in the United States and Europe, where there are currently approximately 1.5 million cases. World-wide there are currently at least 8 million infections, and possibly as many as 20 million infections. In the twelve years since the discovery of the human immunodeficiency virus eleven antiviral drugs have been licenced and scores more have entered clinical trials. Yet in spite of this enormous effort, effective regimens for the prevention of progression to clinical symptoms of AIDS and the treatment of AIDS patients have not yet been achieved.

HIV and related lentiviruses have replication cycles that are typical of all retroviruses. The major biochemical steps in virus growth are shown in Figure 1. It is convenient to think of viral replication as comprising two distinct stages: an early pre-integration stage during which the virus infects the cell, reverse transcription takes place and the proviral genome is translocated to the nucleus and integrated into the host cell genome; and a subsequent post-integration stage, during which the integrated provirus expresses its genes, and virus assembly and maturation takes place. Once integrated, the proviral genome is subject to transcriptional regulation by the host cell, as well as its own transcriptional control mechanisms.

In principle, any step in the virus replication cycle presents a target suitable for drug discovery. However, in practice, most drug discovery programmes have emphasized a few of the HIV enzymes (RT, protease, integrase) in order to take advantage of convenient *in vitro* assays as well as the vast experience of pharmaceutical groups in the design of enzyme inhibitors. Current treatment is restricted to inhibitors of reverse transcriptase and/or the viral protease. Although the available drugs effectively reduce the levels of rapidly replicating virus, the benefit is only short-term because of the emergence of resistant strains. The rate at which drug-resistant mutants emerge following therapy is due to a variety of factors, including the efficacy of the drug itself, the viral replication and recombination rates, and the adverse effects of the resistance mutant on the viral growth rate. Most resistance mutants that are detected in clinical trials pre-exist in HIV infected individuals, but because these mutations reduce virus growth rates, they will only emerge as dominant strains during drug treatment.

Combinations of drugs aimed at multiple targets should be able to reduce virus replication sufficiently to make the emergence of resistant variants highly unlikely. Alternatively, it may be possible to find combinations of drugs which, although directed to the same molecular targets, show a sufficiently different spectrum of escape mutations to prevent cross-resistance. For either strategy to be effective, it is essential to develop new, and increasingly effective, drugs. This paper describes our progress in developing novel inhibitor of HIV replication using an RNA target - an inhibitor of the interaction of the *trans*-activator protein (Tat) and the *trans*-activation response region (TAR) RNA.

## 2. The Tat/TAR RNA Interaction

One of the most surprising results to emerge from the intensive research into the molecular biology of HIV was the discovery that the regulatory proteins interact with specific RNA regulatory sequences, rather than DNA targets. The long terminal repeat (LTR) of HIV acts as an inducible promoter which can be stimulated by the *trans*-activator protein (Tat). Tat is required for viral gene expression both during exponential growth of the virus and during the activation of integrated, but latent proviral genomes. Tat activity requires the *trans*-activation responsive region (TAR) which in HIV-1 is a 59-residue stem-loop RNA found at the 5′-end of all viral transcripts. TAR acts as a binding site for Tat, recognition being centred around a U-rich bulge [1]. The apical loop region is thought to participate in *trans*-activation by acting as a binding site for essential cellular co-factors.

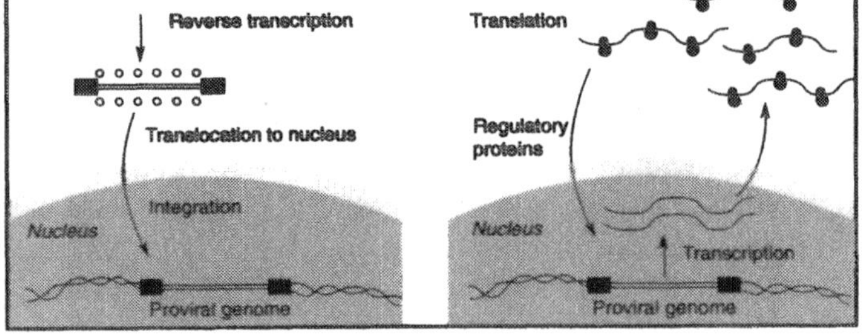


**Figure 1.** Replication cycle of HIV. (Left) Early stages leading to integration of the proviral genome. (Right) Late stages leading to virus assembly. Transcription of the proviral genome requires the HIV regulatory protein Tat, which is synthesized at low levels early during the infection cycle. Fig. from Karn (1995) [18].

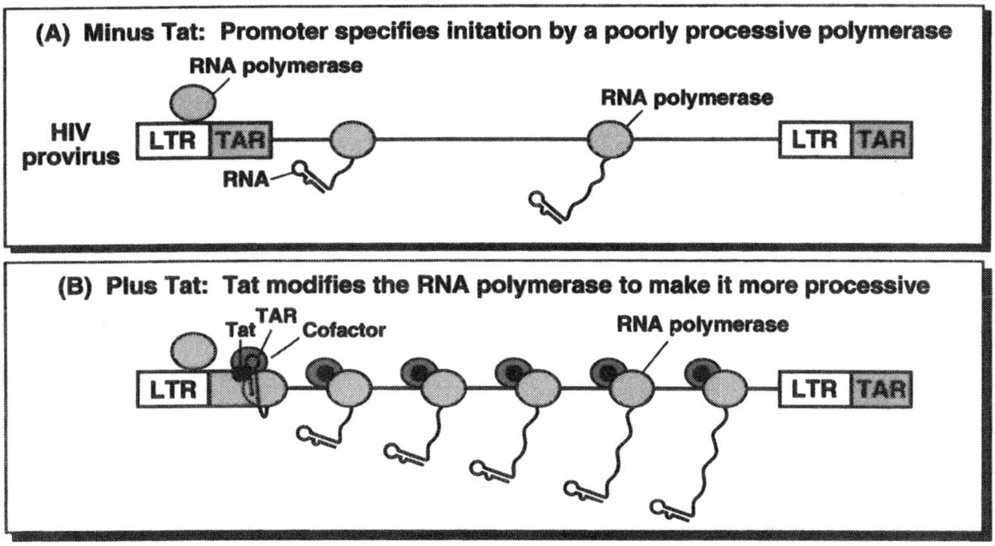


**Figure 2.** Model for the control of transcription elongation by Tat. In a newly infected cell, the binding of cellular transcription factors to the viral LTR DNA sequences stimulates a basal level of transcription. In the absence of Tat the polymerase is unstable and disengages from the template at random sites downstream of TAR. As a result, the majority of the transcripts are short and non-polyadenylated. When Tat is present in the cell, it can bind to nascent TAR RNA sequences and form a complex with RNA polymerase, and cellular co-factors. This new complex is capable of efficient elongation and leads to the production of high levels of the HIV mRNAs. Figure adapted from Karn *et al.* (1996) [2].

In the absence of Tat, the majority of RNA polymerases initiating transcription stall near the promoter [2]. When cells harbouring latent proviruses are stimulated a low level of HIV mRNA synthesis is initiated. As the infection proceeds, the amount of Tat protein in the infected cell increases, and there a dramatic increase in the density of RNA polymerases found downstream of the promoter. Thus, Tat participates in a positive feedback mechanism that ensures high levels of HIV transcription following the activation of cells carrying HIV proviruses.

Our working hypothesis for the mechanism of action of Tat illustrated in Figure 2. According to this proposal, the RNA polymerase engaged by the HIV core promoter is intrinsically unstable. Soon after transcription through TAR, the polymerase either pauses or falls off the template. If Tat is present in the cell, it can associate with TAR RNA and the transcribing RNA polymerase. It seems likely that this reaction involves not only the binding of Tat to TAR RNA but also the recruitment of cellular cofactors, most likely including the TAR RNA loop-binding proteins and cellular elongation factors. The modified transcription complex is then able to transcribe the remainder of the HIV genome efficiently.

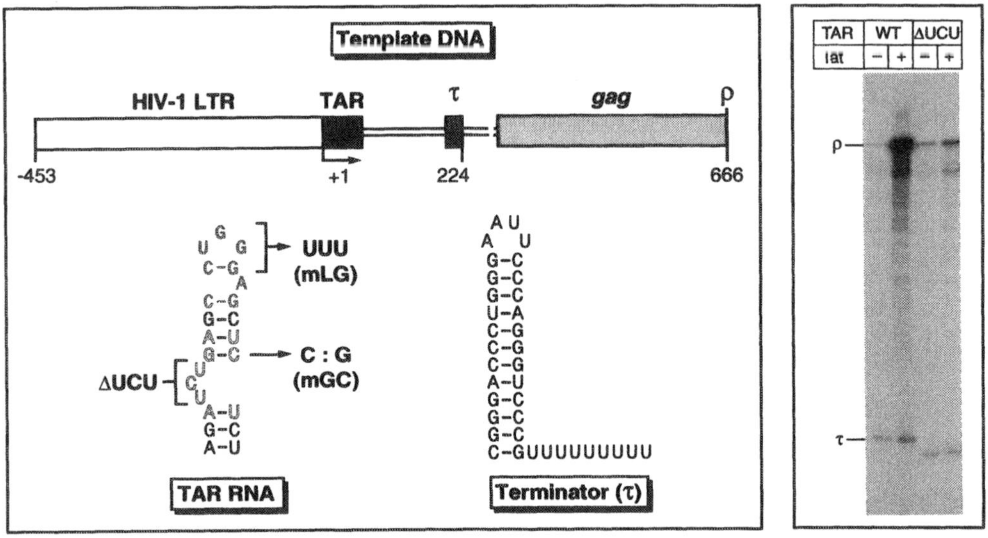


**Figure 3.** Tat stimulates transcriptional elongation but not initiation *in vitro*. (Left) The structure of template DNAs. Each plasmid carried synthetic terminators (τ) inserted 183 nt downstream of the start of transcription. Templates were linearised by cleavage at the *XbaI* site at 668 nt downstream of the start of transcription. (Right) Transcription assay. Transcription reactions were performed in the presence of 0 (-), or 200 ng (+) recombinant Tat protein. Templates carried either an intact TAR element or the ΔU ($\Delta U_{23-25}$) mutations in the TAR element. Addition of Tat protein increased the synthesis of the run-off product (ρ) approximately 25-fold from the wild-type template, but the level of transcription products ending at the terminator (τ) remained relatively constant in the presence or absence of Tat. Figure adapted from Rittner *et al.*, 1995 [5].

## 2.1 Control of transcriptional elongation by Tat

Detailed analysis of the *trans*-activation mechanism is now possible because of the availability of efficient cell-free transcription systems that respond to Tat [3, 4]. The experiment shown in Figure 3 demonstrates that the Tat-stimulated RNA polymerase has an intrinsic anti-termination activity [5]. To measure the elongation capacity of RNA polymerase at a downstream site, a synthetic terminator sequence (τ) consisting of a stable RNA stem-loop structure followed by a tract of 9 uridine residues was placed approximately 200 nts downstream of the start of transcription (Figure 4(A)). The presence of τ caused approximately 30% of the transcribing polymerases to disengage prematurely from the template. Addition of recombinant Tat protein (Figure 4(B)) to the reaction stimulated the production of run-off product (ρ) by more than 25-fold.

Stimulation of transcription by Tat in the cell-free system requires a functional TAR element. Templates carrying the ΔU23-25 mutation in TAR (a mutation that abolishes specific Tat binding *in vitro*) were stimulated less than 3-fold by the addition of Tat (Figure 5(B)). Studies using an extensive series of templates carrying mutations in TAR have shown that Tat-dependent *trans*-activation *in vitro* has the same sequence requirements seen in vivo [3, 5, 6].

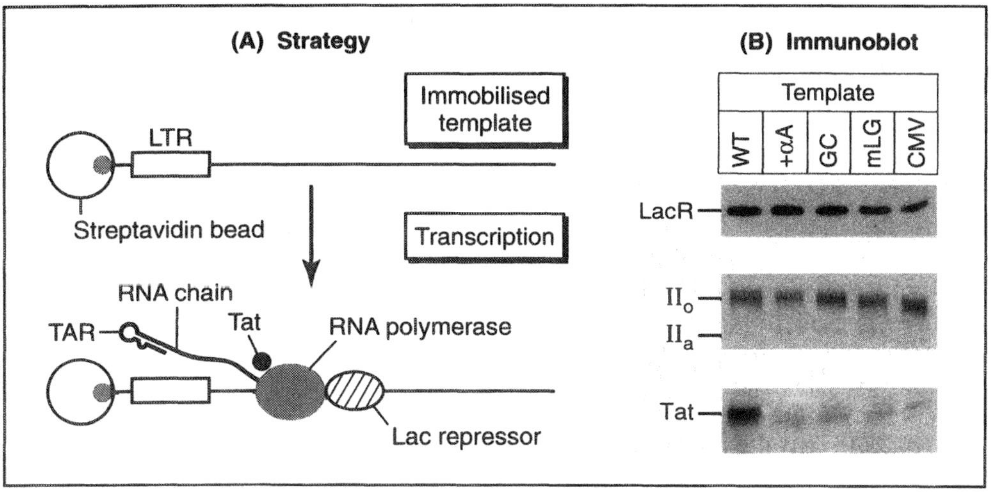


**Figure 4.** A. Experimental strategy for analysing active transcription complexes. Step 1: Cell-free transcription reactions are performed using biotinylated templates in the presence of *lac* repressor protein (LacR). Step 2: Transcription complexes arrested by LacR are purified by binding to streptavidin-coated magnetic beads. The diagram depicts a complex between polymerase and the Tat protein. B. Analysis of the protein composition of the arrested transcription complexes by immunoblotting. Upper panel: LacR. Middle panel: RNA polymerase II large subunit. Only the hyperphosphorylated ($II_O$) form is associated with the template. Lower panel: Tat. Note that Tat is only bound to the transcription complexes when a wild-type TAR element is present. Figure adapated from Keen *et al.*, 1997 [19].

## 2.2 Tat is transferred to the activated transcription-elongation complex

In addition to pausing at RNA stem-loop structures, elongation by RNA polymerase can be blocked by DNA-binding proteins, such as the lac repressor. The block imposed by the *lac* repressor protein can be used to 'trap' actively elongating transcription complexes because, in contrast to polymerases arrested by terminator sequences, the majority of the transcription complexes pausing at the *lac* repressor do not disengage from the template. As outlined in Figure 4(A), we have taken advantage of this property to determine whether Tat becomes attached to transcription complexes that are formed on templates carrying a functional TAR element [7].

Immunoblots of the proteins present in the released transcription complexes are shown in Figure 4(B). Immunoblots for LacR and RNA polymerase provide internal controls for the recovery of proteins bound to the templates following the purification scheme. A strong Tat signal is found in association with transcription complexes purified on templates carrying a wild-type TAR RNA element. Tat association with the transcription complexes is specific and dependent upon a functional TAR RNA element. When TAR was inactivated by either the mGC mutation in the Tat binding site or by the mLG mutation in the apical loop sequence no significant Tat binding to the transcription complex was observed. Similarly, Tat was not acquired when the CMV promoter was to direct transcription instead of the HIV-1 LTR.

## 3. Drug design

### 3.1 RNA binding by the Tat protein

Extensive mutagenesis and chemical probing studies have now defined the key elements required for high affinity binding of Tat to TAR RNA. Tat recognition requires both the presence of the bulged nucleotides and base pairs in the stem [1, 8]. Recent NMR studies of TAR RNA demonstrate that the accessibility of the critical functional groups recognised by Tat is enhanced by a local conformational rearrangement [9-11] (Figure 5). This refolding process involves one of the arginine side chains present in the basic binding domain of the Tat protein. In the presence of the arginine, the stacking of the bulged residues U23 on A22 is disrupted and A22 becomes juxtaposed to G26. This creates a binding pocket where the guanidinium and εNH groups of the arginine are placed within hydrogen-bonding distance of G26–$N^7$ and U23–$O^4$, respectively. The conformational change in TAR RNA also repositions the P22, P23 and P40 phosphates, which provide energetically important contacts with Tat. These phosphates can then be easily contacted by other basic residues found in the TAR RNA binding region. The Tat-TAR interaction therefore provides a clear example of the 'indirect readout' of nucleic acid sequences through recognition of backbone phosphates. The importance of the conformational change in TAR for Tat binding is confirmed by the observation that the mutations that produce the most severe reductions in TAR activity involve G26 and U23 and disrupt the intermolecular interactions that are responsible for the folding transition [12, 13].

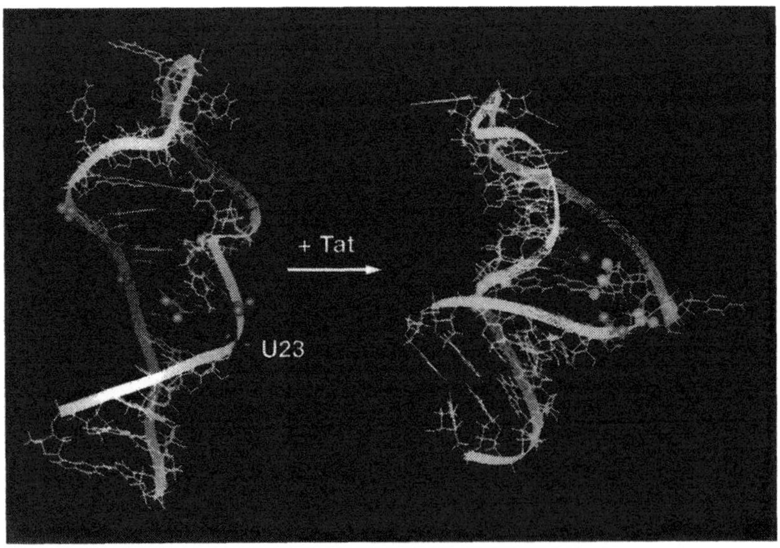


**Figure 5.** Major groove view of the free TAR RNA (Left) and bound TAR RNA (Right) structures. Functional groups which have been identified as critical for Tat binding, including the backbone phosphates, are highlighted by van der Waals spheres. Figure adapted from Aboul-ela *et al.* [11].

## 3.2 Low molecular weight inhibitors of the Tat-TAR interaction

The Tat/TAR RNA interaction is strictly required for HIV replication. A small molecule that is able to bind TAR RNA with high affinity and act as effective competitor inhibitor will consequently, inhibit HIV growth. Inhibitors of this class have great potential for use in combination therapy since they inhibit both rapidly growing virus populations as well as attacking the pool of integrated proviruses that serve as the source of drug-resistant mutants.

In order to discover low molecular weight inhibitors of the Tat/TAR RNA interaction, we adopted a combinatorial chemistry approach based on the synthesis of oligomers containing both peptoid and (all D)-peptide residues [14]. Peptoids are isomers of peptides in which all the side chains are carried by the backbone nitrogens (N-substituted glycines) [15]. Peptoids are more flexible than peptides since intramolecular CO-HN hydrogen bonds are removed and the steric interactions that induce secondary structure are different. For pharmacological applications, peptoids have the advantage of being stabilized against enzymatic degradation.

The strategy for the synthesis and deconvolution of the peptoid library is illustrated in Figure 6. Five positions (residues A to E) were randomized by introducing a set of twenty building blocks carrying a wide range of functional groups. The initial combinatorial library consisted of $3.2 \times 10^6$ ($20^5$) compounds divided into 20 sublibraries of 160,000 compounds each. In order to identify the compound with the highest affinity for TAR RNA in these highly complex mixtures, we followed a deconvolution scheme analogous to that described by Houghten *et al.* [16]. In the first step, 20 sublibraries, each of which contained a unique residue at the first position under investigation (the N-terminal residue, residue A) and randomized sequences at residues B to E. The sublibrary showing the highest TAR RNA binding activity was identified using a gel mobility shift assay. This defined the optimal side chain for residue A. A new set of 20 sublibraries was synthesized each of which contained the optimal residue A, a unique sequence at residue B, and randomized sequences at residues C and D. Continuing with this approach it was possible to progressively limit the complexity of the sublibrary and to identify optimal residues at each position. After the fifth deconvolution cycle, a single compound with high affinity for TAR RNA, CGP64222, was identified.

The selection experiment described above provides very strong evidence that there are strict sequence and steric requirements for the optimal interactions of peptidic compounds with TAR RNA. It is important to note that the best compounds did not arise by indiscriminate insertion of positive charges at each position. For example, D-Arg was never selected in place of the Narg residue, not even as the second best building block.

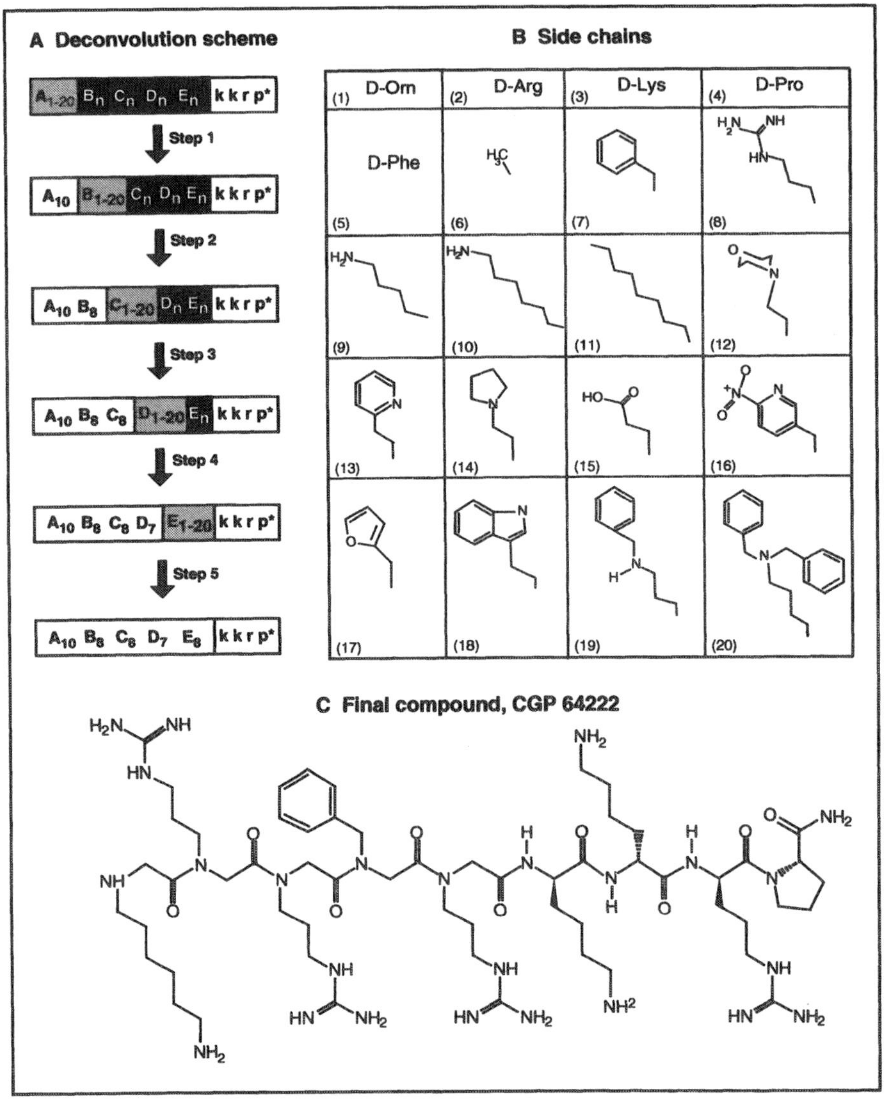


**Figure 6.** (A) Scheme for the deconvolution of combinatorial libraries. Initially 20 sublibraries carrying unique residues in position A ($A_{1\text{-}20}$; grey box; subscripts refer to side chain numbers shown in panel (B)) and a random collection of side chains in positions B, C, D and E ($X_n$; black box) were synthesized. The sublibrary with the greatest inhibitory activity against the Tat/TAR RNA interaction was detected using a gel mobility shift assay. After selecting the optimal residue for position A, another set of 20 sublibraries was prepared with a fixed residue in position A ($A_{10}$, white box), unique residues at position B ($B_{1\text{-}20}$; grey box) and a random collection of side chains in positions C, D and E ($X_n$; black box). This deconvolution process was continued until a single compound was identified. (B) Structure of side chains used to create the combinatorial library. (C) Structure of the final compound, CGP64222. Figure from Hamy *et al.*, 1997, [14].

## 3.3 Molecular modelling of the CGP64222-TAR complex

NMR spectroscopy demonstrates that it is both sterically possible and energetically favourable to fit CGP64222 into the major groove of TAR RNA. First, CGP64222 induces a conformational change in TAR resulting in a structure very similar to that of the ADP-1-bound TAR RNA [9]. When the spectra of the CGP64222-TAR RNA and ADP-1-TAR RNA complexes were compared in detail, no significant differences were observed in the strand containing the bulge (residues A20 to G28), the region where the most significant conformational changes occur upon ligand binding. The conformational change is mediated by direct contacts between an Narg side chain and G26 and U23, as observed for the TAR-ADP-1 complex [9]. Second, chemical shift changes in the pyrimidine-rich strand indicate that side chains from CGP64222 are in proximity to residues C37 to C41. Clear differences between peptide and CGP64222 interactions with TAR RNA were observed in the pyrimidine-rich strand opposite the bulge (C37 to C41). Since CGP64222 should be more flexible than ADP-1, it seems likely that recognition of the pyrimidine-rich strand by interaction with the phosphodiester backbone accounts for the enhanced binding properties of the new compound.

## 3.4 Inhibition of Tat activity in cellular trans-activation assays

The antiviral activity of CGP64222 was tested in cellular assay systems designed to measure its ability to inhibit either viral membrane fusion activity and cell entry or to inhibit HIV-LTR function. Figure 7 shows the activity of CGP64222 and a control peptide carrying 5 L-Arg residues and the sequence of D-amino acids found in CGP64222 (R-R-R-R-R-k-k-r-p* ; Arg5) in a novel assay called the Fusion Induced Gene Stimulation (FIGS) assay [17]. In the FIGS assay, an adherent CD4-positive indicator cell line (SX22-1) is co-cultivated with $HUT_{4\text{-}3}$ lymphocytes that are chronically infected by HIV-1. Mixing the two cell lines results in a rapid development of extended syncytia due to Env-CD4-mediated membrane fusion. The SX22-1 cell line also carries an integrated lacZ reporter gene under the control of the HIV-1 LTR. Blue syncytia, indicating foci of active HIV replication and Tat expression can be easily detected after staining the cells with X-Gal (5-bromo-4-chloro-3-indolyl-ß-D-galactoside). The FIGS assay system is a very useful tool to profile inhibitors of HIV replication since it permits inhibitors of HIV entry and inhibitors of LTR activation to be distinguished [17].

As expected, untreated control cultures contained a high number of blue syncytia. Addition of the Arg5 peptide strongly inhibited cell fusion and resulted in a suppression of syncytia formation. Since Tat protein was not transferred from the infected donor cells to the reporter gene in the indicator cells and there was no induction of ß-galactosidase. In contrast to the non-specific inhibition of cell fusion by the Arg5 peptide, CGP64222 was able to prevent ß-galactosidase synthesis in the extended syncytia. The $IC_{50}$ for CGP64222 in the FIGS assay is 3 to 5 µM. Since Tat protein is readily provided in this system by the donor cell, the failure to synthesize ß-galactosidase must be due to inhibiton of Tat activity by CGP64222.

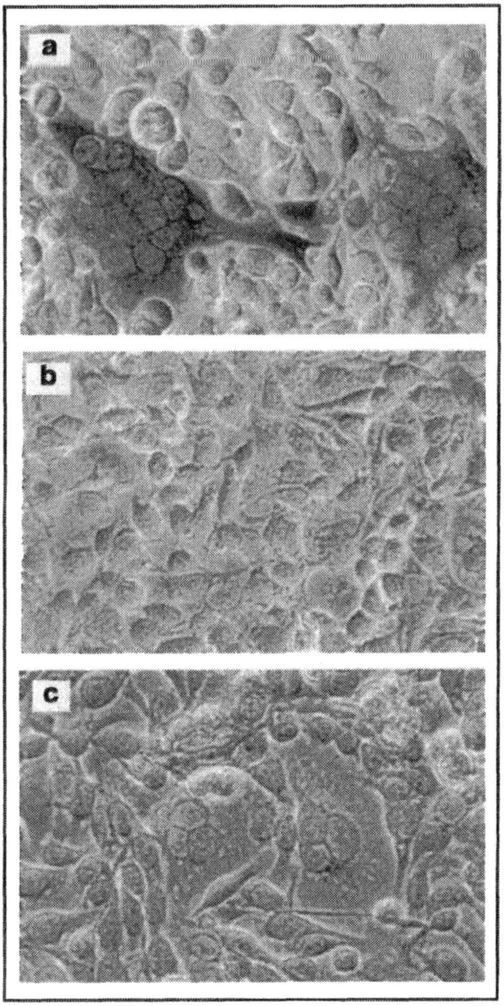

**Figure 7.** Evaluation of CGP64222 in the cellular fusion-induced gene stimulation (FIGS) assay. Tat activity was assessed 24 hrs after co-culture of chronically HIV-1 infected $HUT_{4-3}$ cells with SX22-1 indicator cells. (a) Untreated control cells showing syncytium formation and the subsequent activation of the lacZ gene by Tat. (b) Cells treated with 10 µM Arg-5 peptide. Note that fusion is inhibited. (c) Cells treated with 10 µM CGP64222. Note that generation of syncytia is normal, but Tat activity is completely suppressed. Figure from Hamy *et al.*, 1997, [14].

The control peptide, Arg5, was chosen because it is closely related to a compound (ALX-40C) which has been developed by Allelix Biopharmaceuticals and entered into clinical trials. The Allelix compound is an all-D amino acid peptide containing nine arginine side chains. Although this compound can inhibit Tat binding to TAR RNA in vitro with a $K_i$ similar to CGP64222, we found using the FIGS assay that the primary antiviral action of ALX-40C was to block syncytia formation. This is because the compound fortuitously binds the surface glycoprotein gp120 and does not penetrate well into cells. Thus the antiviral activity of ALX-40C appears to be primarily due to cell surface events. By contrast, the results shown above, demonstrate convincingly that the antiviral effects of CGP64222 are due exclusively to inhibition of Tat activity.

## 4. Conclusions

The HIV *trans*-activator protein, Tat, provides the first example of a protein that regulates transcriptional elongation in eukaryotic cells. Until Tat was discovered, elongation control was a relatively neglected area of transcription regulation. There is still a great deal of ignorance about the mode of action of the key elongation factors, but undoubtedly, identification of the co-factor(s) used by Tat will reveal a great deal about the normal cellular mechanisms of elongation control.

Studies of the interactions of Tat with TAR RNA are also providing new insights into the chemistry of nucleic acid recognition. There are many examples of RNA binding proteins that recognize bases displayed in apical loop and distorted bulge structures. Studies of the Tat-TAR interaction have revealed that the binding reaction involves two distinct steps: First, Tat induces a conformational changes in RNA structure at the bulge. Second, the structural rearrangement in TAR permits Tat to interact with specific functional groups displayed on base pairs in the major groove as well as on adjacent phosphate residues. This new principle of RNA recognition is likely to extend to many other protein-RNA interactions.

Finally, in addition to its academic interest, studies of the Tat-TAR interaction is providing a basis for drug discovery. Using a combinatorial approach we were able to identify a peptidic compound, CGP64222, that is able to effectively compete with Tat binding for TAR RNA. NMR analysis shows that CGP64222 binds directly to TAR RNA at the Tat binding site. Thus, CGP64222 is the first example of an antiviral compound that selectively inhibits and RNA/protein interaction. We are optimistic that by using modern structure-based design approaches this early lead can be developed into a drug that will make a useful contribution to the management of AIDS.

## 5. References

1. Gait, M. J. & Karn, J. (1993) *TIBS* 18, 255-259.
2. Karn, J., Churcher, M. J., Rittner, K., Keen, N. J. & Gait, M. J. (1996) in *Control of transcriptional elongation by the human immunodeficiency virus Tat protein*, ed. Goodbourn, S. (Oxford University Press, Oxford), pp. 256-288.
3. Graeble, M. A., Churcher, M. J., Lowe, A. D., Gait, M. J. & Karn, J. (1993) *Proc. Natl. Acad. Sci.* USA 90, 6184-6188.
4. Marciniak, R. A., Calnan, B. J., Frankel, A. D. & Sharp, P. A. (1990) *Cell* **63**, 791-802.
5. Rittner, K., Churcher, M. J., Gait, M. J. & Karn, J. (1995) *J. Mol. Biol.* **248,** 562-580.
6. Churcher, M. J., Lowe, A. D., Gait, M. J. & Karn, J. (1995) *Proc. Natl. Acad. Sci. USA* **92,** 2408-2412.

7. Keen, N. J., Gait, M. J. & Karn, J. (1996) *Proc. Natl. Acad. Sci. USA* **93,** 2505-2510.
8. Karn, J., Gait, M. J., Churcher, M. J., Mann, D. A., Mikaélian, I. & Pritchard, C. (1995) in *Control of human immunodeficiency virus gene expression by the RNA-binding proteins Tat and Rev, eds.* Nagai, K. & Mattaj, I. (Oxford University Press, Oxford), pp. 192-220.
9. Aboul-ela, F., Karn, J. & Varani, G. (1995) *J. Mol. Biol.* **253,** 313-332.
10. Puglisi, J. D., Tan, R., Calnan, B. J., Frankel, A. D. & Williamson, J. R. (1992) *Science* **257,** 76-80.
11. Aboul-ela, F., Karn, J. & Varani, G. (1996) *Nucl. Acids Res.* **24,** 3974-3981.
12. Churcher, M., Lamont, C., Hamy, F., Dingwall, C., Green, S. M., Lowe, A. D., Butler, P. J. G., Gait, M. J. & Karn, J. (1993) *J. Mol. Biol.* **230,** 90-110.
13. Weeks, K. M. & Crothers, D. M. (1991) *Cell* **66,** 577-588.
14. Hamy, F., Felder, E., Heizmann, G., Lazdins, J., Aboul-ela, F., Varani, G., Karn, J. & Klimkait, T. (1997) *Proc. Natl. Acad. Sci. USA* **94,** 3548-3553.
15. Simon, R. J., Kania, R. S., Zuckerman, R. N., Huebner, V. D., Jewell, D. A., Banville, S., Ng, S., Wang, L., Rosenberg, S., Marlowe, C. K., Spellmeyer, D. C., Tan, R., Frankel, A. D., Santi, D. V., Cohen, F. E. & Bartlett, P. A. (1992) *Proc. Natl. Acad. Sci. USA* **89,** 9367-9371.
16. Houghten, R. A., Pinilla, C., Blondelle, S. E., Appel, J. R., Dooley, C. T. & Cuervo, J. H. (1991) *Nature* **354,** 84-86.
17. Wyatt, J. R., Vickers, T. A., Roberson, J. L., Buckheit, R. W. J., Klimkait, T., DeBaets, E., Davis, P. W., Rayner, B., Imbach, J. L. & Ecker, D. J. (1994) *Proc. Natl. Acad. Sci. USA* **91,** 1356-1360.
18. Karn, J. (1995) in *An introduction to the growth cycle of human immunodeficiency virus,* ed. Karn, J. (Oxford University Press, Oxford), Vol. II, pp. 3-14.
19. Keen, N. J., Churcher, M. J. & Karn, J. (1997) *EMBO J.* submitted.

H. van der Goot, (Editor)
TRENDS IN DRUG RESEARCH II



# Diverse Approaches to Combinatorial Library Design

Eric J. Martin, Roger E. Critchlow, David C. Spellmeyer, Steven Rosenberg, Kerry L. Spear, and Jeffrey M. Blaney

Chiron Corp., 4560 Horton St., Emeryville, CA 94608, USA

Combinatorial libraries for drug discovery can be produced with several different goals: maximum diversity, "drug like" properties, bias toward particular receptor families, lead optimization, or structure based drug design. The most suitable approach to designing a library depends on its specific purpose. Never the less, some general tools such as the creation of a property space, statistical experimental design, and stratified sampling from candidate bins can be useful in a wide variety of library design problems.

This paper will compare the designs of 4 different libraries, intended for 4 different purposes: a pure diversity design, a design biased toward the physicochemical properties of orally available drugs, a library biased by docking into the active site of thrombin, and a lead optimization library designed to increase the potency and decrease the molecular weight of an initial hit that bound to Urokinase Plasminogen Activation Receptor (UPAR).

## 1. INTRODUCTION

Combinatorial libraries are used to search for drugs. When almost nothing is known about the requirements of the target, the search must be very broad. Here, it is beneficial to maximize library's diversity. When more is known, such as a requirement for oral bioavailability, one can constrain the library to emphasize compounds with physicochemical properties typically found in existing orally available drugs. If the 3-D structure of the target is known, methods of structure based drug design can be combined with diversity optimization to create a target biased diversity library. Once a promising hit is found, optimization libraries can be used to improve the potency, solubility, stability, or other properties of that lead. All of these approaches rely on computing the similarity or dissimilarity between potential candidate molecules or fragments. Some also benefit from the ability to compute additional properties such as binding, physicochemical properties, metabolic stability, or toxicity. Due to the limitations of even the most sophisticated computational methods, however, all of these approaches further benefit from a mechanism to incorporate the intuition of experienced medicinal chemists.

A general method useful in all of these problems is to create a multidimensional property space, where each molecule or combinatorial building block is represented by a point, and proximity reflects similarity. Mathematical sampling methods can then be used to select representative subsets from the larger candidate set that fill the property space as efficiently as possible. The candidates can also be categorized by other additional properties, from molecules that could be docked into a 3-D receptor structure, to groups that an experienced scientist has "a good feeling about". Stratified sampling can then be used to pick a certain fraction of the substituents from each category, while still attempting to fill space as completely as possible. In this way, diversity calculations can be combined with virtually any other kind of information to design combinatorial libraries that are as general or as targeted as desired.

This paper presents four typical combinatorial library designs, tempered by various types and degrees of bias. It describes a pure diversity design, a design biased by the physicochemical properties typical of orally available drugs, a library biased by docking candidates into the active site of thrombin, and finally, a lead optimization library designed to increase the potency and decrease the molecular weight of an initial hit that bound to Urokinase Plasminogen Activation Receptor (UPAR).

## 2. METHODS

Selection of substituents for a combinatorial library entails two phases: the creation of a "property space" in which proximity between substituents reflects structural similarity, and the subsequent selection of points that are well distributed throughout that space.

### 2.1. Property Calculations

It is difficult to know, *a priori*, which properties will be relevant for any given drug discovery program. However, one might anticipate that molecular shape, bulk physical properties such as membrane permeability or pKa, constituent heterocyclic rings or chemical functional groups, metabolic stability, and the geometrical arrangement of potential sites of receptor interaction such as charge, H-bonding, aromatics or hydrophobics, might all be important. Thus, many kinds of structural characterization should be included in the property space. These can include estimated atomic or molecular properties, such as topological indices, octanol/water partition coefficient (Kow), or volume. They can also include 2-D or 3-D measures of similarity. Multidimensional scaling can be used to convert the similarities to Cartesian coordinates (latent properties), so they can be combined with other properties into a single Cartesian property space.

#### 2-D Similarity and properties

The calculation of a 2-D property space for library design has been previously described.[1] We routinely compute the octanol/water partition coefficient, 81 topological indices, "chemical functionality" similarities based on Tanimoto similarities between Daylight 2-D substructure fingerprints, and "receptor recognition" similarities from comparing "atom layer tables", which give the distribution along the substituent of atomic radii, hydrogen bonding groups, charged groups, and aromatic groups. These properties were chosen to characterize similarity and diversity with respect to lipophilicity, shape, chemical functionality, and distribution of key receptor binding features. The properties are automatically calculated using the program MAKESPACE, which takes lists of commercially available reagents or candidate substituents as input. It normalizes the structures by removing counter ions, standardizing resonance forms, etc. It then finds the best commercial source for each reagent, computes the properties, and loads the structures and properties into a Daylight THOR[2] database for searching with MERLIN[3].

**3-D similarity**

The properties just described do not explicitly include 3-D information. It is difficult to know, *a priori*, which conformations will be relevant in a drug screening. However, it is possible to compute properties for most of the accessible conformations of each substituent, then compare these ensembles of properties to produce *flexible* 3-D similarities. We compute flexible 3-D shape and 3-point pharmacophore (3PP) similarities as follows.

Three dimensional shape similarities are calculated by first generating a basis set of "polyominoes", which are all the shapes that can be made by joining cubes at their faces. We use all 3-D polyominoes of up to 7 cubes, with a unique cube identified for the site of attachment to orient the substituent. For 7 cubes, this yields 3736 shapes. For computational convenience, the cubes are replaced with overlapping spheres. Distance geometry is then used to flexibly dock each substituent into each of the shapes. The docking results for each substituent are stored as a bit string, where each shape is assigned to a bit, and set bits indicate those shapes that the substituent could achieve. The strings are compared by the Tanimoto metric, yielding similarities in the ensembles of 3-D shapes achievable by each pair of substituents.

Three dimensional 3PP similarity is analogous. The Chem-X software is used to flexibly identify all 3PPs that can be achieved by each substituent.[4] The pharmacophoric feature types include H-bond donors and acceptors, charge, Aromatic center, and a special site-of-attachment dummy atom. Rather than using the software to generate an accumulated fingerprint for collections of molecules, we generate a bit string indicating the accessible 3PPs for all conformations of each individual substituent, and output the logical AND and logical OR for each pair of substituents. The quotient of these is the Tanimoto flexible 3PP similarity between substituents.

## 2.2. Creating "Property Space"

Most of the calculations yield matrices of pairwise similarities rather than tables of properties. Furthermore, many of the properties are redundant. The MAKESPACE program uses principal components analysis (PCA) to reduce the 81 topological indices to 5 latent shape properties. The 2D and/or 3D similarity matrices are converted to latent properties by non-linear MDS. The numbers of MDS dimensions are chosen as the fewest required to reproduce the pairwise dissimilarities within an average relative standard deviation of 10%. MAKESPACE then scales the data and combines them into one property space, using PCA again to remove any additional redundancies. The final result is a property space of typically from 15 to 25 dimensions. MAKESPACE stores the substituent structures and various physical properties, including log P, pKa, MW, and the number of rotatable bonds, in a THOR database, so they can be searched by structural features or sorted by properties.

## 2.3. Experimental Design

The program TAILOR performs substituent selection by D-optimal design. A "diverse" set of substituents should avoid redundancy and fill property space. The D-optimal algorithm picks a "design set" of points from the larger "candidate set" of available reagents, such that they sample a wide range of properties (*i.e.* are not redundant), and are nearly orthogonal (*i.e.* they fill space). Orthogonality tries to sample the full dimensionality of the space, even if the number of points is as few as the number of dimensions. A pre-selected "core set" can be augmented to a larger size while maximizing the overall diversity. TAILOR can perform up to 16 sequential augmentations, drawing from different candidate sets, allowing the constitution of a design to be carefully "tailored". TAILOR also generates standard bench mark sets to calibrate the diversity score for evaluating the loss of diversity due to biasing a design toward a particular biological target or distribution of physicochemical properties.

In addition to specifying the size of the design, the core set, and the candidate set, D-optimal design also requires the specification of a model. If the number of points to be selected is much larger than the number of dimensions, the D-optimal algorithm may suggest sampling some points twice, indicating that a smaller design would be sufficient to model a linear response surface. TAILOR can automatically add quadratic terms, cross terms, or higher order terms to remove the excess degrees of freedom.

## 2.4. Docking

The procedure GADOCK uses a genetic algorithm (GA), combined with distance geometry docking, to identify combinatorial library products that compliment the shape and charge of a 3-D protein binding site. GADOCK is comprised of two perl scripts. The first script takes as input the smiles codes for the candidate substituents of a proposed combinatorial synthesis. It automatically generates bit strings to encode each substituent, and manages the input parameters needed to run the public domain genetic algorithm GA162[5]. The second script provides the GA with the "fitness" score for each member of the new population. It decodes the bit string for each molecule returned by the GA, generates 3D structures with RUBICON,[6] docks each structure with the program DGEOMDOCK,[7] and assigns the fitness score. Partial atomic charges used for the docking calculations are assigned with the program ASSIGN_CHARGE. The fitness score is calculated as GASCORE = DOCKSCORE - NUMROT - 0.01*MW, where DOCKSCORE is the intermolecular force field energy returned from the docking calculation, NUMROT is the number of rotatable bonds,[8] and MW is the molecular weight.

The DGEOMDOCK calculation can be broken down into two parts: a shape-fitting procedure, and a scoring procedure. The shape fitting procedure is comprised of flexibly docking a molecule using distance geometry methods into a set of spheres which fill a protein active site, yielding an approximate conformation and orientation. The active-site spheres are generated with the C program MakeSpheres.[9] Minor editing of sphere placement was performed by hand. The crudely docked geometry is then refined by conjugate-gradient minimization, using distance geometry for the intramolecular constraints and an approximate molecular mechanics force field for intermolecular interactions. The final fitness score for each molecule is the best from 100 random starting trials, generated from random atomic coordinates. Scores of previously docked molecules are stored and re-used to avoid duplicate docking calculations.

## 3. RESULTS

Four libraries were designed to demonstrate a range of approaches.

### 3.1. Pure D-optimal Diversity Library

The first library was designed purely to maximize diversity. The combinatorial synthetic scheme required aliphatic amine starting materials for two diversity positions and phenols for a third. Of the approximately 8000 aliphatic amines and 9000 phenols in the Available Chemicals Directory (ACD)[10], only 943 amines and 750 phenols were found to be suitable after eliminating those that were too large (MW > 250), too expensive (cost > $250/g), too unreactive, too unstable, contained toxic groups, or contained other functionality that would have interfered with the synthesis. A property space was computed with 5 topological shape descriptors, 7 chemical functionality descriptors, and 6 receptor interaction descriptors, and log P, for a total of 19 dimensions. PCA showed that the two smallest PCs explained only 0.5% of the variance, so these were discarded, leaving a 17 dimensional property space.

Fifty substituents were needed for the first amine site. At this point a handful of additional problematic reagents were eliminated, leaving a candidate set of 762 amines. The D-optimal set of 50 points was computed for a saturated quadratic model. This most diverse set of 50 substituents had a D-score of 136. This score has no absolute scale, so benchmarks were computed for calibration. The D-optimal set of 50 from a candidate set of the 346 most rigid amines, *i.e.* those with 2 or fewer rotatable bonds, was 105. The D-optimal set of 50 from the 229 smallest amines, with MW < 130, was 66. The most diverse set of 50 that could be made from only the pure hydrocarbon substituents, *i.e.* compounds with shape diversity but no functional diversity, had a D-score of 61. Random sets of 50 substituents from the full set of 762 had scores of only 50. The 50 nearest analogs to tyramine, an extremely homologous set, had a score of -148.

### 3.2. Property-biased Broad-screening Library

Examining the structures of this most diverse set revealed that many of the substituents were too large, too lipophilic, or too flexible to be likely to yield orally available drug candidates. In order to understand the tradeoff between diversity and the distribution of physical properties, the 762 candidates were subsetted into overlapping categories, or "bins", according to several properties. For example, a "seed" bin contained 4-methoxybenzylamine, which was the validated point nearest the centroid of the property space, as well as 4-hydroxyphenethylamine

Table 1. "Bin" profiles used in the property-biased amine design.

| Name | Description | No. | Range | Tries |
|---|---|---|---|---|
| seed | center + 2 pharmas | 3 | 3 | 1 |
| LoRgPlV | Low MW, rigid, polar, validated | 13 | -10-14 | 3 |
| boc | Available with boc protected amine | 6 | 1 | 1 |
| tbu | Available with t-butoxy protected acid | 9 | 2 | 2 |
| LoPlrV | Low MW, polar, validated | 21 | -15-19 | 3 |
| LoRgPlr | Low MW, rigid, polar | 77 | 5-6 | 2 |
| LoPlr | Low MW, polar | 126 | 4-5 | 2 |
| DrgPlr | Drug like and polar | 88 | -30-34 | 3 |
| OKamine | Diamines that need no protection | 46 | 1 | 2 |
| DrugV | Drug-like and validated | 11 | 2-4 | 2 |
| valid | Reaction has been validated | 76 | 2 | 2 |
| Drugish | Contain pieces from 100 top drugs | 151 | 1 | 1 |
| LowMW | Molecular Weight < 130 | 229 | 2 | 2 |
| polar | H-bond acceptor and log P < 1.5 | 378 | -46 | 3 |
| FF_xtrm | union of FF and extreme bins | 57 | 2 | 2 |
| good | Chemistry is expected to work well | 699 | 2 | 2 |

Column "No." is the number of candidate substituents in that category. Column "Range" gives the acceptable numbers of substituents to be taken from that bin. A range starting with "-" indicates taking enough substituents from that bin to bring the total to that number. Column "Tries" shows the number of numerical attempts to find the D-optimal compound set.

and benzhydrylamine, whose corresponding side chains were previously found in potent ligands for the α1-adrenergic and μ-opiate receptors.[11] Bin "LoRgPlV" contained the 13 compounds which were low MW (< 130), rigid (fewer than 2 rotatable bonds), polar (log P < 1.5), and whose chemistry had already been validated. Descriptions of the bins are summarized in Table 1. Ranges of bin membership, also shown in Table 1, were then chosen for each category to bias the distribution of properties toward polar, rigid, low MW compounds with fragments found in known drugs and whose chemistry had already been validated. Separate designs were made for each of the 1320 profiles consistent with these ranges, and with a total design size of 50 amines. I.e., each design of 50 was generated from a profile starting with the "seed" set, then augmenting by D-optimal design with the specified number of substituents from bin "LoRgPlV", then further augmented with the "boc" set, and so on, until all bins had been appropriately sampled. The entire calculation took about 1 day elapsed time on an SGI Indigo 2 with a 150 MHz R4400 CPU.

Some notable designs are presented in Table 2, in order of decreasing diversity. The scores ranged from 102 down to 83. Included are the 2 most diverse designs, the most diverse design with maximum bias in each of the 5 categories, the least diverse design, and design #35 which was chosen as the best compromise between property bias and diversity.

Table 2. Scores and profiles of selected designs from Table 1.

| Rank | Score | Drug | Valid | Polar | Low | Rigid |
|---|---|---|---|---|---|---|
| Range | 83-101 | 2-14 | 11-17 | 30-32 | 18-24 | 12-17 |
| 1 | **101.6** | 12 | 12 | 31 | 18 | 13 |
| 2 | **100.6** | 6 | 12 | **32** | 20 | 15 |
| ***35*** | ***97.1*** | ***11*** | ***14*** | ***30*** | ***21*** | ***16*** |
| 67 | 96.1 | 6 | 15 | 30 | 22 | **17** |
| 74 | 95.9 | **14** | 13 | 30 | 18 | 14 |
| 178 | 94.7 | 9 | **17** | 30 | 23 | 14 |
| 470 | 92.4 | 4 | 17 | 30 | **24** | 14 |
| **1320** | 82.8 | 9 | 15 | 31 | 22 | 14 |

Column 1 is the rank order. Column 2 is the Diversity score. Columns 3-7 are total numbers drawn from bins that include each property. *E.g.*, the "drug" column is the sum of the "Drug-polar" and "Drug-valid" columns.

Figures 1-4, present histograms and quantile box plots for 3 sets of substituents: the full set of 762 amines, the final property-biased design of 50 compounds, and the maximally diverse D-optimal design of 50 compounds with no bias at all. The full candidate set gives the distributions expected of random sets. Distributions are presented for 4 properties: molecular weight, CLOGP, number of rotatable bonds, and distance from the center of property space. Diamonds in the box plots show means and standard deviations. The boxes indicate the 25, 50 and 75 percentiles. Other tick marks are at 0%, 0.5%, 2.5%, 10%, 90%, 97.5%, 99.5%, and 100%.

### 3.3. Ligand Biased Diversity Library

The previous example shows how much detailed bias can be incorporated even into broad screening libraries. Stronger bias yet is employed in a lead optimization library. In this example, about 150,000 "peptoids" were screened as mixtures of over 500 compounds in a UPAR binding assay. Several NSG ligands were found. Figure 5 shows the most potent, CHIR5585, which was a 250 nM trimer with a molecular weight of 708. Our goal was a lower MW ligand with low nanomolar affinity. Toward that end, a series of about 15 individual peptoids were synthesized to get a rough SAR. This small series suggested minimal tolerance to variation in the C-terminal position, moderate tolerance in the middle position, and fairly loose requirements at the N-terminal position. An optimization library was then designed, containing 15 substituents at the N-terminal position, 11 indane analogs in the middle position, and 8 glycinamides in the C-terminal position. The substituents were chosen by a blend of computational and "art-based" methods. Searching property space in the vicinity of the lead's sidechains produced large lists of analogs from commercially available reagents. A combination of medicinal chemistry intuition and database searching expanded these lists. D-optimal design then chose a diverse sampling of from 8 to 11 members from each of these three fairly homologous subsets.

Finally, in the more tolerant N-terminal position, the 10 analogs were further augmented by a second D-optimal design step with 5 additional, low MW, diverse substituents.

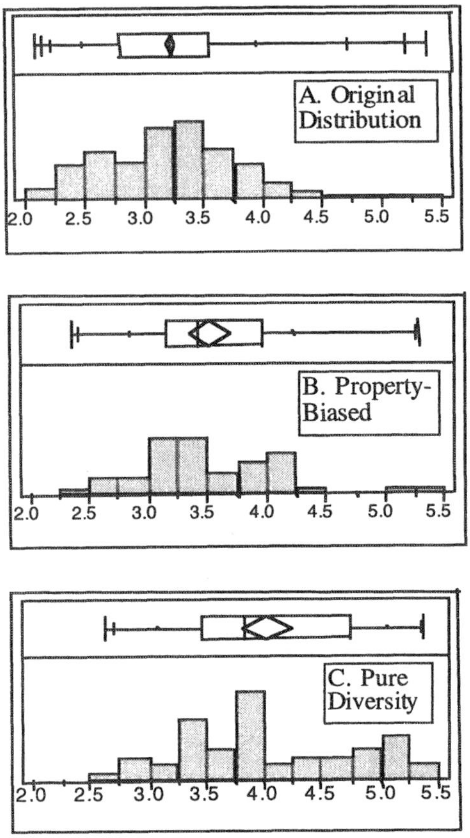


Figure 1. Histograms and quantile box plots showing that the radial distribution in tailored designs does not emphasize the extremes of property space as does the simple D-optimal design of 50 candidates.

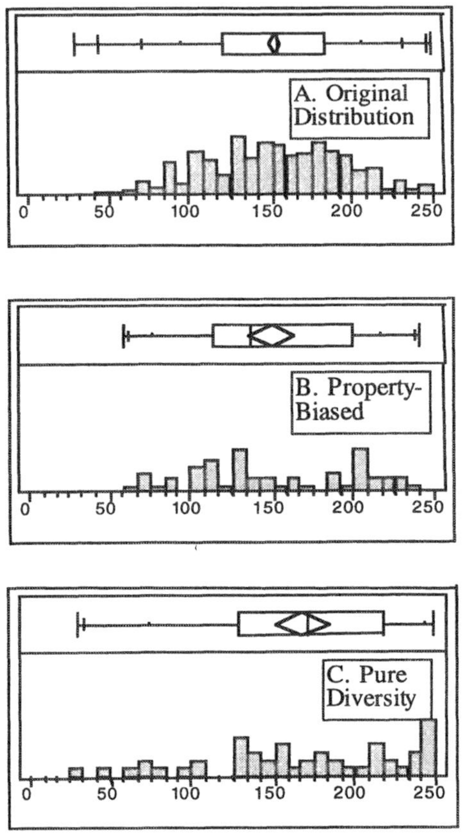


Figure 2. Histograms and quantile box plots show that the molecular weight distribution of tailored designs does not emphasize large compounds as does the simple D-optimal design of 50 candidates.

The optimization library was synthesized in a single robot run by the "split and mix" method as 15 pools of 88 members each for a total of 1320 compounds. It was screened by affinity selection mass spectroscopy. The most potent ligand, CHIR11509, had a molecular weight of 636 and a potency of 15 nM, achieving our activity and MW goals.[12] Figure 5 shows the format of the library and SAR.

The published crystal structure of NAPAP bound to thrombin[13] was used to demonstrate how the 3-D structure of a target protein can be employed to bias a screening library. Figure 6 outlines a virtual dipeptide library of over 1,000,000 NAPAP analogues, constructed from 64 sulfonyl chloride N-terminal capping groups, 32 amino acids, and 16 secondary amine C-terminal capping groups. The commercially available candidate reagents were chosen both for diversity, and for similarity to the side chains found in NAPAP. An initial population of 40 molecules was generated at random, docked, and scored as above. Using the "microga" algorithm with uniform cross-over, the molecules were propagated to give a new population, and the procedure was repeated for 40 generations. A total of 2,420 compounds were docked and scored.

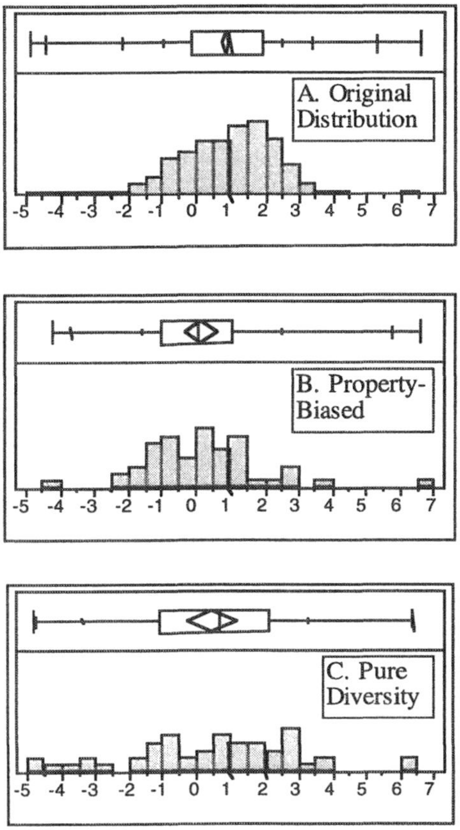


Figure 3. Histograms and quantile box plots showing that the log Kow distribution of tailored designs does not emphasize the extremes of high and low lipophilicity as does the simple D-optimal design of 50 candidates.

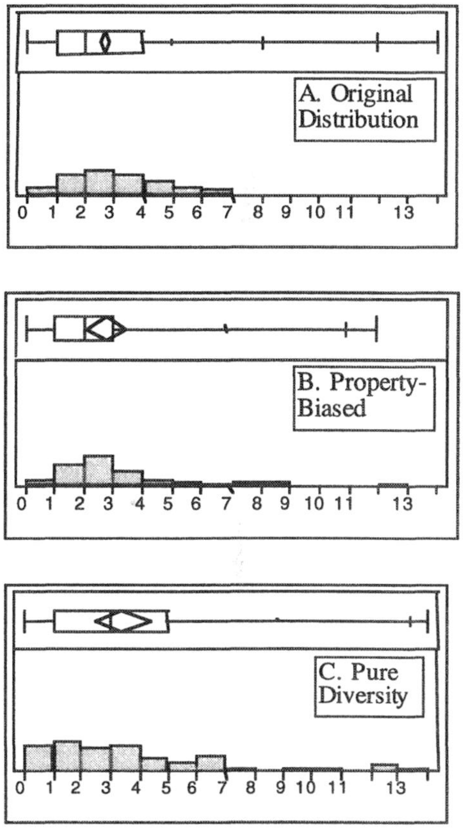


Figure 4. Histograms and quantile box plots showing that the distribution of the number of rotatable bonds for tailored designs does not emphasize flexible substituents as does the simple D-optimal design of 50 candidates.

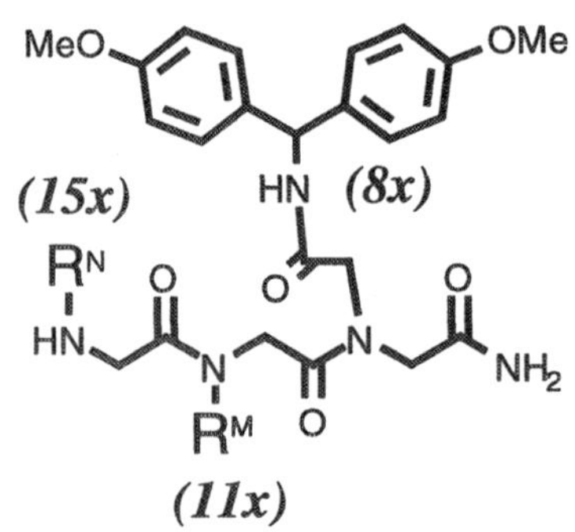


| Description | $R^N$ | $R^M$ | $K_i$ (nM) |
|---|---|---|---|
| CHIR5585 (Original Hit) | OH | | 250 |
| CHIR11648 (Change $R^M$) | OH | | 250 |
| CHIR5952 (Change $R^N$) | | | 700 |
| CHIR11509 (Best Ligand) | | | 15 |

Figure 5. Structures of the UPAR antagonists. The optimization library was synthesized as 15 pools of 88 compounds each. The SAR showed strong interactions between the N-terminal and middle substituents.

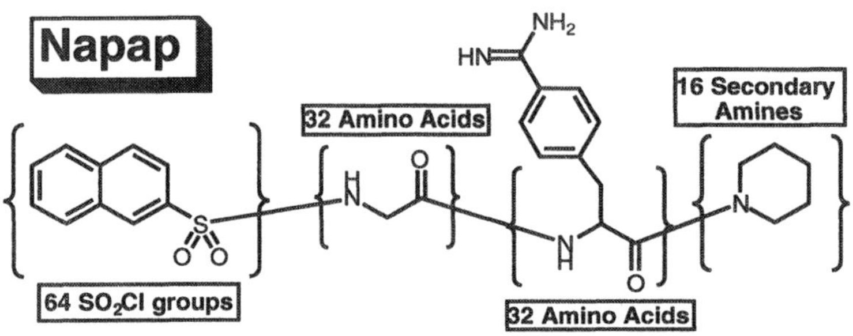


Figure 6. Diagram of building blocks for virtual library of NAPAP analogs.

The 1000 top-scoring molecules were surveyed. The 5 most frequently recurring side-chains for each position are shown in Figure 7. The number by each structure shows in how many molecules that substituent occurred; an indication of that substituent's suitability for a thrombin biased library. The best scoring of the thrombin dockings fell into several "families" or modes of docking. This result is typical of genetic algorithms. Thus, the best scoring side chains for each position are not always similar to each other, since they often match different receptor features. These substituents should all compliment the shape and electrostatics of the thrombin active site.

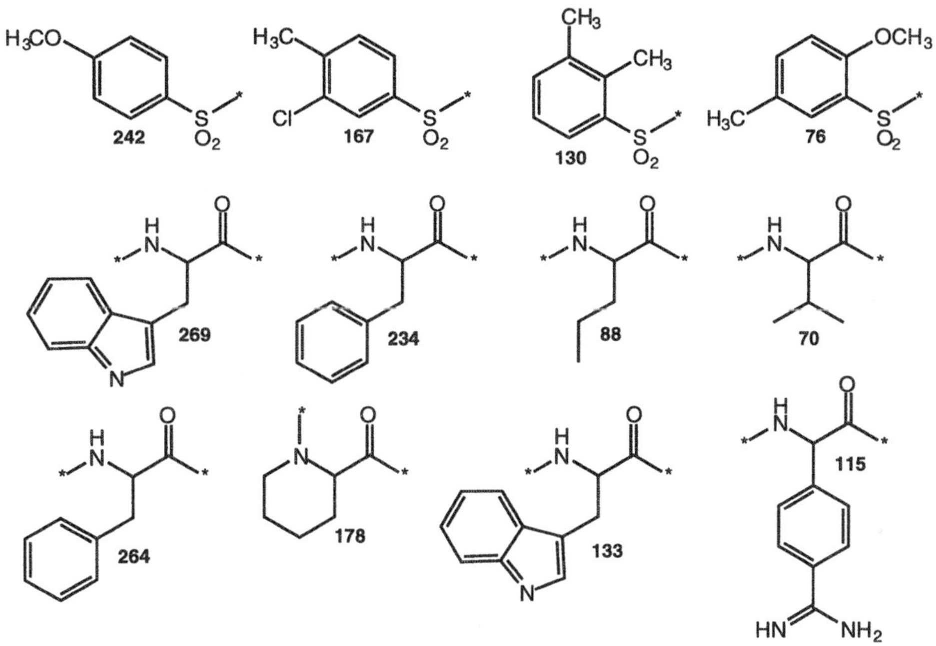


Figure 7. Substituents most frequently used by the GA at each synthetic site.

## 4. DISCUSSION

### 4.1. Comparison of diverse and property-biased libraries.

The maximally diverse D-optimal design, with a score of 136, was significantly more diverse than any of the property-biased designs, with scores ranging from 83 to 102. This decrease in diversity was the price paid for tailoring the distribution of physicochemical properties. Still, even the most biased of these designs scored much higher than random libraries, with averages scores of 50 +/- 5. Thus, even strongly constrained D-optimal libraries have more diversity than an arbitrary choice of substituents. Surprisingly, the most diverse design, #1, had an unusually high drug-like bias, having 12 of a possible 14 members from the drug-like bins. However, its members tend to be high molecular weight and flexible, and very few had been validated. Design #2, with a score of 101, had only 6 drug-like substituents, but had better low MW and rigid bias. Design #35, with a score of 97, was chosen as a good compromise between diversity, property bias, and synthetic ease.

Concern has been raised that pure D-optimal designs (as well as other pure diversity designs) sample only the "outer edges" of property space. The radial distributions in Figure 1 show that, for better or worse, the pure D-optimal diversity set does indeed over-sample the extremes of property space relative to a random distribution. The property-biased design shows only a modest outward shift relative to the candidates. Apparently, the additional constraints of stratified sampling from bins reduces the D-optimal algorithms propensity to mainly sample remote regions of property space.

The other 3 property histograms in Figures 2-4 show that the pure diversity design emphasized large, flexible groups with either extremely high or extremely low lipophilicity. Orally available drugs tend to be smaller, more rigid compounds with intermediate lipophilicity, so criticisms that pure diversity designs bias libraries away from the ideal properties for small molecule drugs also appear justified. Since only a few members from the extreme bins were permitted, the property-biased library's distributions have wide tails, but are not as flat and extreme as those of the pure diversity designs. It is less hydrophobic than the original distribution, and includes a few extreme values, but it concentrates most of the members in the desirable moderately hydrophilic region. 75% of the substituents in the property-biased set have 3 or fewer rotatable bonds, compared to 4 in the original distribution and 5 in the pure diversity set, showing that property bias has limited the fraction of flexible substituents. The median (50%) molecular weight in the property-biased design is lower than the original distribution and much lower than the pure diversity design, but there is a curious bimodal distribution with peaks at about 130 and 200. Despite the low original frequency of the highest MW slice, it is the most frequent value in the pure diversity design. This suggests that structural diversity might require complexity, and complexity might require high MW. The low molecular weight bins, with a cut-off at 130, were strongly emphasized in the property-biased profile. The sequential D-optimal algorithm apparently emphasized the heaviest members available in each candidate bin, giving one peak near 130 from sampling the low MW bins, and another near the highest value of 250 from sampling the bins with no specified MW restrictions.

### 4.2. Ligand biased library

The UPAR optimization library, made in a single robot run, with fewer than 1% as many compounds as the original screening set, produced a ligand with 15 fold greater potency while dropping the molecular weight by 72 units. Clearly, the rapid exploration of a small region of property space around an initial lead can efficiently ascend a local peak in the activity response surface.

More importantly, the SAR shown in Figure 5 demonstrated the particular strength of combinatorial optimization. The best substituent for the C-terminal position was the same as the initial lead. In the middle position, the optimized substituent was very similar in shape and lipophilicity to the initial lead. However, in the N-terminal position, the p-hydroxyphenethyl substituent was replaced by propargyl; a dramatic change. Furthermore, replacing the middle position alone gave no increase in potency, and the N-terminal replacement alone actually decreased potency by a factor of 3. It is only when these two were simultaneously replaced that the large improvement was seen. Such interactions are very difficult to predict, and were unlikely to have been discovered without making and testing such a large combinatorial library.

### 4.3. Structure biased library

Structure based combinatorial library design differs from typical *de novo* drug design problems in two important ways: first, all of the compounds suggested by the GA are readily accessible, and second, the goal of the calculation is not to design an individual potent ligand, or even a handful of potent ligands, but rather to identify substituents with good electrostatic and shape complementarity to the target receptor. We know that our molecular mechanics scoring function has limitations. In particular, it does not adequately treat hydrophobic interactions. Hence, the specific compounds suggested by the GA may not be very potent ligands. Furthermore, tight binding is not the only requirement for an effective drug. Thus, rather than suggesting a library using only the compounds found by the docking exercise, the favored substituents are used to form an elite "docked" bin in a property biased design. A structure biased thrombin library would contain many members sampled from the docked bin, but also additional members chosen to maximize diversity under the constraints of a good physical property distribution as described above. This gives serendipity a good opportunity to compensate for the known limitations of the computational model.

## 5. SUMMARY

Broad screening libraries are often referred to as "random libraries". This designation might appropriately indicate that the library is equally suited for any arbitrary target. However, this exercise showed that random selection of compounds was poor both in structural diversity and in distribution of physicochemical properties. More recently, such libraries have been referred to as "diversity libraries". Pure diversity designs, however, were found to be systematically biased towards heavy, flexible compounds with very high or very low lipophilicity. These properties are poorly suited to yield bioavailable drugs. Contrary to what these names suggest, designing optimal libraries for broad screening requires a combination of property calculations, structural diversity calculations, experience, and good medicinal chemical intuition. Stratified D-optimal sampling from bins provides machinery to combine all of these means.

Conversely, focused combinatorial libraries can be used for lead optimization or structure-based library design. Yet even in these cases, since our understanding is always imperfect, these approaches can benefit from an additional diversity component incorporated by stratified D-optimal sampling. Finally, medicinal chemists have invaluable intuition, developed from years of experience, that cannot be quantified or encoded algorithmically. Stratified sampling from bin profiles allows experienced drug discovery teams to design well tailored libraries that are diverse, but satisfy these additional, often nebulous, factors as well.

## ACKNOWLEDGMENTS

Special thanks to Steven Cato at Chemical Design for help with computing substituent 3PPs and corresponding Tanimoto similarities.

## REFERENCES


1. Martin, E. J.; Blaney, J. M.; Siani, M. A.; Spellmeyer, D. C.; Wong, A. K.; Moos, W. H., Measuring Diversity: Experimental Design of Combinatorial Libraries for Drug Discovery. *J. Med. Chem* , **1995**, *38*, 1431-1436.
2. Weininger, D., *Thor*; Daylight Chemical Information Systems: Irvine, CA, 1993.
3. Weininger, D., *Merlin*; Daylight Chemical Information Systems: Irvine, CA, 1993.
4. Chemical Design Ltd., *Chem-X Combinatorial Chemistry User Guide*, JUL96; Jantec Printing Ltd: Oxon, England, 1996.
5. Carroll, D. L., *GA162*, v1.6.2; http://www.staff.uiuc.edu/~carroll/ga.html: Urbana, IL, 1997.
6. Weininger, D., *Rubicon Reference Manual, v4.51*; Daylight Chemical Information Systems, Inc.: Irvine, CA, 1997.
7. Blaney, J. M.; Dixon, J. S., *DGEOMDOCK*, Chiron Corp.: Emeryville, CA, 1992.
8. Jan, D., *rotomers*, Chiron Corp.: Emeryville, 1996.
9. Siani, M., *MakeSpheres*, Chiron Corp.: Emeryville, CA, 1995.
10. MDL, *Available Chemicals Directory*, Molecular Design Limited: San Leandro, CA, 1993.
11. Zuckermann, R. N.; Martin, E. J.; Spellmeyer, D. C.; Stauber, G. B.; Shoemaker, K. R.; Kerr, J. M.; Figliozzi, G. M.; Siani, M. A.; Simon, R. J.; Banville, S. C.; Brown, E. G.; Wang, L.; Moos, W. H., Discovery of Nanomolar Ligands for 7-Transmembrane G-Protein Coupled Receptors from a Diverse (N-Substituted)Glycine Peptoid Library. *J. Med. Chem.* , **1994**, *37*, 2678-2685.
12. Rosenberg, S.; Spear, K. L.; Martin, E. J., *Preparation of peptide ligands of the urokinase plasminogen activator receptor*; Patent No. WO 9640747 A1, Chiron Corporation, USA; Rosenberg, Steve; Spear, Kerry L.; Martin, Eric J.: PCT Int. Appl., 1996.
13. Brandstetter, H.; Turk, D.; Hoeffken, H. W.; Grosse, D.; Stuerzebecher, J.; Martin, P. D.; Edwards, B. F. P.; Bode, W., Refined 2.3 Angstroms X-Ray Crystal Structure of Bovine Thrombin Complexes Formed With the Benzamidine and Arginine-Based Thrombin Inhibitors Napap, 4-Tapap and Mqpa: A Starting Point for Improving Antithrombotics. *J.Mol.Biol.*, **1992**, *226*, 1085.

H. van der Goot, (Editor)
TRENDS IN DRUG RESEARCH II



# Heterocyclic mixture-based combinatorial libraries: synthesis and analysis of composition

J. S. Kiely and Y. Pei

Department of Exploratory and Combinatorial Chemistry, Trega Biosciences, Inc., 3550 General Atomics Court, San Diego, CA 92121, USA

## 1. INTRODUCTION

A driving force for developing and investing in combinatorial chemistry technologies is the realization that within preclinical lead discovery one of the rate limiting tasks is the synthesis of new chemical entities[1,2]. This rate limiting effect of chemical synthesis is irrespective of the intended use of the synthesized molecules, whether new lead discovery or lead optimization. The historical method of "one at a time" synthesis of new compounds is wholly inadequate for the capacity of even modest throughput 96 well microtiter plate-based assays. Existing historical collections will only fill part of the foreseeable need for total numbers chemical entities for screening and provide little in the way of new chemical entities compared to demand. This is because existing methods for creation of historical collections have limited growth potential. Additionally, historical collections suffer from prior depletion of patentable structures. From this point in time and into the future there will be an increasing and likely massive demand for new small molecule structures for screening in new targets developed from genomics projects. Thus this emerging demand for new compounds for screening purposes is one of the needs that pharmaceutical researchers have identified as requiring new methodologies[3].

This current and future need is beginning to be met through various methods for the production of thousands to hundreds of thousands of new chemical entities as single compound arrays or small pool size controlled mixtures. Collectively these have become known as chemical libraries. These libraries are organized collections of new compounds built from subsets of smaller building blocks through the systematic synthesis of a majority of the possible combinations. The inventories of new chemical libraries are based on a substantial and continually growing number of well-controlled optimized chemistries that yield medicinally relevant heterocycles and small molecule acyclic structures[2,4].

Screening efficiency has become an important consideration for preclinical drug discovery. The drive for greater efficiency has the paired goals of maximizing the number of compounds screened in order to uncover high value leads and minimizing the costs per assay point. This goal of economy has driven the industry towards high throughput screens wherever possible. These high throughput assay benefit from array-based combinatorial libraries. There remains, however, a significant number of assays that for various reasons are not true high throughput screens. The reasons include both cost of biological reagent and slow signal production (long incubation times). These factors precludes these 'non'-high throughput assays from benefiting from large arrays of compounds. In keeping with the goal of maximizing the numbers of compounds in these modest throughput assays, the combinatorial library utilized to provide the large numbers of compounds must be available as a mixture-based libraries. This type of library is organized into small subsets(pools) each having a defined structural parameter and with the number of compounds in each pools being low. For modest throughput assays this achieves the goal of tailoring the library format to the physical requirements of the assay and to the maximizing the number of compounds screened. Herein we described the examples of the production and analysis of mixture-based heterocyclic libraries intended for lead discovery in assays where single compounds arrays are inappropriate.

It is our belief that the chemical libraries utilized for lead discovery must be based on robust and highly optimized chemistries. The development of the reaction sequence described below have been carried out using solid phase organic synthesis techniques. These techniques are then adapted to the well established but closely held tea-bag technology [5,6] and to new robotic synthesis systems to produce the combinatorial libraries. We has focused on solid phase techniques for several reasons. First for libraries of tens of thousands and greater numbers of compounds, it is far easier to assemble heterocyclic and acyclic systems with three or more diversity sites on solid phase than with solution-based methods where one is usually limited to two or fewer diversity sites. This allows the solid phase based methods to more readily create medicinally relevant structures where multi-point pharmacophore structures are usually present. Secondly, in accessing these more complex structures, it is common to require 4 or more (usually <7) chemical steps. With the ability to drive reactions to completion on solid phase with large excesses of soluble reagents and easily remove by-products and excess reagents, it is far easier to accomplish these longer syntheses on a solid support than it is using solution phase methods{7,8].

We believe that the quality of the combinatorial libraries produced must be at the highest attainable level. Stringent and well defined criteria by which to judge library's composition need to be put in place. Mass spectrometry(MS) and HPLC-MS are our current tools of choice for the analysis of combinatorial library quality. For mixture-based combinatorial libraries quality is judged from several different pieces of data, which when taken together provide the data to make a determination of the quality of the library.

First, the chemistry to produce the library must be highly optimized and shown to work with the most sterically and electronically demanding examples anticipated within the full library. Secondly array syntheses are carried out in which every anticipated building block of the library is tested with at least one combination of the other building blocks. This is done to ensure that all building blocks undergo the anticipated reactions as planned. One proceeds to the synthesis of the library when those that fail to undergo the reaction have been identified and eliminated from further consideration and when a sufficient number of building blocks work successfully in the library to achieve the library size desired.

Thirdly during actual production of a mixture-based libraries, every library pool must have associated with it at every reaction step at least one control. A control is a single representative compound that is synthesized simultaneously with the pools. The controls utilized during the intermediate steps are analyzed before proceeding to the next reaction. This will assure that up to that point in the reaction sequence the chemistry being carried out on the pool has been accomplished as planned. At the completion of the synthesis of the mixture-based library, the final pools also require that a single compound control which is one of the compounds within the pool be associated with it. This control and the pool itself must be analyzed before use of the pool in any assays. Introduction of these highly important controls is facilitated a great deal by the use of the tea-bag technology, that so readily allows segregation and sorting of control and pools samples.

## 2. LIBRARY SYNTHESIS

The reaction sequence of a three diversity site heterocyclic library of 3,4-dihydro-2(1H)-quinolinones prepared by solid phase synthesis is shown in Figure 1. This was recently disclosed by us in preliminary form[9]. The final quinolone was assembled utilizing three building block sets. The sequence begins with coupling of amino acids to resin followed by condensation of a resin-bound amine with a 2-nitro-benzaldehyde under conditions previously optimized in the context of a different library[10]. The resultant imine was reacted with *in situ* generated ketenes in a [2+2] cycloaddition reaction. The 2-nitro group, a masked amine functionality, is reduced to the amine with tin dichloride and undergoes β-lactam ring opening to the 2,3-dihydro-2-quinolone[11]. This was subsequently cleaved from the resin to yield the desired heterocyclic compounds.

The optimization of the ketene cycloaddition-reduction-ring opening reaction sequence was carried out on Wang resin. A three dimensional matrix was designed to probe the limitations of each set of building blocks, the resin-bound amines, the ketene component (an acid chloride), and the 2-nitro-benzaldehydes. In this optimization each ketene component was tested with all pairwise combinations of amines and 2-nitro-benzaldehydes as shown in Figure 2. The amino acid were selected to probe the consequences of increased steric hinderance as previous work in our laboratory had shown this to be an important factor in cyclization of some imines. The benzaldehyde selections probed steric hinderance (6-chloro-2-nitrobenzaldehyde), increased and decreased electron density (4,5-dimethoxy- and 4-nitro- 2-nitrobenzaldehydes) and the effects of a free hydroxyl (5-hydroxy-2-nitro benzaldehyde). The acid chlorides chosen provided information on the reactivity of the ketenes with varying α–substitution. Table 1 lists the yields for the various test compounds.

Figure 1.

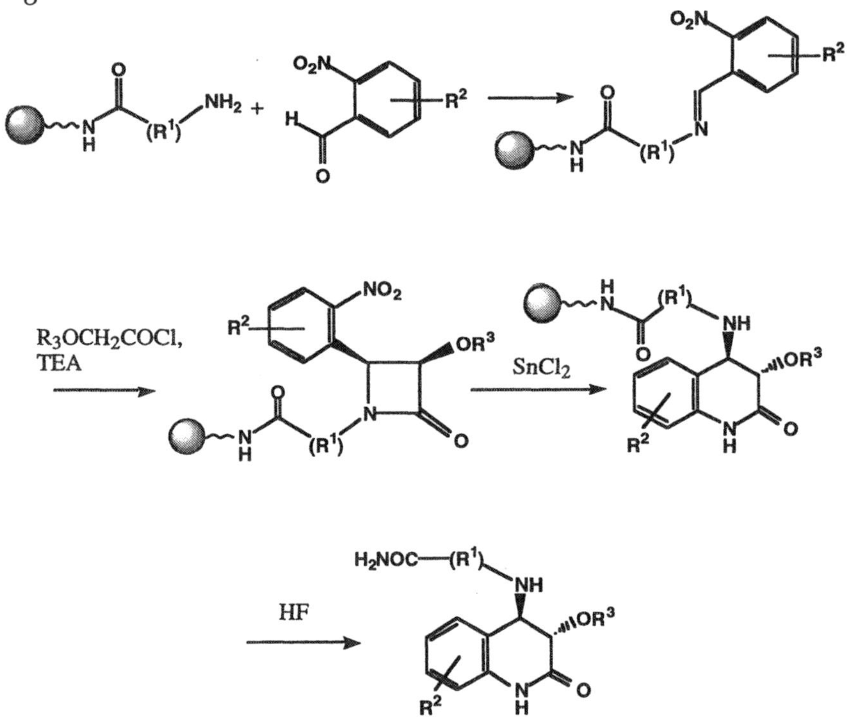

Figure 2.

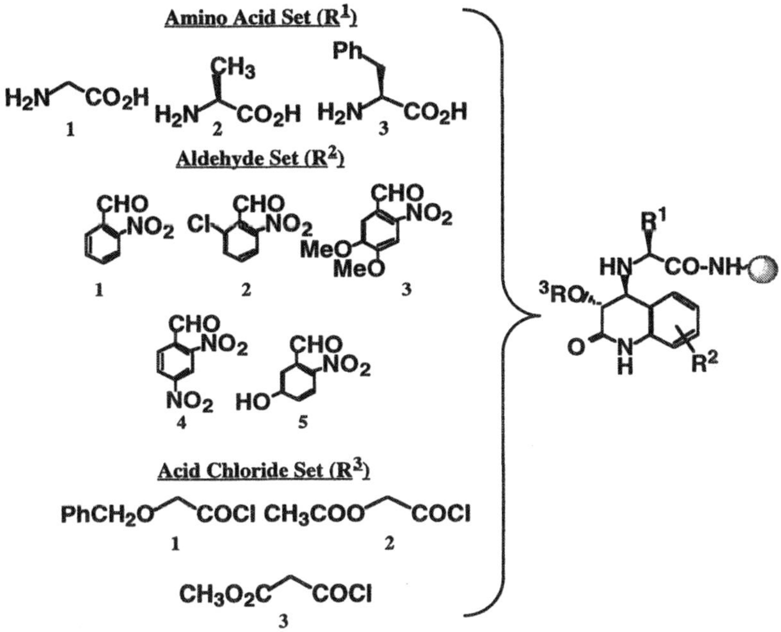


Table 1.
Optimization Matrix. Ketene reaction with pairwise amines and benzaldehydes

| Benzaldehyde | 1 | 2 | 3 | 4 | 5 |
|---|---|---|---|---|---|
| Amine | | | | | |
| KETENE 1 | | | | | |
| 1 | 0% | 0% | 50% | 0% | 0% |
| 2 | 68% | 0% | 83% | 0% | 40% |
| 3 | 65% | 0% | 82% | 0% | 70% |
| KETENE 2 | | | | | |
| 1 | 0% | 0% | 55% | 0% | 0% |
| 2 | 69% | 0% | 92% | 0% | 65% |
| 3 | 90% | 0% | 89% | 0% | 86% |
| KETENE 3 | | | | | |
| 1 | 0% | 0% | 0% | 0% | 0% |
| 2 | 0% | 0% | 80 | 0% | 0% |
| 3 | 0% | 0% | 0% | 0% | 0% |

Building block numbers taken from Figure 2.

As can be ascertained from Table 1, sterically hindered or electron deficient benzaldehydes and carbomethoxy substituted ketene (#3) did not work in the reaction sequence. These data dictated that from commercially available sources there were six 2-nitro benzaldehydes and five ketene precursors that were likely to be successful under the reaction conditions employed in the optimization phase. In addition to 2-nitro benzaldehyde itself, these included 5-chloro, 3-methoxy, 4,5-dimethoxy, 5-benzyloxy (under the cleavage conditions this converts to a free hydroxy substituent), and 4,5-methylenedioxy. The ketenes that could be generated were derived from the following substituted acetylchlorides: methoxy, phenoxy, 4-chlorophenoxy, acetoxy, and benzyloxy. Additionally, non-alpha substituted amino acids only work with electron rich benzaldehydes. This last observation could be due to the use of Wang (acid labile) resin in a highly Lewis acidic environment during the $SnCl_2$ reduction. The glycine test case was repeated using MBHA resin and found to proceed well. All subsequent chemistry was carried out on the more acid stable MBHA resin.

The optimization chemistry provided the information to proceed with the preparation of a mixture-based library based upon this reaction sequence. The library was constructed using the Houghten tea-bag technology[5,6] to carry out the divide, couple, and recombine (split and mix)[12,13,14] procedure for generating the mixtures. In planning the library, it was decided to keep the library mixture (pools) as small as reasonable (under 150 compounds) in order to facilitate effective testing. With the limited number of benzaldehydes and acetylchlorides available it was possible to format the library as 30 subsets of the pairwise combinations of the benzaldehydes and acetylchlorides ($R^2$ and $R^3$, in the Figure) and limit the mixture to all the possible amino acids making up the $R^1$ position. A selection of sixty nine different α-amino acids were utilized for $R^1$ spanning as diverse a set of substitutents as was available. All appropriate single compound controls were included in all phases of the preparation of this library. This provided a final library comprising 4140 3,4-dihydro-2(1H)-quinolinones as summarized in Figure 3.

Figure 3.

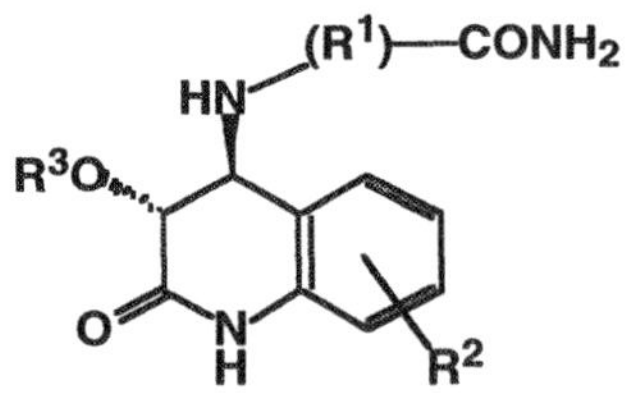


$R^1$ derived from 69 amino acids

$R^2$ derived from 6 ortho-nitrobenzaldehydes

$R^3$ derived from 5 acid chlorides

Library Size = 30 pools of 69 enantiomeric pairs

= 4140 Compounds

## 3. LIBRARY ANALYSIS

Upon completion of the library synthesis mass spectroscopy was utilized to analyze the components of the library. For mixture-based pools every pool and its associated control was analyzed by mass spectroscopy. We believe that due to the added complexity brought on when pools are being prepared that nothing less than complete analysis of the library is sufficient to provide confidence that the library synthesis proceeded as planned. Thus, every pool and its associated control is compared, a comparison of various pools to one another is performed to check for anticipated shifts in mass, and to the predicted mass pattern of the pool to assure that the mixture is as anticipated. This adherence to very high quality standards provides the necessary data to have full confidence in the library fidelity.

Figure 4 shows a representative pool mass spectrum obtained by infusion delivery atmospheric pressure chemical ionization using a Finnegan LCQ quadrapole mass spectrometer. The spectrum was obtained under conditions selected to minimize fragmentation and in a TFA buffer to eliminate sodium adducts. Panel A displays the actual pool data while Panel B is the predicted spectrum based on spreadsheet-based plot of expected molecular ions. The comparison of the two is clearly quite good. The anomalous peak observed at low m/z is due to a common fragmentation pathway for all components of the pools (Figure 5.). For this particular library, we were unsuccessful in completely suppressing this fragmentation. However, it is noninterfering with analysis of the pool(s) and clearly a mass spectrometer artifact.

Figure 4.

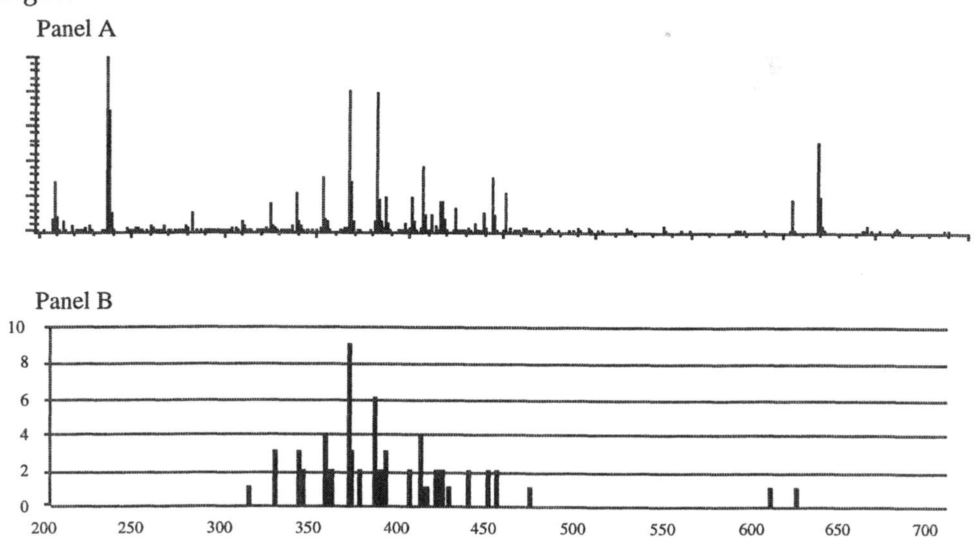

Figure 5.

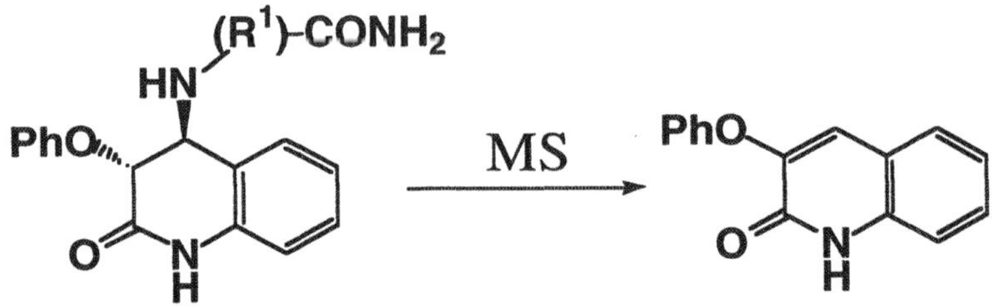


An HPLC-MS analysis of the corresponding control compound that was associated with the pool described above is shown below in Figure 6. The left hand trace of the Figure is the HPLC trace as monitored by total ion current (TIC) and the right hand spectrum is the mass spectrum for the major peak. The minor peak at longer retention time showed the same spectrum as the major peak and is attributed to a diastereomer, probably due to epimerization of one of the stereocenters. In the mass spectrum the high molecular weight peaks are attributed to dimer minus a proton and the ion-molecule pair of the m/z238 fragment with parent molecule.

Figure 6.

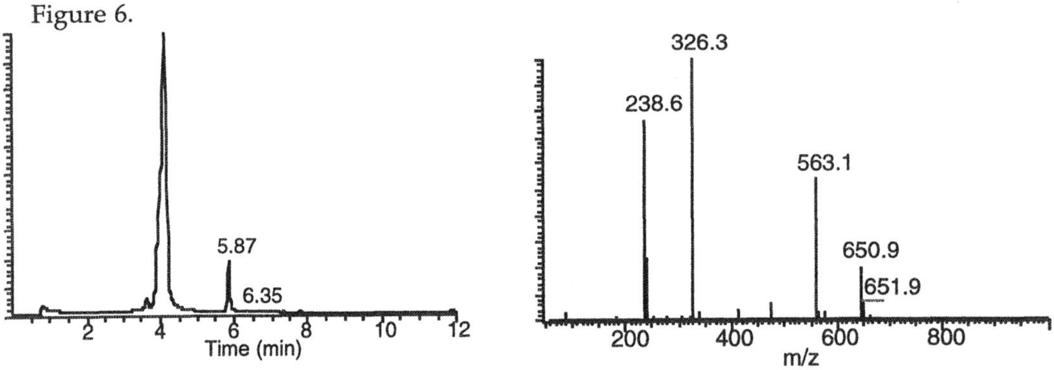


For the individual pools it is also possible to employ HPLC-mass spectroscopy to probe whether the intended masses are actually present within the pool. This is accomplished by analyzing the pool by HPLC and detecting the individual compounds by mass as they are eluted. Figure 7 shows a sampling of the peaks at individual mass ranges from the HPLC-MS analysis of the pool from Figure 5. The Figure shows the total ion current(TIC) trace for the elution of the pool from the HPLC and then the traces for the expected molecular ions for selected members of the pool. For example, within the pool were two compounds both with molecular weights of m/z368 these are shown in the first trace below the TIC trace. As can be seen no other material is present with the same molecular weight and two peaks are seen as expected. For molecular weight 326 two peaks are seen as expected along with two minor components which are likely due to epimerization as was seen for the control. Finally, at molecular weight 354 four peaks are observed as expected.

Figure 7.

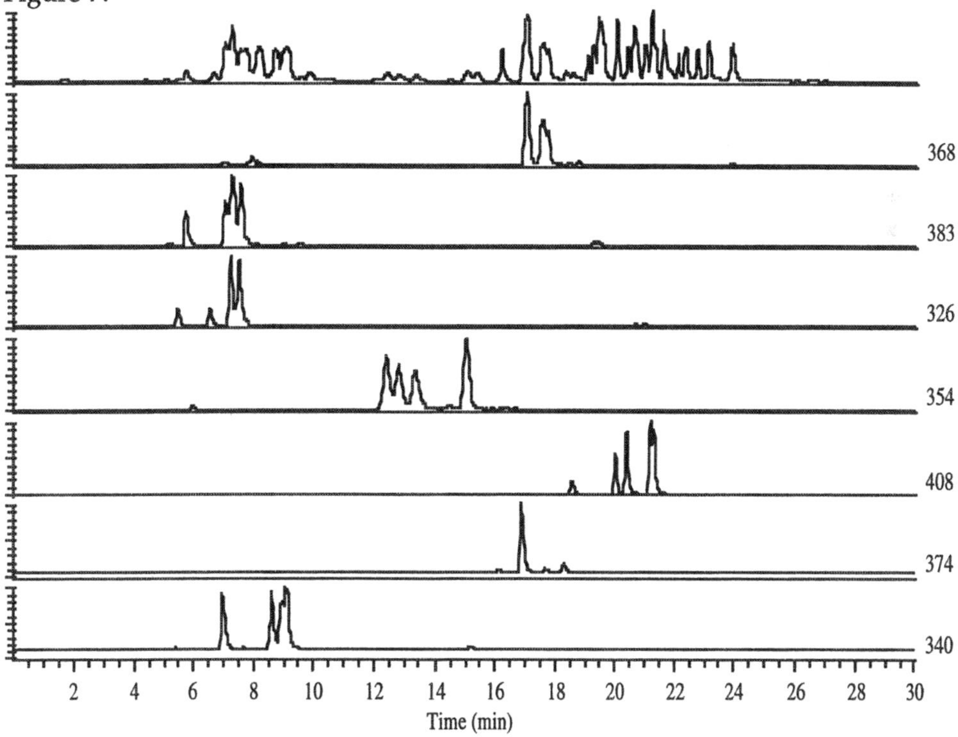


In summary, as shown by the results present, current chemical capabilities exist for the production of combinatorial libraries of substantial size based on well-controlled chemistries with very stringent analytical control of the components of the library. For this library of 4170 3,4-dihydro-2(1H)-quinolinones, the data presented is representative of all pools of the library.

This type of heterocyclic mixture-based combinatorial library can be made and employed for screening for new leads (data to be reported in due course). To effectively carry out the preparation of a combinatorial library of this type, it is very important to control the chemistry each of the steps of the sequence through rigorous optimization and to incorporate controls at every step in the reactions sequence and for every final pool produced. The current generation of HPLC-mass spectrometer systems are capable of generating sufficiently precise data for the individual pools of the library, that when coupled with data regarding control compounds synthesized simultaneously, one can obtain reliable data on which to base a decision on a libraries quality and suitability for use in lead discovery assays.

## 4. REFERENCES

1. Gallop, M.A., Barrett, R.W., Dower, W.J., Fodor, S.P.A. and Gordon, E.M., J. Med. Chem., **37** (1994) 1233-1251.
2. Terrett, N.K., Gardner, M., Gordon, D.W., Kobylecki, R.J. and Steele, J., Tetrahedron, **51** (1995) 8135-8173.
3. Moos, W.H., Pavia, M.R., Ellington, A.D., and Kay, B.K.,(eds.), Ann. Rep. Combin. Chem. Mol. Div., ESCOM, Leiden, 1997.
4. James, I.W. in Ann. Rep. Combin. Chem. Mol. Div., ESCOM, Leiden, 1997, pp. 326.
5. Houghten, R.A., Proc. Natl. Acad. Sci USA, **82** (1985) 5131-5135.
6. Houghten, R.A., US Patent No. 4,631,211 (1986).
7. Coe, D.M. and Storer, R. in Ann. Rep. Combin. Chem. Mol. Div., ESCOM, Leiden, 1997, pp. 50.
8. Hall, S.E. in Ann. Rep. Combin. Chem. Mol. Div., ESCOM, Leiden, 1997, pp. 326.
9. Pei, Y, Houghten, R.A., and Kiely, J.S., Tet. Lett. (1997) in press.
10. Griffith, M.C., Dooley, C.T., Houghten, R.A., and Kiely, J.S. in Chaiken, I. and Janda, K.D., eds., Molecular Diversity and Combinatorial Chemistry, American Chemical Society, Washington, D. C.,1996, pp. 50
11. Meyers, H.V., Dilley, G.J., Durgin, T.L., Powers, T.S., Winssinger, N.A., Zhu, H. And Pavia, M.R., Molecular Diversity, **1** (1995), 13.
12. Houghten, R.A., Pinilla, C., Blondelle, S.E., Appel, J.R., Dooley, C.T. and Cuervo, J.H., Nature, **354** (1991) 84-86.
13. Lam, K.S., Salmon, S.E., Hersh, E.M., Hruby, V.J., Kazmierski, W.M. and Knapp, R.J., Nature, **354** (1991) 82-84.
14. Furka, A., Sebestyen, F., Asgedom, M. and Dido, G., Intl. J. Pep. Prot. Res., **37** (1991) 487-492.

H. van der Goot, (Editor)
TRENDS IN DRUG RESEARCH II



# The Quest for New Chemical Entities to Treat the Gastric Pathogen *Helicobacter pylori*

Thomas C. Kühler*

Astra Research Center Boston, Inc.
128 Sidney St
Cambridge, Ma 02139, USA

Flurofamide and proton pump inhibitors as urease inhibitors *in vitro* and their efficacy *in vivo* is reviewed. The *in vitro* and *in vivo* profiles of both proton pump inhibitors and their corresponding sulfides and sulfones as anti *H. pylori* agents is discussed. The prospect of using genomics to identify new intervention therapies is analyzed.

## 1. INTRODUCTION

*Helicobacter pylori* plays an important role in peptic ulcer disease [1,2] and eradication of the bacterium in patients with peptic ulcers dramatically decreases the risk of ulcer recurrence [3,4,5]. The obvious remedy for treating a bacterial infection with antimicrobials, however, has often proven futile [6] and only a few combination regimen have reached clinical practice [7,8,9,10]. The need for alternative or novel treatments is thus still evident. In addition, the emergence of strains resistant to treatment with *e.g.* metronidazole [11,12] or clarithromycin [13] is another compelling argument for trying to identify new drugs directed against *H. pylori*.

## 2. UREASE AS A TARGET

*H. pylori* allocates a considerable proportion of its anabolic activity towards the production of urease, EC 3.5.1.5. [14], which early on became a conspicuous biochemical marker as well as a target for intervention therapy [15,16]. Both flurofamide (**1**), an earlier developed potent urease inhibitor, as well as proton pump inhibitors (*cf.* **2** for the generic structure), more recently conceived as potential urease inhibitors, have been tested for in vivo efficacy.

* Internet: thomas.kuhler@arcb.us.astra.com

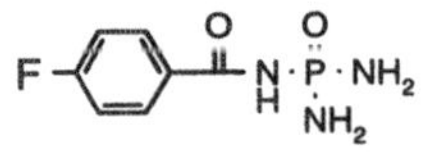

1

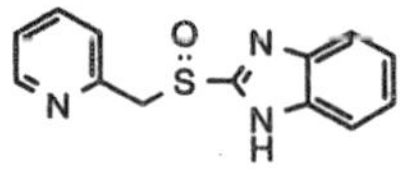

2

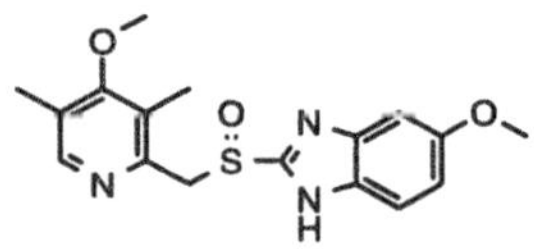

3

Chart 1. Structures of the urease inhibitor flurofamide (**1**), a generic proton pump inhibitor (**2**), and the proton pump inhibitor omeprazole (**3**).

Flurofamide has been tested in both the *Helicobacter mustele* infected ferret [15] and the *Helicobacter felis* infected mouse [16] models whereas omeprazole (**3**) has been tested only in the *H. felis* infected mouse [16] model. Both compounds do potently inhibit urease activity *in vitro* and flurofamide has been shown to inhibit urease activity *in vivo* by some 95-100% for 24 h after a single oral dose of 23 mmol/kg in the ferret model [15]. Despite good activities at the enzyme level (both flurofamide and omeprazole) and excellent pharmaco dynamic and pharmaco kinetic properties (flurofamide) neither treatment regimen proved efficacious in either of the animal models tested as shown in Figure 1.

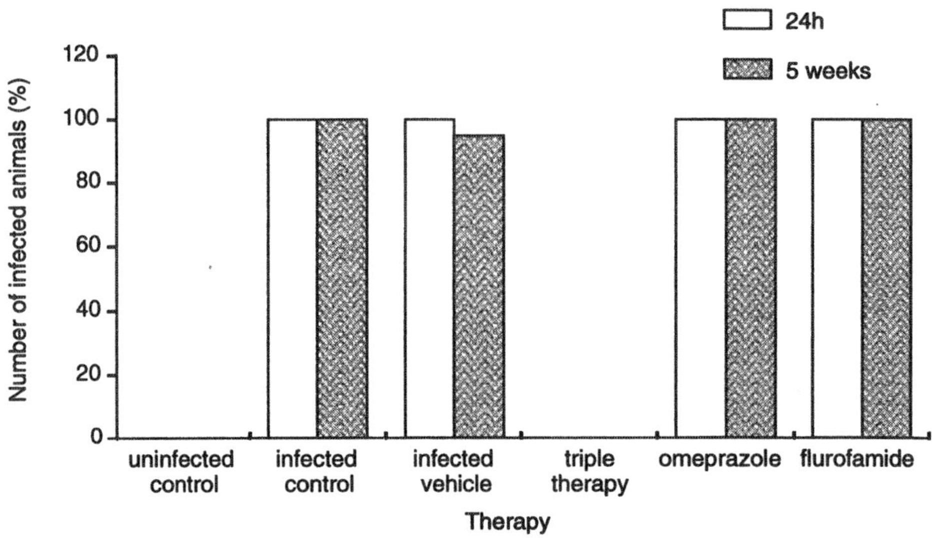


Figure 1. *H. felis* status in various control and treated SPF mice 24 h and 5 weeks after stopping treatment, respectively. The suppression and eradication rates are expressed as the number of urease positive animals divided by the number of animals checked x 100% (n > 9). Triple therapy was made up by 0.185 mg bismuth, 0.675 mg metronidazol, and 1.500 mg tetracycline and was administered once daily

for two weeks followed by bismuth alone once daily for another two weeks. In this group of animals, vehicle alone was administered at the second daily dosing occasion. Omeprazole (1.295 mg corresponding 125 mmol/kg) and flurofamide (1.250 mg corresponding to 230 mmol/kg) were each dosed twice daily for four weeks. The amounts stated are per mouse and day. From reference 16.

Although urease activity is required during the colonization of gnotobiotic piglets [17] and SPF mice [18] urease inhibition seemingly does not provide a route to clear an ongoing *Helicobacter*-infection. This is in good agreement with urease negative mutants being viable [19,24].

## 3. ANTIBACTERIAL ACTIVITY OF OMEPRAZOLE

Omeprazole and structural analogues were first developed as potent proton pump inhibitors, *i.e.* inhibitors of the $H^+/K^+$-ATPase of the parietal cell in the oxyntic mucosa of the stomach [20]. Beside the inhibitory effect on the proton pump (and as discussed above the urease) omeprazole also exerts an antibacterial effect on *H. pylori* [21,22]. Omeprazole is a prodrug and the question was posed whether it was the intact prodrug **3**, the activated species **4**, or both that were responsible for the antibacterial activity.

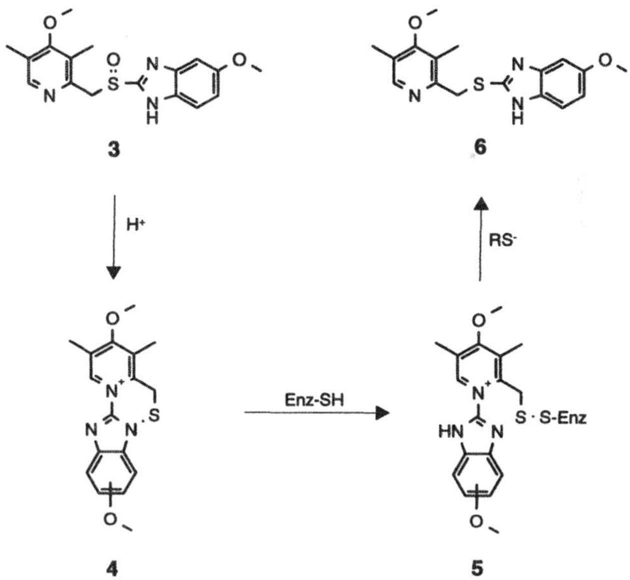


Scheme 1. Structures of the prodrug omeprazole (**3**), the activated species (**4**), the enzyme-inhibitor disulfide complex (**5**), and the corresponding sulfide (**6**). For details of this chemistry please check Reference 23.

When a sterically hindered omeprazole analogue (7), incapable of transforming to the activated species corresponding to structure **4** above, was studied *in vitro,* the antibacterial activity was found to be moderate but selective for *Helicobacters* [24].

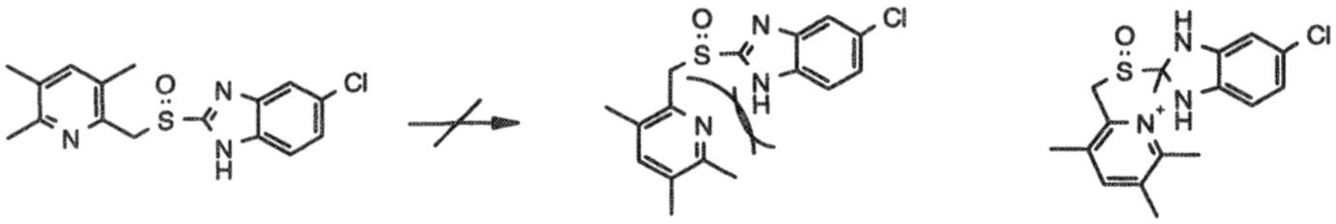

7 8 9

Scheme 2. A proton pump inhibitor analogue **7** carrying a methyl group in the 6 position of the pyridine moiety. The methyl group incurs an unfavorable steric interaction as indicated in **8** preventing the formation of the spiro compound **9**. Compound **9** would have been the first intermediate in a cascade eventually leading to the activated species corresponding to structure **4** in Scheme 1. For details of this chemistry please check Reference 23.

When omeprazole was studied *in vitro* [22] under conditions of low pH when it readily is converted to the activated species **4** the antibacterial activity was augmented but not selective for *Helicobacters* any more. Under these conditions several other Gram-negative as well as several Gram-positive genera were susceptible to killing. The bactericidal activity correlated well with the degree of covalent binding of omeprazole to bacteria cell proteins (as evidenced with tritium labeled omeprazole on *H. pylori*) and killing was hypothesized to be due to modifications of accessible sulfhydryl groups. This notion was substantiated by the finding that the addition of β-mercaptoethanol, a known scavenger of the active compound **4**, almost entirely abolished bactericidal activity.

When additional omeprazole analogues were studied, a correlation between the rate of conversion to the active species and the antibacterial activity was observed such that a more rapid conversion resulted in an increased antibacterial activity. As expected, increased antibacterial activity was associated with a decreased selectivity for the genus *Helicobacter*.

Since there was an inverse correlation between potency and selectivity (the opposite would have been desired) and a direct correlation between potency and acid secretion inhibition (again the opposite would have been desired) the proton pump inhibitors were abandoned as anti *H. pylori* agents.

## 4. ANTIBACTERIAL ACTIVITY OF SULFIDES

The same selective antibacterial activity as observed for compound **7** was early on also observed for sulfones of the generic structure **10** [24], the sulfide **6** [25], as well as some other sulfides with the same basic frame work, *cf.* **11**, n = 0 [21].

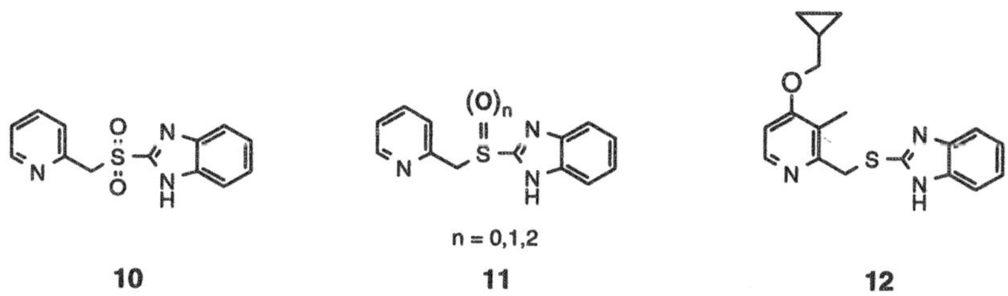

Chart 2. A generic sulfone **10** (corresponding to the generic proton pump inhibitor **2** as shown in Chart 1), generic structure of compounds claimed as anti *H. pylori* agents (**11**), and structure of a sulfide **12** that has been tested for *in vivo* efficacy.

These results indicated that there was an intrinsic antibacterial activity connected with the framework of these compounds that did not require prior activation to an active species. This, of course, attracted considerable interest and attempts to optimize the antibacterial activity has resulted in close to 30 patents being filed to date claiming compounds of the generic structure **11** as anti *H. pylori* agents. Only one study, however, has discussed the *in vivo* efficacy of such sulfides, *e.g.* compound **12**, and evidenced the lack of a pronounced effect [26].

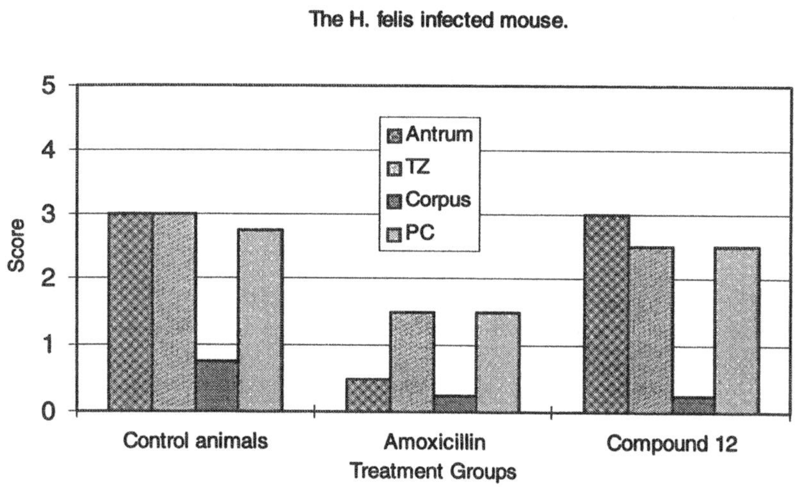


Figure 2. The *H. felis* infected mouse. Effects of amoxicillin and test compound **12**

(minimum bactericidal concentration = 0.5 μg/mL against *H. pylori* strain ATCC 43504) on stomach *H. felis* cell counts. The degree of infection was assessed by microscopically examining stained sections of the stomach and estimating the bacterial density on a scale from 0 to 5. Amoxicillin was given orally at 15 mmol/kg (6.25 mg/kg) bid for four days and compound **12** was given orally at 125 μmol/kg (41 mg/kg) bid for four days, respectively. From reference 26.

Compound **12** did, however, reduce gastric acid secretion and the conclusion was that any sulfide of the generic structure **11**, n = 0, *in vivo* would partly be oxidized to the corresponding sulfoxide **11**, n = 1, yielding a proton pump inhibitor.

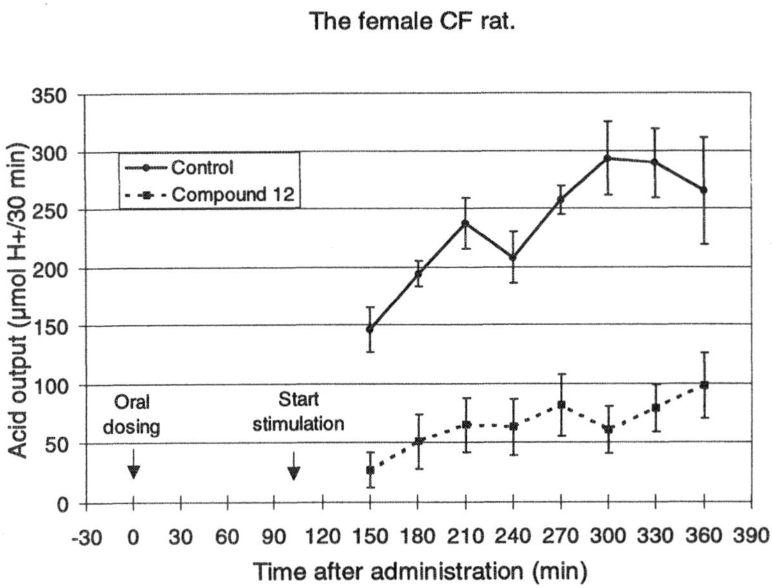


Figure 3. Stimulated gastric acid secretion in the female chronic fistula rat model. Acid output after a single oral dose of compound **12** (95 μmol/kg, 31 mg/kg). From reference 26.

The investigators suggested that whereas the antibacterial activity obviously had the potential of decreasing *H. felis* cell counts *in vivo* the proton pump inhibitory effect (less acid secretion) promoted (*H. felis*) cell growth rate *in vivo* very much the same way as observed for the proton pump inhibitors and *H. pylori* in man [27].

Hence, sulfides of the generic structure **11**, n = 0, are compromised by their propensity to become potent proton pump inhibitors upon metabolic oxidation *in vivo* and therefore carry structural features that disqualifies them for development as new anti *H. pylori* therapeutics.

## 5. THE PROSPECTS OF GENOMICS

Proton pump inhibitors, proton pump inhibitor derivatives such as the corresponding sulfides or sulfones, and numerous other (diverse) structures found in the patent literature have displayed, if not excellent then good, anti *H. pylori* activity *in vitro*. Yet no new chemical entities have emerged for which the *in vitro* activity successfully could be translated into the *in vivo* situation. On the contrary, the number of failures could make one speculate that *H. pylori* when grown *in vitro* is extremely troubled and any compound (with true cidal activity or not) could inflict irreparable damage rendering a false hit.

The traditional drug discovery paradigm starting with *in vitro* and/or *in vivo* models mimicking a disease, and a compound with some activity in those models, followed by "black box" optimization, does obviously not work to well with *H. pylori*. Access to the (complete) genome, and thereby the possibility of mapping the entire anabolic/metabolic/catabolic pathways of the organism, provides a unique knowledge and thereby a possible advantage. Essentiallity of key processes can be identified and established by the gene knock-out technique after which relevant proteins could be over expressed. High throughput screening, structural based design, or mechanism based design would then furnish starting points for rational optimization of new compound classes acting on novel and true targets.

Figure 4. The genomics to drugs approach. Access to complete genomic information allows the mapping of anabolic/metabolic/catabolic pathways. After identifying targets for intervention they can be over expressed and be subjected to high throughput screening. When hits are identified they can be validated as leads and optimized to furnish Candidate Drugs.

The outlined strategy is not necessarily a unique or new approach in any of its aspects, but combined with knowledge of the entire genome, as opposed to single genes or single cloned proteins, it certainly provides unprecedented possibilities.

The genomics to drugs approach to identify new chemical entities directed towards the eradication of *H. pylori* infections is currently pursued at Astra Research Center in Boston, Ma.

## ACKNOWLEDGEMENTS

All former and current colleges and co-workers at Astra Hässle AB, Sweden and at Astra Research Center Boston, Inc., Ma., respectively, are recognized for their contributions. I especially would like to mention Dr. Nils-Åke Bergman and Dr. Vladimir Shcherbuchin for their contributions on the SAR:s, Dr. Håkan Larsson and Dr. Björn Mellgård for the pharmacology studies, Dr. Jan-Eric Sjöström for the microbiology work, Dr. Jan Fryklund for providing the urease assay, and Marianne Swanson for excellent synthetic help.

## REFERENCES

1. Working Party Reports, *36-45 World Congress on Gastroenterology*, Sydney **1990**.
2. *Helicobacter pylori in Peptic Ulcer Disease. NIH Consensus Statement.* **1994**, *Feb 7-9;12(1)*, 1-17.
3. Forbes, M. F.; Glaser, M. G.; Cullen, D. J. E.; Warren, J. R.; Christianssen, K. J.; Marshall. B. J. Duodenal ulcer treated with *Helicobacter pylori* eradication: seven year follow up. *Lancet* **1994**, *343*, 258-60.
4. Labenz, J.; Rühl. G. H.; Bertrams, J.; Börsch, G. Clinical course of duodenal ulcer disease one year after omeprazole plus amoxycillin or triple therapy plus ranitidine for cure of *Helicobacter pylori* infection. *European Journal of Gastroenterology & Hepatology*. **1994**, *6*, 293-297.
5. Tytgat, G. N. J. Long-Term Consequences of *Helicobacter pylori* Eradication. *Scand. J. Gastroenterol. Suppl.* **1994.** *205*, 38-44.
6. Bayerdörffer, E.; Ottenjahn, R. The Role of Antibiotics in *Campylobacter pylori* Associated Peptic Ulcer Disease. *Scand J Gastroenterol.* **1988,** *23 (suppl 142)*, 93-100.
7. Unge, P.; Gad, A.; Eriksson, K.; Bergman, B.; Carling, L.; Ekström, P.; Glise, H.; Gnarpe, H.; Junghard, O.; Lindholmer, C.; Sandzén, B.; Strandberg, L.; Stubberöd, A.; Weywadt, L. Amoxicillin added to omeprazole prevents relapse in the treatment of duodenal ulcer patients. *E. J. Gastroenterol & Hepat.* **1993,** *5*, 325-331.
8. Logan, R. P. H.; Gummett, P. A.; Hegarty, B. T.; Walker, M. M.; Baron, J. H.; Misiewicz, J. J. Clarithromycin and omeprazole for *Helicobacter pylori. Lancet.* **1992,** *340*, 239.
9. Tytgat, G.N., *Review article: treatments that impact favorably upon the eradication of* Helicobacter pylori *and ulcer recurrence. Alimentary Pharmacology & Therapeutics.* **1994,** *8*, 359-368.
10. Axon, A.T. The Role of Omeprazole and Antibiotic Combinations in the Eradication of Helicobacter pylori - An update. *Scand. J. Gastroenterol.* **1994** *29 (Suppl 205)* 31-37.
11. Haas, C.E.; Nix, D.E.; Schentag, J.J. In Vitro Selection of Resistant *Helicobacter pylori. Antimicrob. Agents Chemother.* **1990,** *34*, 1637-1641.
12. van Zwet, A.A.; Thijs, J.C.; Shievink-de Vries, W.; Shiphuis, J.; Snijder, J.A.M. In Vitro Studies on Stability and Development of Metronidazole Resistance in *Helicobacter pylori. Antimicrob. Agents Chemother.* **1994,** *38*, 360-362.
13. Debets-Ossenkopp, Y.J.; Sparrius, M.; Kusters, J.G.; Kolkman, J.J.; Vandenbroucke-Grauls, C.M.J.E. Mechanism of clarithromycin resistance in clinical isolates of *Helicobacter pylori. FEMS Microbiol. Lett.* **1996,** *142*, 37-42.

14. Mobley, H. L. T.; Hausinger, R. P. Microbial Ureases: Significance, Regulation and Molecular Characterization. *Microbiol. Rev.* **1989**, *53*, 85-108.
15. McColm, A. A.; Bagshaw, J.; O'Malley, C.; McLaren, A. Is urease a lethal target for therapy of *Helicobacter pylori? Microbial Ecology in Health and Disease (Proceedings from The 6th International Workshop On Campylobacter, Helicobacter & Related Organisms).* **1991**, *4*, 145 (abstract H4-10).
16. Kühler, T.C.; Fryklund, J.; Bergman N-Å.; Weilitz J.; Larsson H. Structure-Activity Relationship of Omeprazole and Analogues as *Helicobacter pylori* Urease Inhibitors. *J. Med. Chem.* **1995**, *38*, 4906-4916.
17. Eaton, K. Y.; Brooks, C. L.; Morgan, D. R.; Krakowka, S. Essential Role of Urease in Pathogenesis of Gastritis Induced by *Helicobacter pylori* in Gnotobiotic Piglets. *Infect. and Immun.* **1991**, *59*, 2470-2475.
18. Ferrero, R. L.; Lee, A. The importance of urease in acid-protection for the gastric-colonizing bacteria *Helicobacter-pylori* and *Helicobacter-felis* new-species. *Microbial. Ecol. Health Dis.* **1991**, *4*, 121-134.
19. Ferrero, R. L.; Cussac, V.; Courcoux, P.; Labigne, A. Construction of isogenic urease negative mutants of *Helicobacter pylori* by allelic exchange. *J. Bacteriol.* **1992**, *174*, 4212-4217.
20. Fellenius, E.; Berglindh, T.; Sachs, G.; Olbe, L.; Elander, B.; Sjöstrand, S.E.; Wallmark, B. Substituted benzimidazoles inhibit acid secretion by blocking $H^+/K^+$-ATPase. *Nature.* **1991**, *290*, 159-161.
21. Iwahi, T.; Satoh, H.; Nakao, M.; Iwasaki, T.; Yamazaki, T.; Kubo, K.; Tamura, T.; Imada, A. Lanzoprazole, a novel benzimidazole proton pump inhibitor, and related compounds have selective activity against *Helicobacter pylori. Agents Chemother.* **1991**, *28*, 490-496.
22. Sjöström, J-E.; Fryklund, J.; Kühler, T.; Larsson, H. In Vitro Antibacterial Activity of Omeprazole and Its Selectivity for *Helicobacter* spp. Are Dependent on Incubation Conditions. *Antimicrob. Agents Chemother.* **1996**, *40*, 621-626.
23. Lindberg, P.; Nordberg, P.; Alminger, T.; Brändström, A.; Wallmark, B. The Mechanism of Action of the Gastric Acid Secretion Inhibitor Omeprazole. *J. Med. Chem.* **1986**, *29*, 1327-1329.
24. Unpublished results.
25. Sjöström, J-E; Kühler, T.; Larsson, H. BASIS FOR THE SELECTIVE ANTIBACTERIAL ACTIVITY IN VITRO OF PROTON PUMP INHIBITORS AGAINST HELICOBACTER SPP. Accepted for publication in *Antimicrob. Agents Chemother.*
26. Kühler, T.C.; Swanson, M.; Shcherbuchin, V.; Larsson, H.; Mellgård B.; Sjöström, J-E. Structure Activity Relationship of 2-[[(2-pyridyl)methyl]thio]-1H-benzimidazoles as Anti Helicobacter pylori Agents in vitro and Evaluation of their in vivo Efficacy. Submitted for publication in *J. Med. Chem.*
27. Lee, A.; Dixon, M.F.; Danon, S.J.; Kuiper, E.; Megraud, F.; Larsson, H.; Mellgård, B. Local acid production and *Helicobacter pylori*: a unifying hypothesis of gastroduodenal disease. *E. J. Gastroenterol. & Hepat.* **1995**, *7*, 461-465.

H. van der Goot, (Editor)
TRENDS IN DRUG RESEARCH II



# Chemistry and SAR of prokinetic motilides: non-peptidic motilin receptor agonists

P. A. Lartey

Pharmaceutical Products Division, Abbott Laboratories, Abbott Park, IL 60064, USA*

The advent of the motilide class of prokinetic agents commenced with the recognition that an often reported side effect of erythromycin (**1**) on the gastrointestinal (GI) tract, 'stomach upset', is due to a pharmacological action of **1**, as a motilin receptor agonist [1]. Motilin **2** is a 22-aminoacid peptide, elaborated by endocrine (Mo) cells in the gastrointestinal mucosa [2] and secreted cyclically during interdigestive periods [3]. The release of this hormone is associated with induction of caudally migrating propulsive contractions with well-characterized patterns, referred to as the migrating myoelectric complex (MMC). Exogenously (intravenous) administered **2** induces coordinated GI contractions in man, resulting in gastric emptying [4].

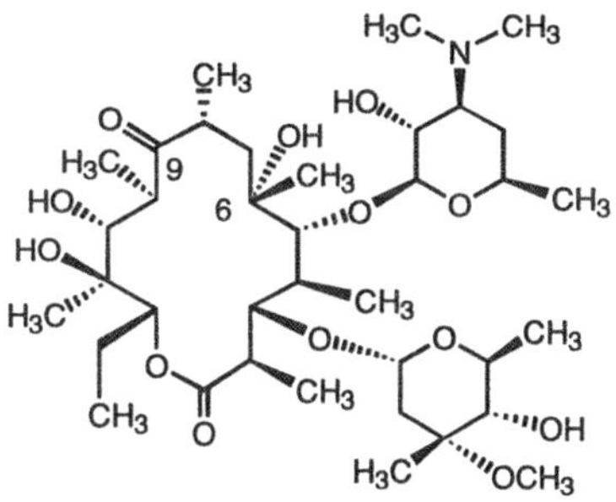


Phe-Val-Pro-Ile-Phe-Thr-Tyr-Gly-Glu-Leu-Gln
Gln-Gly-Lys-Asn-Arg-Glu-Lys-Glu-Glu-Met-arg

**1** **2**

* Current address:
Infectious Diseases Research
Pharmacia and Upjohn
7000 Portage Rd.
Kalamazoo, MI 49001
USA

Erythromycin binds to motilin receptors isolated from the GI tract of rabbit and man, and displaces bound [$^{121}$I]-motilin [5]. It induces contractions similar to those induced by **2** in man [6]. Like motilin, **1** also increases lower esophageal sphincter (LES) pressure and motility.

The prokinetic activity of **1** is independent of its antibacterial action. The latter depends on selective binding of **1** to the 50*S* ribosome of bacteria and inhibition of bacterial protein synthesis. Hence it was proposed that compounds derived from erythromycin with improved prokinetic action but no antibacterial activity, dubbed motilides [7], will be useful as drugs for the treatment of GI motility disorders such as gastroesophageal reflux disease (GERD), diabetic gastroparesis, functional dyspepsia and irritable bowel syndrome.

## 1. MOTILIDES

The first such compound to be described as a potential drug candidate was EM-523 (**3**) [8]. Compound **3** is an 8,9-dehydro derivative of erythromycin A-6,9-hemiacetal. Compound **3** is much more potent ($pED_{50}$ = 8.5) than **1** ($pED_{50}$ = 5.8) in a rabbit duodenal smooth muscle contractility assay, but has little or no antibacterial activity (MIC > 50 μg/mL vs. *Streptomyces pyogenes*) compared to **1** (MIC = 0.01 μg/mL). However, as in the case of similar derivatives, the enol ether moiety of **3** is highly unstable under aqueous acidic conditions [9] such as occurs in the stomach.

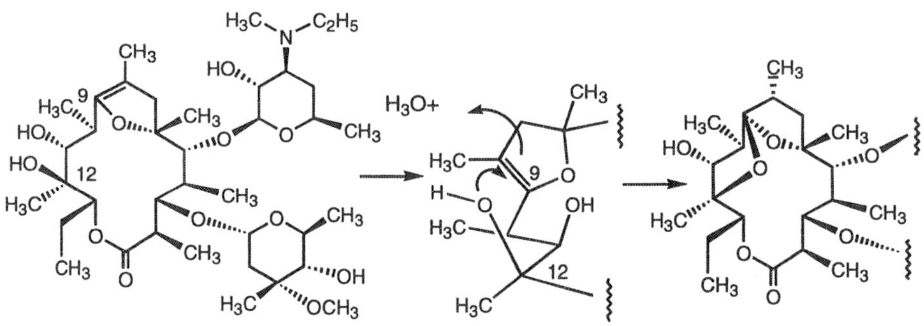


**3** **4**

Scheme 1. Acid catalyzed spiro-acetalation

Under these conditions, **3** undergoes intramolecular ketalization to afford **4** (Scheme 1), which lacks prokinetic activity. Hence **3** has no oral efficacy.

### 1.1. Motilides in clinical development

Currently four new motilides, reported to be orally active, are under investigation as potential prokinetic drugs [10, 11]. These are EM-574 (**5**), GM-611 (**6**), KC-11458 (**7**) and ABT-229 (**8**).

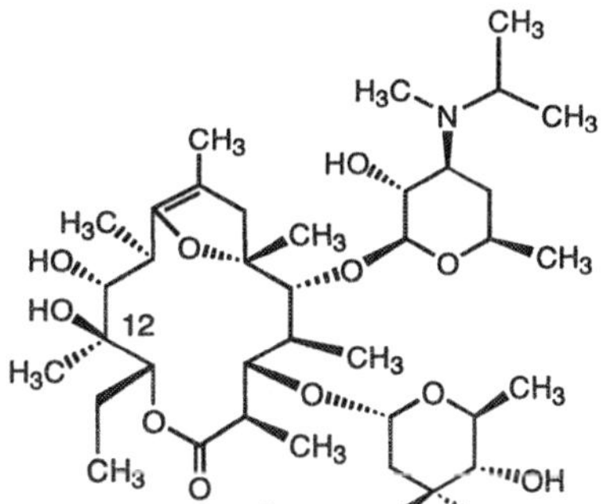

5

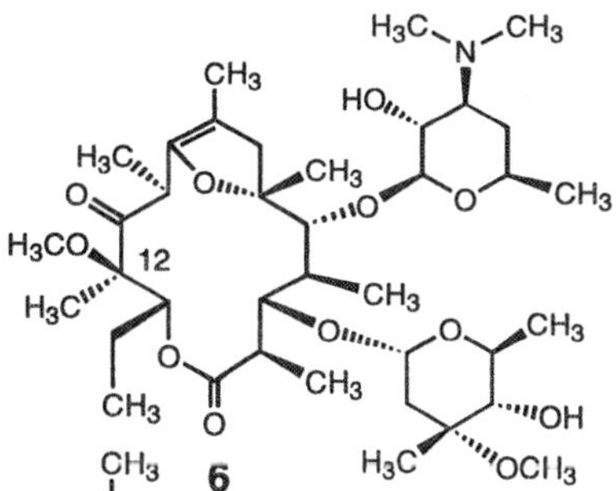

6

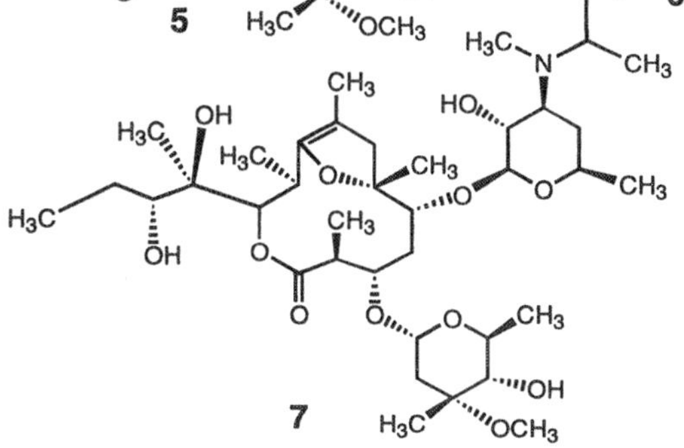

7

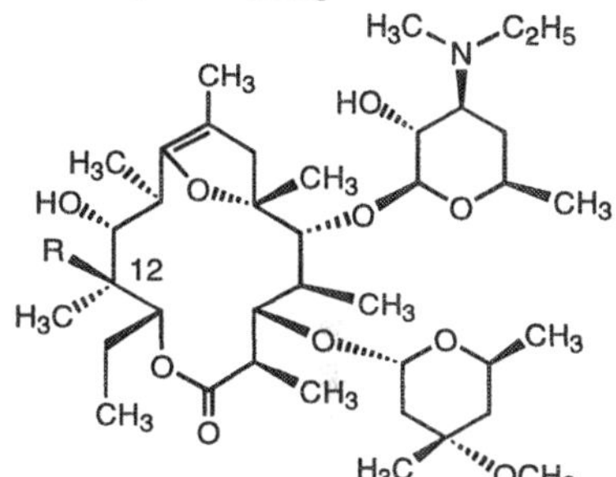

**8** R = H
**9** R = OH

### 1.2. ABT-229

Compound **8** is a derivative of erythromycin B which lacks the C-12 hydroxy group and is not subject to the intramolecular ketalization described in Scheme 1. Compound **8** is therefore relatively more acid stable [12] than either its 12-hydroxy congener **9** or **3**. The relative acid stability of **8** is reflected in its oral bioavailability (39%) compared to 2.7% for **9** and 1.4% for **3** (Table 1).

Table 1
Comparative activities and pharmacokinetics

| Compound | *In Vitro* Activity[1] (pED$_{50}$) | Pharmacokinetics | |
|---|---|---|---|
| | | $t_{1/2}$ (h) | F (%) |
| **3** | 8.45 | 2.4 | 1.4 |
| **8** | >12 | 5.5 | 39 |
| **9** | 11.15 | 2.6 | 2.7 |

1. pED$_{50}$ = -log concentration for half maximal response in isolated rabbit duodenal smooth muscle.

The deoxygenation of the 4"-position also dramatically improved potency, making **8** >300,000 times more potent than **1** in inducing contractility in rabbit duodenal smooth muscle. Compound **8** does not bind to bacterial ribosomes and hence lacks antibacterial activity. It is currently in phase II clinical trials for the treatment of diabetic gastroparesis and GERD.

## 2. SAR OF MOTILIDES

### 2.1. 6,9-ether

An important structural feature for potency of motilides is the 6,9-oxygen bridge, which in compounds **5** to **9** is in the form of an enol ether.

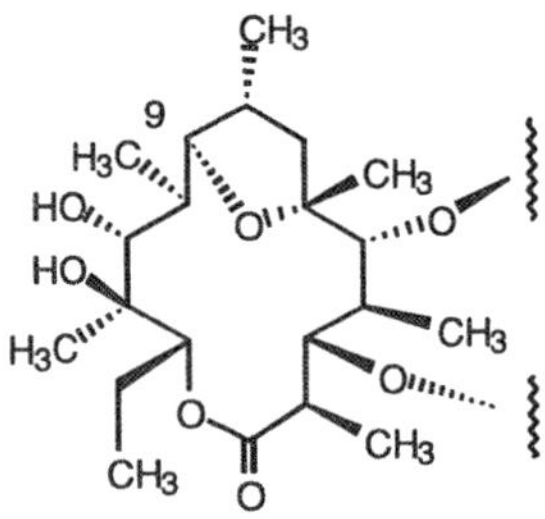

**10**

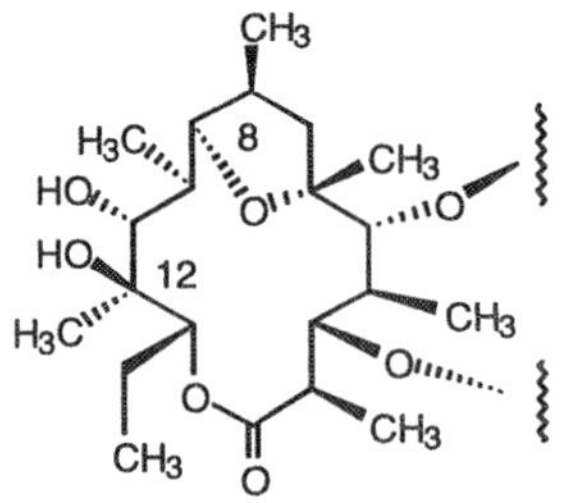

**11**

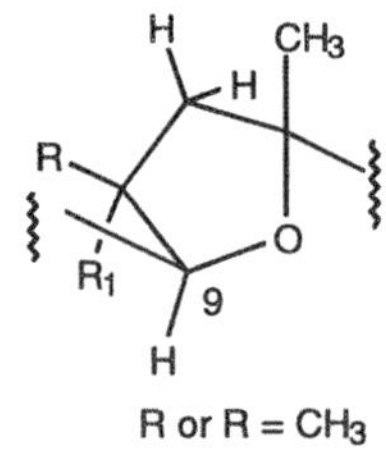


**12**

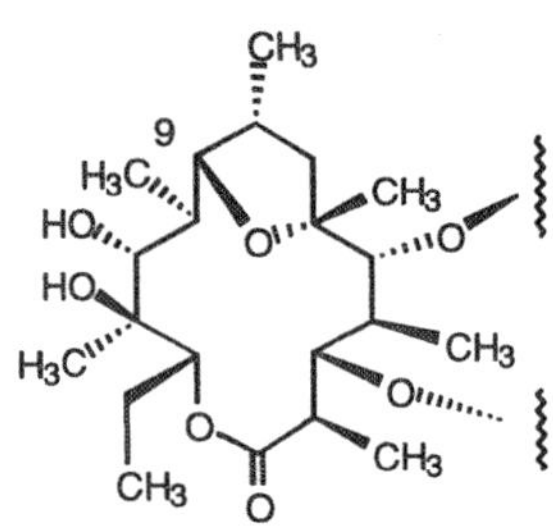

**13**

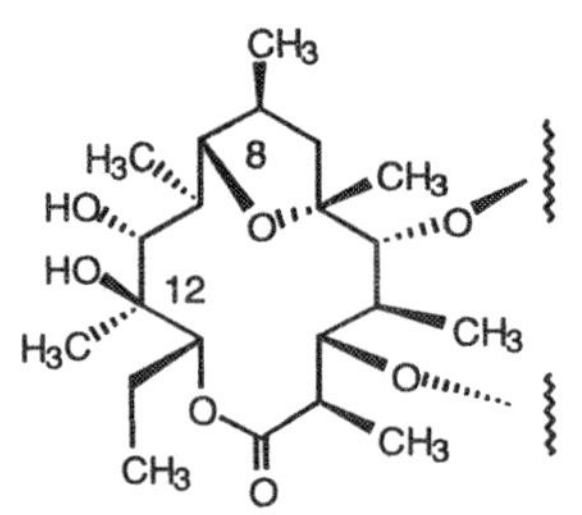

**14**

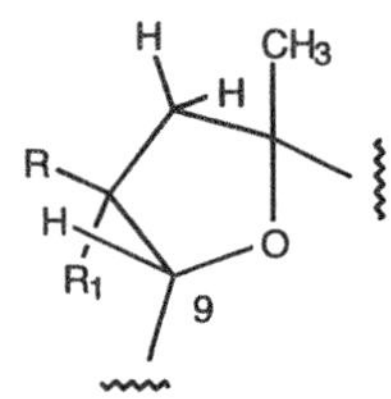

**15**

A logical structural modification to retain potency and increase acid stability would be to reduce the double bond of the enol ether moiety. Four possible diastereomers are possible from such a transformation; **10** and **11** with the 9*R* configuration around the nascent tetrahydrofuranyl ring system **12,** as well as **13** and **14** with the 9*S* configuration **15**. Recent work in our laboratory focused on relating the structures of these diastereomers to their prokinetic potencies.

The stereoselective synthesis of **10** has been previously described [13]. In Scheme 2, we give conditions for the stereoselective reduction of the enol ether in 8,9-anhydroerythromycin-6,9-hemiacetal (**16**) to afford **11** and **13** in a 4:1 ratio of a mixture of diastereomers.

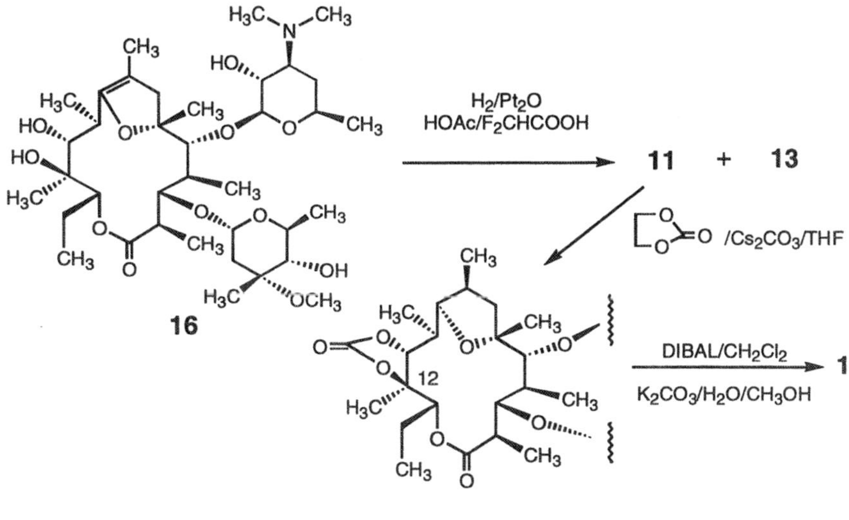


17

Scheme 2. A practical route for the preparation of an 8*S*, 9*R* analog

The epimeric configurations at C-9 gave rise to two different conformations for **11** and **13** around the C-10 to C-12 region. In compound **11**, the OH-groups at C-11 and C-12 are *syn* on the macrolactone ring, while in **13**, they are orthogonal. These differences in the relative orientations of the hydroxy groups enabled us to develop a practical separation method, wherein treatment of the mixture with ethylene carbonate resulted in selective derivatization of **11** as its 11,12-di-*O*-cyclic carbonate **17**, which could then be separated from **13** and subsequently deprotected to afford **11**. Compound **14** has thus far been elusive.

The 9*R* congeners **10** and **11** had good prokinetic activities ($pED_{50}$ = 6.80 and 7.15, respectively) while the 9*S* analog **13** was weaker in activity ($pED_{50}$ = 5.71) than erythromycin. This suggests that the 9*R* configuration is preferred and that the elusive compound **14** will most likely have weak activity.

**2.2. C-12 epimers and macrolactone conformation**

Synthesis of the C-12 epimer (12*S*) **18** of the erythromycin B derivative **19** has been described [14]. Interestingly, the conformation of **18** differed considerably from that of **19** around the C-10 to C-13 region. Based on Nuclear Overhauser effect (NOE) experiments, the conformation of **19** (**20**), around that region, was shown to be identical to the X-ray conformation of **10** [13]. Compound **18**, on the other hand, has conformation **21**, in which H-11, H-12 and H-13 are on the same side of the macrolactone ring and hence show NOEs to each other.

This latter conformation for **17** may be preferred presumably because the configuration at C-12 would have yielded an unfavorable 1,3-diaxial C-10:C-12-methyl group interaction if the conformation were **21**. While both compounds have good activities, the conformation of **19** appears to be preferred, the former being considerably more potent ($pED_{50}$ = 8.7) than **18** ($pED_{50}$ = 7.0).

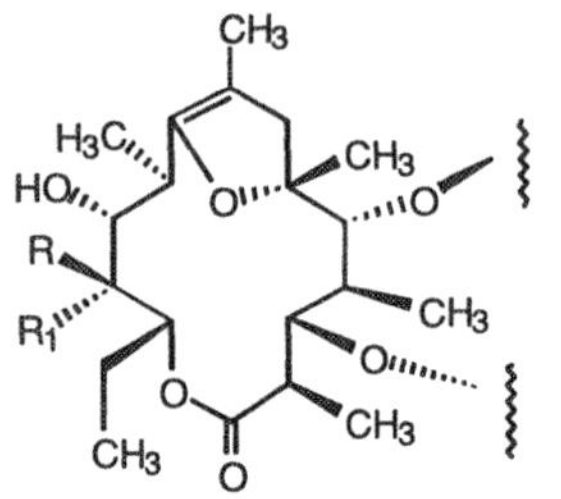


**18** $R = CH_3$, $R_1 = H$
**19** $R = CH_3$, $R_1 = H$

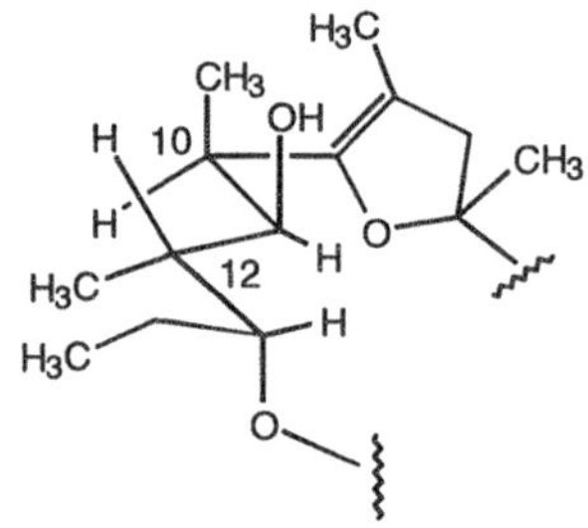


**20**

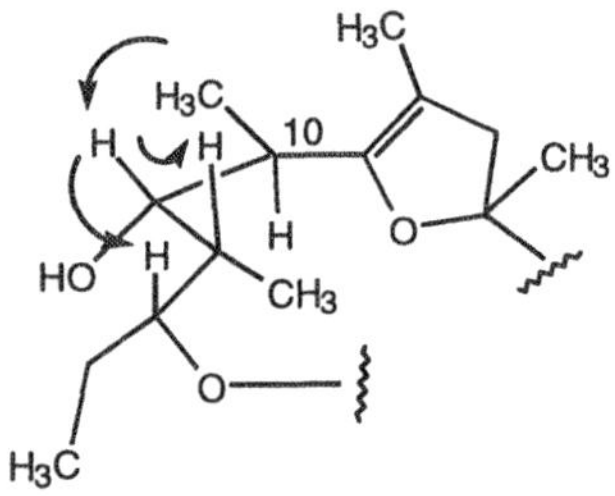


**21**

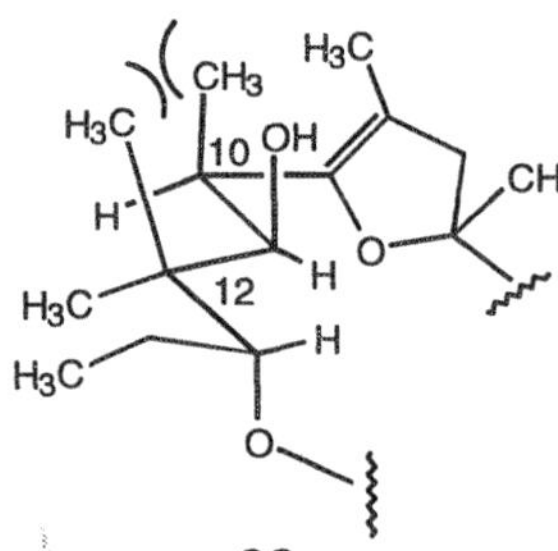


**22**

### 2.3. Macrolactam derivatives

The base catalyzed ring contraction of the 14-membered macrolactone 8,9-anhydro erythromycin-6,9-hemiacetal **16** to give 12-membered macrolactone **23** has been described [15]. The unraveling of the former macrolactone oxygen at C-13 and its stereochemical relationship to the vicinal 12-OH provided an opportunity for the conversion of **23** to macrolactam **26**.

Thus, as shown in Scheme 3, because the 12-OH group is relatively more hindered, epoxidation of **23** was stereoselective to afford **24** (Scheme 3). Epoxide ring opening of **24** is apparently controlled by the same steric factors, hence treatment with ammonia led to regio- and stereoselective opening at C-13 to give amine **25**. Subsequent ring-closure, usually under the conditions of the epoxide ring opening, afforded **26**. Compounds **10** and **11**, in which the enol ether moiety had been reduced, were likewise converted to their macrolactam congeners **27** and **29** respectively [16].

Compound **26** retained good (p$ED_{50}$ = 7.10 ) prokinetic activity. The more acid stable derivatives **27** and **29** were likewise potent.

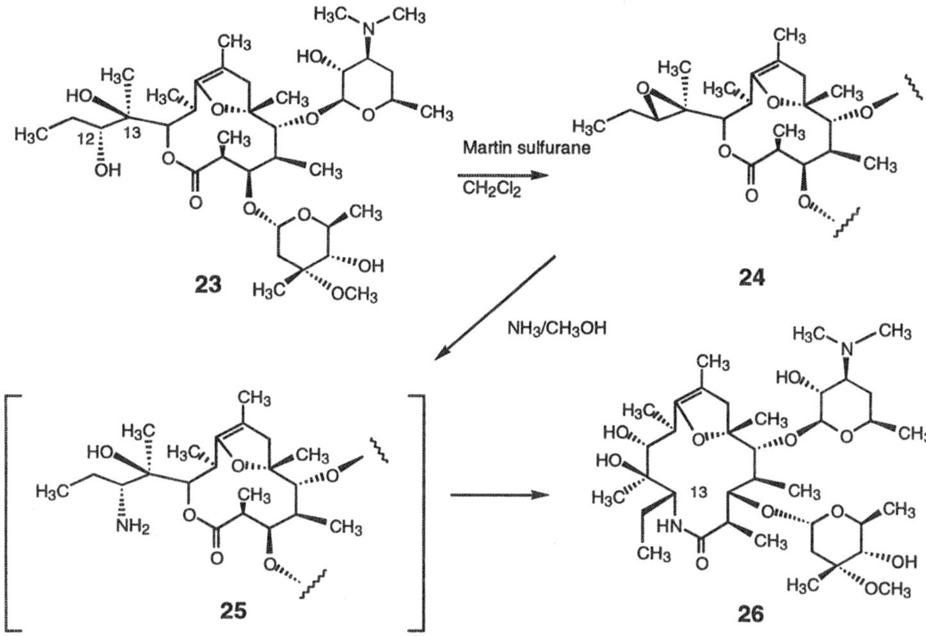


Scheme 3. Synthesis of macrolactams

Compound **28**, a 4"-deoxy-3'-N-isopropyl analog, is a highly acid stable compound, with oral bioavailability of 60%. This compound is currently undergoing further preclinical evaluation.

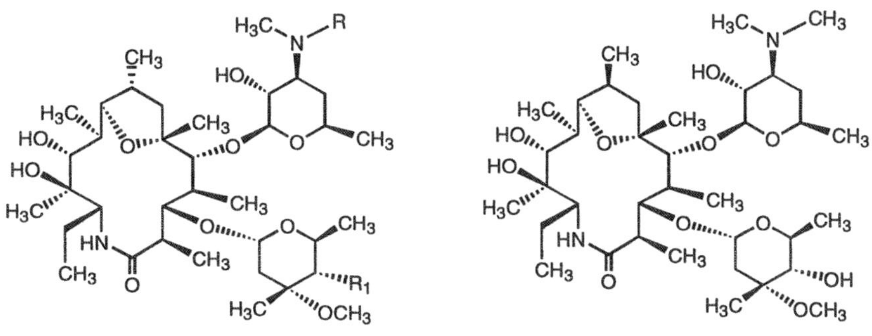

**27** R = $CH_3$, $R_1$ = OH
**28** R = $(CH_3)_2CH$, $R_1$ = H

**29**

### 2.3.4. Macrolactone ring size

Thus far, only 14- and 12-membered macrolides have been reported to have any prokinetic activity. The 15-membered ring system as in azithromycin and the 16-membered macrolides such as josamycin appear not to induce any GI motility. In an earlier communication [17], we reported synthesis of the novel 13-membered macrolactone derivative **30**. Subsequent studies have shown that **30** and its congeners retain prokinetic activity. As illustrated by the selected examples in Table 2, members of this novel series exhibit good prokinetic activities albeit weaker than the corresponding 14-membered counterparts.

Table 2
Activities of novel 13-membered lactones

| | 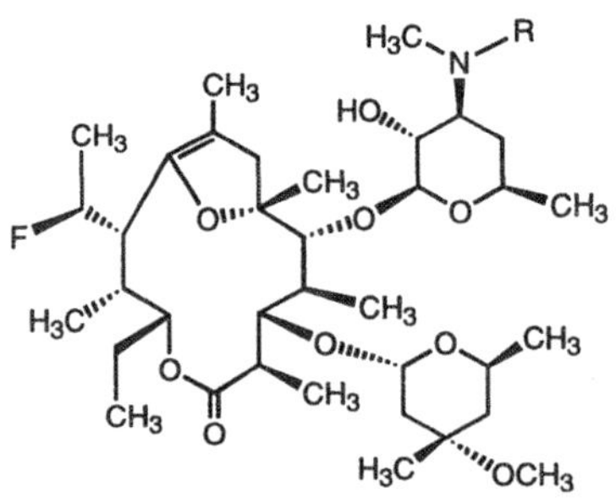 | | 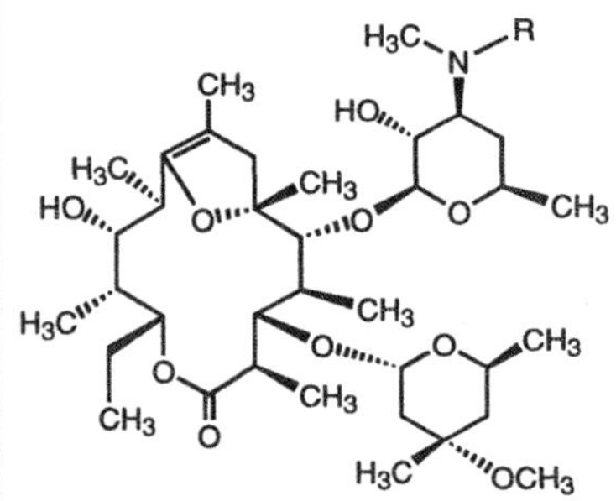 | |
|---|---|---|---|---|
| **R** | Compound | $pED_{50}$ | Compound | $pED_{50}$ |
| H | **30** | 7.43 | **31** | 9.78 |
| $CH_3$ | **32** | 7.35 | **33** | 9.78 |
| $C_2H_5$ | **34** | 7.94 | **8** | >12.0 |

### 2.4. Effect of desosamine sugar modification

The desosamine sugar with its 3'-amino functionality appears to be required for prokinetic activity. The effects of modifying the C-3'-amino group have been extensively studied [7, 18, 19]. Omura *et al.* have shown that a secondary, tertiary or quarternary amine is required for activity. In a study of relative activity in dog, quarternary ammonium derivative **37** was shown to have potent activity (Table 3), while primary amine **35** was even weaker in activity than erythromycin. Other alkyl substituents in the tertiary amine series improved potency, however large groups tended to reduce activity.

Table 3
Effect of desosamine modification on prokinetic activity[1]

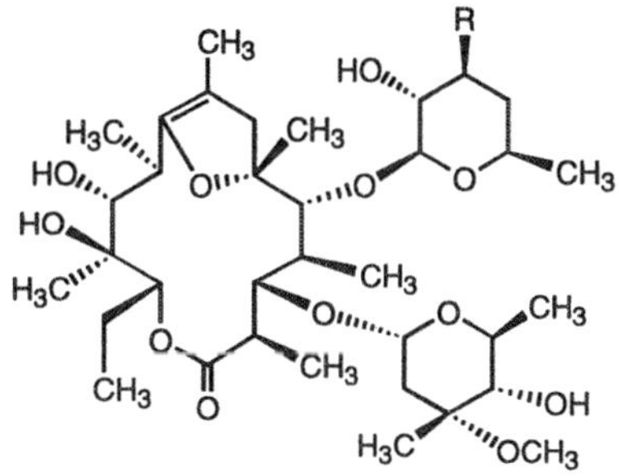


| Compound | R | pED50 |
|---|---|---|
| **35** | 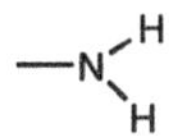 | <1.0 |
| **36** | 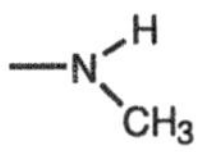 | 14.9 |
| **5** | 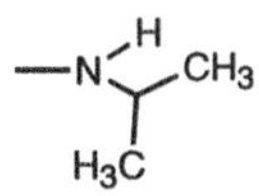 | 248 |
| **37** | 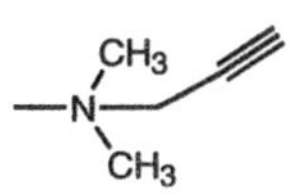 | 202 |

1. Mean of GI contractility in anesthetized dog relative to **1**.

### 2.5. Importance of the cladinose sugar

Earlier studies indicated that acylations and carbamoylations of the 4"-OH of the cladinose sugar led to reduction in prokinetic potency [7]. Also in some examples, removal of the 4"-OH did not afford any improvements in potency [20]. In our hands, 4"-deoxygenation of the cladinose sugar greatly potentiated activity [12]. Availability of the more acid stable 6,9-ether **10**, enabled us to cleave the cladinose sugar cleanly, so as to study the activities of various descladinosyl analogs modified in the macrolactone.

Thus as shown in Scheme 4, the descladinosyl derivative **38** was selectively protected for versatility, either as cyclic carbonate **39** or base stable di-*O*-isopropylidene intermediate **40**. Compound **39** was converted to 3-*O*-methylthiomethyl analog **41**, Raney Nickel reduction of which provided 3-*O*-methyl congener **42**.

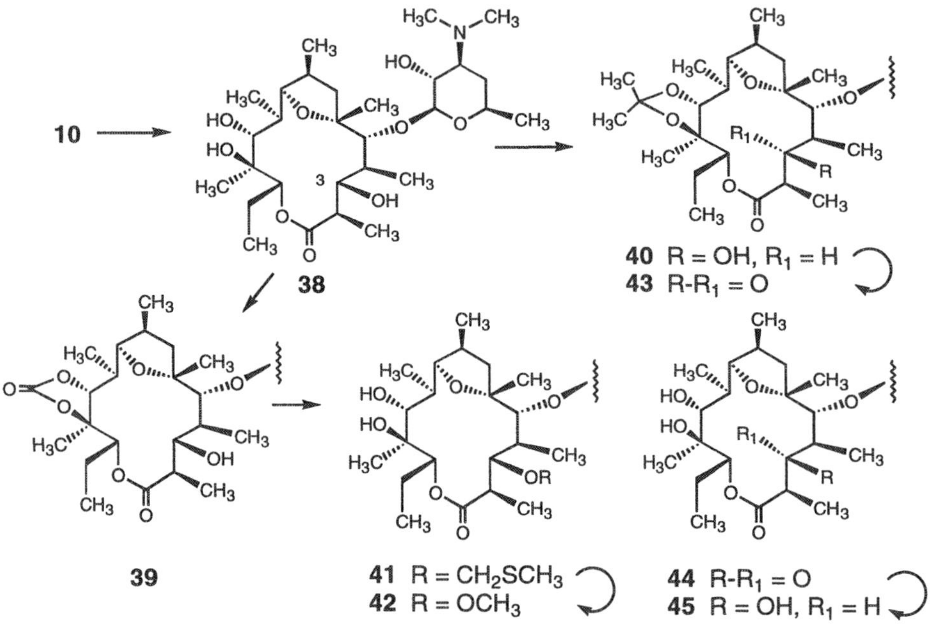


Scheme 4. Preparation of descladinosyl analogs modified at C-3

Similarly, Swern oxidation of **40** gave protected ketone **43**, which was deprotected to give ketone **44** or subjected to a Wolf-Kishner reduction to give 3-deoxy congener **45**. These transformations allowed for exploration of issues regarding polarity, simple alkylation, lack of substitution and C-3 geometry. By all intents and purposes, these compounds were devoid of useful activity ($pED_{50}$ = 4). Thus, the presence of the cladinose sugar is essential for activity.

## 3. SUMMARY

The preceding discussion gives an account of the discovery, chemistry and SAR of the motilide class of GI prokinetic agents. The medicinal chemistry of this class is far from complete. As members of this series progress to become useful drugs for the treatment of GI disorders, more chemistry will likely be reported, thus contributing to knowledge of the SAR.

## REFERENCES


1. Z. Itoh, M. Nakaya, T. Suzuki, H. Arai and K. Wakabayashi, Am. J. Physiol., 274 (1984a) G688.
2. A.G.E. Pearse, J.M. Polak, S.R. Bloom, C. Adams, J.R. Dryburgh and J.C. Brown, Virchows Arch. B. Cell Pathol., 16 (1974) 111.
3. J.E.T. Fox in Z. Itoh (ed.) p73, Motilin, Academic Press, San Diego, 1990.
4. J.E.T. Fox, E.E. Daniel, J. Jury, A.E. Fox and S.M. Collin, Life Sci, 33 (1983), 819.
5. T.L. Peeters, G. Matthijs, I. Depoortere, T. Cachet, J. Hoogmartens and G. Vantrappen, Am. J. Physiol., 257 (1989) G470.
6. T. Tomomasa, T. Karoume, H. Arai, K. wakabayashi and Z. Itoh, Dig. Dis. Sci., 31 (1986) 157.
7. K. Tsuzuki, T. Sunazuka, S. Marui, H. Toyoda, S. Omura, N. Inatomi and Z. Itoh, Chem Pharm. Bull (Tokyo), 37 (1989) 2687.
8. T. Satoh, N. Inatomi, H. Satoh, S. Marui, Z. Itoh ans S. Omura, J. Pharmacol. Exp. Ther., 254 (1990) 940.
9. P. Kurath, P.H. Jones, R.S. Egan and T.J. Perun, Experentia, 27 (1971) 362.
10. O. Kawamura, T.Sekiguchi, T. Horikoshi, M. Miyakazi and K. Obokata, Gastroenterology, 106 (1994) A522.
11. H. Takanashi, K. Yogo, K. Ozaki, M. Ikuta, H. Koga and H. Nabata, Gastroenterology, 106 (1994) A575.
12. P.A. Lartey, H.N. Nellans, R. Faghih, A. Petersen, C.M. Edwards, L. freiberg, S. Quigley, K. Marsh, L.L. Klein and J.J. Plattner, J. Med. Chem., 38 (1995) 1793.
13. L.L. Klein, L.A. Freiberg, P. Kurath and R.F. Henry, J. Org. Chem., 58 (1993) 3209.
14. P.A. Lartey, R. Faghih, T. Pagano, H.N. Nellans, A. Petersen and J.J. Plattner, J. antibiot., 48 (1995) 730.
15. I.O. Kibwage, R. Busson, G. Janssen, J. Hoogmartens, H. Vanderhaege and J. Bracke, J. Org. Chem., 52 (1987) 990.
16. R. Faghih, L. Freiberg, K. Marsh, T.L. Peeters, A. Petersen, H.N. Nellans, J.J. Plattner and P.A. Lartey, Abst. P-29, 11th. Noordwijkerhout-Camerino Symposium, Noordwijkerhout (1997)
17. C. Burnell-Curty, R. Faghih, T. Pagano, R.F. Henry and P.A. Lartey, J. Org. Chem., 61 (1996) 5153.
18. S. Omura, K. Tsuzuki, T. Sunazuka, S. Marui, H. Toyoda, N. Inatomi and Z. Itoh, J. med. Chem., 30 (1987) 1941.
19. T. Sunazuka, K.. Tsuzuki, S. Marui, H. Toyoda, S. Omura, N. Inatomi and Z. Itoh, Chem. Pharm. Bull, 37 (1989) 2701.

20. H. Koga, T. Sato, K. Isuzuki and H. Takanashi, Bioorg. Med. Chem. Lett., 4 (1994) 1649.

## ACKNOWLEDGMENTS

I am greatly indebted to the scientists from Abbott Laboratories and to our collaborators (listed below) for their contributions to much of the work reported in these proceedings. My gratitude also goes to Dr. Jacob J. Plattner for his support and vision throughout the studies leading to the discovery of ABT-229.

| Chemistry | Prokinetic Activity | Pharmacokinetics |
|---|---|---|
| Hui-Ju Chen | Hugh N. Nellans | Kenan Marsh |
| Cynthia Curty-Burnell | Albert Petersen* | |
| Leslie Freiberg* | | |
| Ramin Faghih | **Structural Chemistry** | **Motilin Receptor Binding** |
| Carla Edwards | Rodger Henry | Theo L Peeters |
| Larry Klein | Ky Kim | (Catholique Univ. Leuven |
| Ekow Yankah | Thomas Pagano | Belgium) |

* Deceased

H. van der Goot, (Editor)
TRENDS IN DRUG RESEARCH II



# Modulators of 5-HT functions in the treatment of Gastrointestinal Disorders.

F.D. King*, L.M. Gaster, G.J. Sanger and K.A. Wardle

SmithKline Beecham Pharmaceuticals, Harlow CM19 5AD, United Kingdom

## 1. INTRODUCTION

Ninety percent of the body's 5-HT is found in the GI tract, mainly in enterochromaffin cells, and there is considerable evidence to link 5-HT with the aetiology of a number of gastrointestinal disorders [1-4]. A major physiological role of 5-HT within the gut is to mediate various defensive reactions (nausea, vomiting, diarrhoea, abdominal pain, etc) which, when uncontrolled, leads to serious disorders and reductions in the quality of life. These disorders are clearly potential targets for 5-HT receptor antagonists. Conversely, partial agonists at the 5-$HT_4$ receptor, such as cisapride and metoclopramide, are used to restore gut motility levels back to normal and are used for disorders which include those associated with gastro-oesophageal reflux (oesophagitis, heartburn), non-ulcer dyspepsia, gastric stasis (diabetic neuropathy) and functional constipation. Indeed, the relative commercial success of cisapride has re-stimulated much interest in the area of 5-$HT_4$ receptor agonists [2].

Normal physiological stimuli, such as eating, release 5-HT. However, excessive amounts of 5-HT are also released by certain pathological stimuli including distension, ischaemia, stress and by cytotoxic treatments given to cancer patients. From discoveries made within our laboratories, investigation of the the latter led to the now widespread use of selective 5-$HT_3$ receptor antagonists such as granisetron to control nausea and vomiting in cancer patients.

The aim of this paper is to describe our efforts in the identification of a potential treatment for irritable bowel syndrome (IBS). IBS manifests itself as abdominal pain in association with disturbed bowel habits, normally diarrhoea but sometimes diarrhoea alternating with constipation or constipation alone. It is one of the most common of the gastrointestinal disorders and it has been estimated that between 8 and 19% of adults in the Western world have symptoms consistent with IBS. Whilst most people do not seek health care, it is estimated that functional bowel disorders account for between 20-50% of referrals to gastroenterology clinics.

Current treatments of functional bowel disorders are unsatisfactory. High fibre diets offer some long term benefit for certain types of patients but do little for the pain, whereas non-specific muscle relaxants may produce a small reduction in the incidence of abdominal pain, but have little beneficial effect on any of the other symptoms. A novel, effective treatment for IBS is therefore needed.

## 2. IDENTIFICATION OF A NOVEL 5-HT-MEDIATED APPROACH TO THE TREATMENT OF IBS

The symptoms of IBS, that of abdominal pain, diarrhoea and flatulence, can be mimicked by administration of 5-HT and raised numbers of 5-HT-containing enterochromaffin cells have been detected in the rectal mucosa of IBS patients. Certain of the selective 5-$HT_3$ receptor antagonists have been shown to reduce 5-HT-sensitised visceral pain, but this is not a consistent finding for all of the antagonists. These compounds may also reduce colonic propulsive activity and thereby induce constipation (a possible benefit with diarrhoea-evoking anti-cancer treatments). In consequence, a number of selective 5-$HT_3$ receptor antagonists have been, and possibly still are being investigated in the clinic as potential treatments for IBS. However, as these compounds may induce constipation, the level of enthusiasm is not high.

In the 1970's we identified a large number of compounds which had low affinity for dopamine receptors but were potent stimulants of gastric motility, two of which are BRL-20627 and renzapride [5]. We used these compounds to link gastric prokinetic activity to a 5-HT-mediated facilitation of release of acetylcholine within the enteric nervous system, which acts on the gut longitudinal muscle to cause contraction. Using these and other related compounds, Bockaert & colleagues characterised this receptor and designated it as the 5-$HT_4$ receptor subtype [6]. Thus in the upper gut, 5-$HT_4$ receptor activation appears to be an important pharmacological method by which gastrointestinal motility can be increased in a coordinated manner. Interestingly, these compounds and others such as cisapride, have since been shown to act as only partial 5-$HT_4$ receptor agonists. This finding may explain both their therapeutic selectivity for the gut, where the 5-$HT_4$ receptor is usually well coupled to its effector mechanisms, as well as the reason why such drugs mimic some, but not necessarily all, of the effects of endogenous 5-HT within the gut.

Thus, 5-$HT_4$ receptor activation stimulates colonic propulsion and fluid secretion, and the side effects of even the partial 5-$HT_4$ receptor agonists are abdominal cramps and diarrhoea. Based on these findings, one could envisage that a selective 5-$HT_4$ receptor antagonist would normalise the gut by blocking the inappropriate 5-HT induced sensitisation, colonic propulsion and fluid secretion, without necessarily affecting normal gut function and producing constipation. We therefore believed that 5-$HT_4$ receptor blockade had the potential to be an effective treatment for IBS.

To identify selective antagonists to test this hypothesis, neither the cloned receptor nor a binding assay was available. Therefore, for our primary 5-$HT_4$ receptor activity screen we used an isolated tissue preparation from the guinea-pig. However, to be robust enough to handle a regular flow of new compounds and to provide reliable SAR, it was necessary to develop a new assay. We began by recognising that in guinea-pig isolated intestine, both 5-$HT_3$ and 5-$HT_4$ mediated contractions had been demonstrated, reflected in a biphasic response to 5-HT. The smaller contractile response at low 5-HT concentrations was mimicked by the prokinetic 5-$HT_4$ receptor partial agonists, whilst the larger contractile response at higher 5-HT concentrations was blocked by selective 5-$HT_3$ receptor antagonists. We then found that, although both the ileum and colon showed the same biphasic 5-$HT_4$/5-$HT_3$ receptor-mediated contractions in response to 5-HT, the 5-$HT_4$ receptor-mediated response in the colon was greater [7]. This encouraged us to further develop this model and it did provide us with a robust primary screen. The screen was set up so that initially the viability of the tissue was tested by application of acetylcholine, and then a sub-maximal response to 5-HT was identified in each tissue, which acted as its own control. The test compound was then incubated, the effect on the response to 5-HT measured and the antagonist potency expressed as a $pIC_{50}$. For compounds of interest, a full $pA_2$ was measured. In order to validate the model, the potency of the only known, non-selective antagonist, tropisetron was estimated as $pIC_{50}$ 5.5, with a $pA_2$ of 6.4.

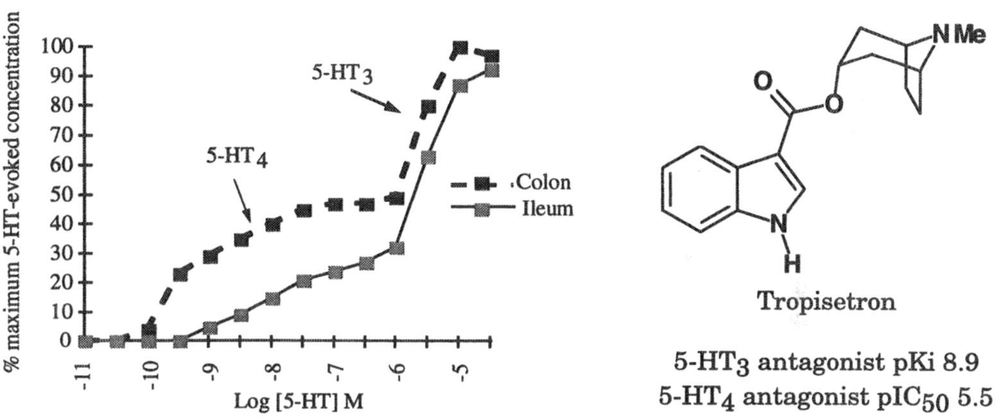


Figure 1. Contractile effect of 5-HT on guinea-pig isolated ileum and colon and the structure of Tropisetron (ICS 209-530)

## 3. IDENTIFICATION OF BENZAMIDE 5-$HT_4$ RECEPTOR ANTAGONIST

Isolated tissue screens do not have the high throughput necessary for screening large compound banks. We therefore used our smaller bank of 5-$HT_4$ receptor partial agonists from our upper gut motility enhancing work to identify potent, low intrinsic activity leads. The strategy we intended to adopt was to convert high potency partial agonists to high potency antagonists. From our initial screening we identified BRL-49332 as a suitable lead partial agonist, with a $pIC_{50}$ of 8.0, over 100x more potent than tropisetron and metoclopramide. BRL-49332 has the 4-amino-5-chloro-2-methoxybenzamide core structure that was optimal for agonist potency and efficacy. Studies on the aromatic substitution pattern confirmed a parallel SAR for the antagonist aspect of its function, but with little evidence of reduced efficacy.

Partial agonist leads: Antagonist:

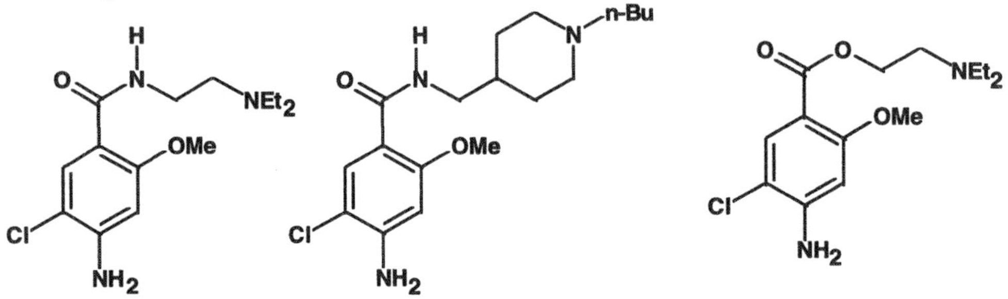


Metoclopramide $pIC_{50}$ 5.7 BRL-49332 $pIC_{50}$ 8.0 SDZ 205-557 $pIC_{50}$ 6.7

Figure 2. Structures of metoclopramide, BRL-49332 and SDZ 205-557.

However, Sandoz reported that the ester analogue of metoclopramide, SDZ 205-570 was an antagonist at the 5-$HT_4$ receptor [8]. In our model we found the compound had a $pIC_{50}$ of 6.7 and a $pA_2$ of 7.8, about 10 fold more potent than metoclopramide as an antagonist. The ester analogue of BRL-49332, SB-203904, showed a similar 10 fold increase in potency, with a $pIC_{50}$ of 9.0, and in our isolated colon model was also a 5-$HT_4$ receptor antagonist. We now had the potency and efficacy that we required, but the ester was too unstable to be suitable for a drug candidate. We therefore took the strategic decision on synthesis grounds to optimise potency as esters, and then investigate stable ester bioisosteres.

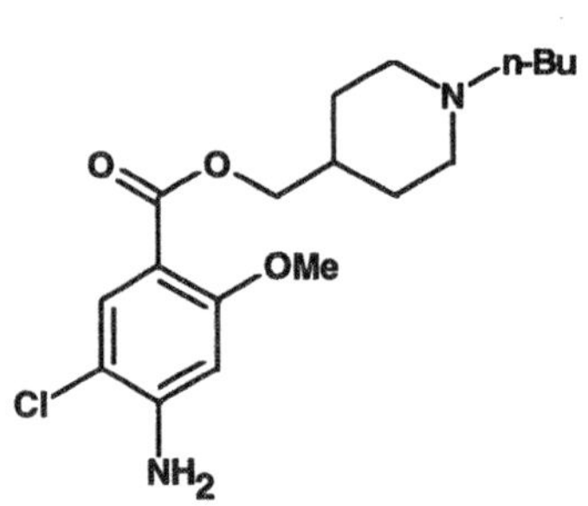


SB-203904 $pIC_{50}$ 9.0

N-substituent: $H<C_1<C_2<C_3<C_{4-6}>C_{7-12}$

ring size: 6>7~5>4

O - N distance:

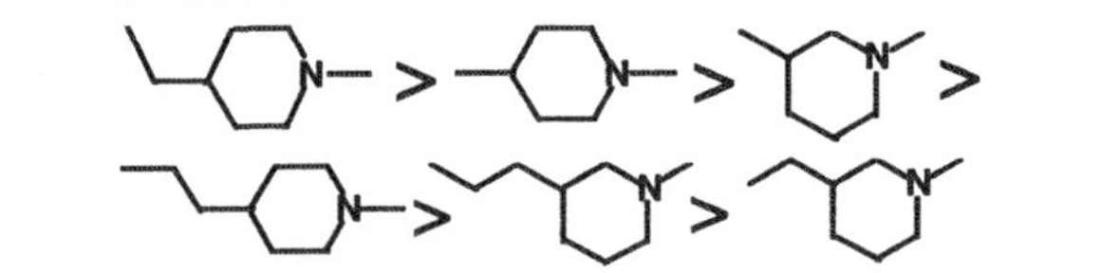

cyclisation:

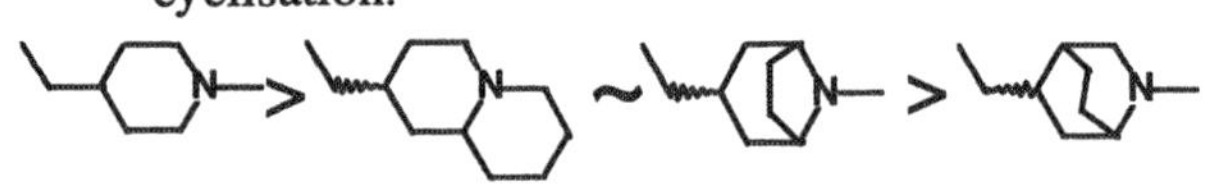

Figure 3. Structure of SB-203904 and summary of SAR.

For side chain optimisation we investigated the N-substituent, the O to N distance, ring size and cyclisation. For the aromatic, we investigated substituent replacement and cyclisation in SB-203904, but as tropisetron had antagonist activity, heterocycles were also investigated. For the side chain we found that there was a progressive increase in potency with increasing chain length up to $C_{4-6}$, and then a very gradual decline in potency up to $C_{12}$, beyond which we felt it was not worth pursuing. O to N distance was optimal at a 4-carbon distance with the 4-methylenepiperidine, and the piperidine was more potent than the azepine, pyrrolidine or azetidine. Further cyclisation of the N-alkyl group retained potency but was unnecessarily complex. Bridging across the piperidine ring showed reduced potency. Thus, by this means we could not improve on the potency of our original lead.

In parallel with this, we also investigated the aromatic and this proved to be more successful. Again we confirmed that for simple benzoates, the 4-amino-5-chloro-2-methoxy substitution was optimal. However, initial investigations in cyclising the methoxy group to give fused dioxa-aromatics were encouraging. Work on the simple unsubstituted systems showed that the benzodioxan ring size was optimal. We then introduced the preferred substituents and identified SB-204070 as our most potent ester to date. SB-204070 had a $pIC_{50}$ of 10.1 which translated into a $pA_2$ of 10.8. It was an antagonist in our colon model and was nearly 1,000 fold selective over other 5-HT, dopamine and adrenergic receptors investigated [9].

Because of its high potency and selectivity, SB-204070 was evaluated *in vivo*. Since there were no confirmed animal models of IBS, we could only test our disease hypothesis against specific 5-HT-mediated effects which form part of the complex of IBS. For this purpose we used the precursor to 5-HT, 5-HTP to induce an increase in faecal pellet output from conscious mice [10].

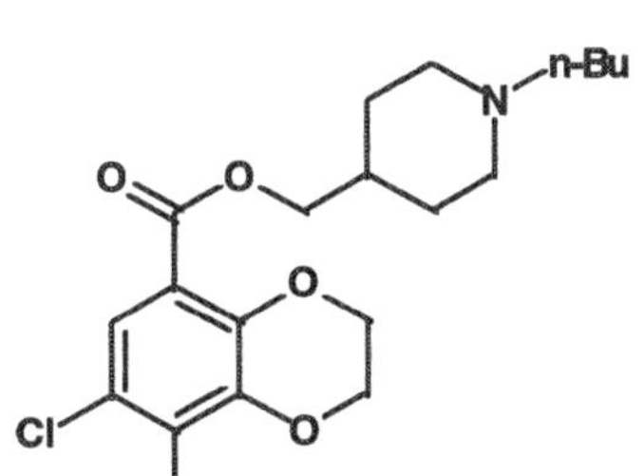


SB-204070 $pA_2$ 10.8

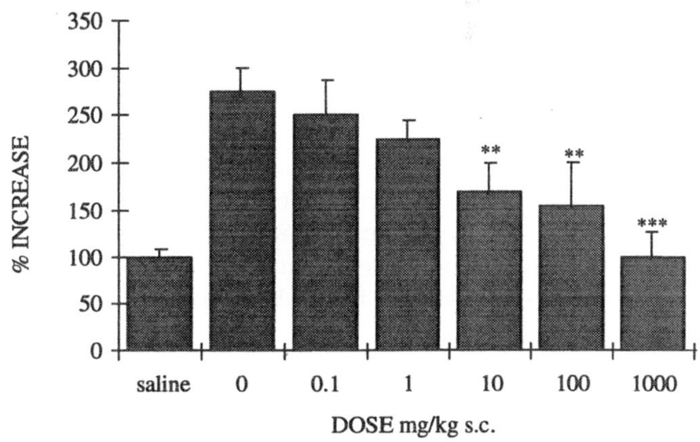


Figure 4. Structure of SB-204070 and its effect on faecal pellet output in 5-HTP treated mice: ** $P < 0.01$; *** $P < 0.001$

SB-204070 dose-dependently reduced the 5-HTP-induced increase in the number of pellets with an estimated $ED_{50}$ of 3.9 µg/kg s.c.. However no activity was obtained on oral dosing in the range 1-1,000µg/kg. Of key importance, SB-204070 did not affect normal patterns of defecation, highlighting an important difference between this approach and the constipation-inducing 5-$HT_3$ receptor antagonist approach for the treatment of IBS.

The lack of oral activity with SB-204070 suggested poor oral bioavailability, presumably due to ester hydrolysis. Therefore, in line with our original strategy, we investigated the amide analogue of SB-204070 in the belief that we would achieve greater metabolic stability, more consistent *in vivo* activity and hopefully retain antagonist activity. The amide, SB-205800 did, in fact, prove to be a pure 5-$HT_4$ receptor antagonist in the guinea-pig isolated colon with a $pA_2$ of 10.0. Similar to the ester, SB-205800 also showed nearly 1000 fold or greater selectivity over a range of 5-HT and dopamine receptors used as our initial selectivity package. For the 5-$HT_4$ receptor affinity we used a ligand binding assay using rat caudate tissue and a radioligand [$^{125}$I]SB-207710, the iodo-analogue of SB-204070, which is now commercially available from Amersham [11].

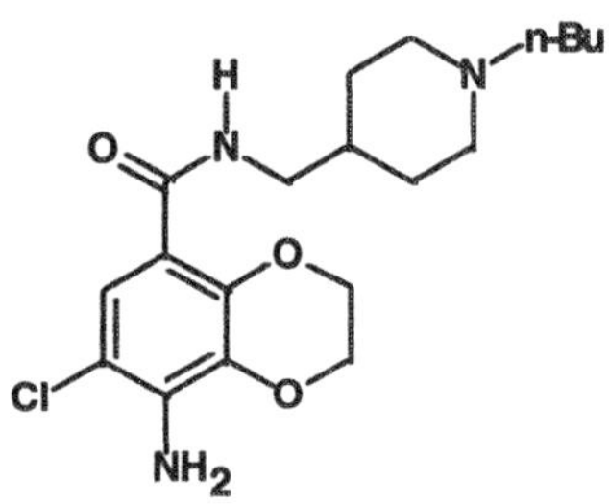


SB-205800 $pA_2$ 10.0

| Receptor | Affinity pKi |
|---|---|
| 5-$HT_{1A}$ | <5 |
| 5-$HT_{1D\alpha}$ | <6 |
| 5-$HT_{1D\beta}$ | <6 |
| 5-$HT_{1E}$ | <5 |
| 5-$HT_{2A}$ | <6 |
| 5-$HT_{2C}$ | <6 |
| 5-$HT_3$ | <6 |
| 5-$HT_4$ | 8.9 |
| $D_2$ | 6.0 |
| $D_3$ | <5 |

Figure 5. Structure and selectivity of SB-205800

SB-205800 therefore satisfied our preliminary criteria for a potential development candidate and the synthesis is shown in Figure 6.

Starting from the commercially available 3-hydroxysalicyclic acid, reaction with dibromoethane gives a reasonable yield of the benzodioxan-5-carboxylic acid. In order to direct nitration to the 8-position, the acid is first dibrominated in the more reactive 6 and 7 positions. Catalytic hydrogenation reduces the nitro group to the amino, and also reductively debrominates. The amino group is acylated in order to moderate the chlorination reaction and to direct chlorination to the 7-position. De-acetylation followed by amide coupling using carbonyldiimidazole gives SB-205800 in an overall yield of 26%. The amine is readily prepared from isonipecotamide by alkylation with n-butyl bromide followed by reduction of the amide with lithium aluminium hydride.

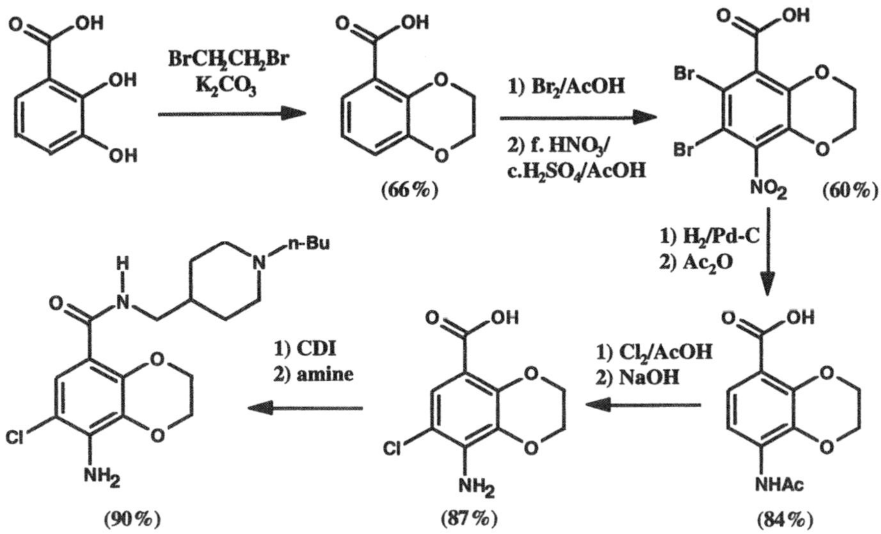


Figure 6. Synthesis of SB-205800

SB-205800 was further characterised in an *in vivo,* conscious animal pharmacodynamic model of 5-$HT_4$ function developed within the group, the Heidenhain pouch dog, which was originally used to characterise our 5-$HT_4$ receptor agonists. In this model, a section of the stomach is extrinsicaly dennervated and a cannula inserted to measure pressure changes within the pouch. As the pouch is separate from the main stomach, drugs can be dosed orally. Pressure increases can be evoked by administration of 5-HT, its precursor 5-HTP and by selective 5-$HT_4$ agonists such as renzapride. The pressure changes evoked by 5-HT are insensitive to standard 5-$HT_1$, 5-$HT_2$ and 5-$HT_3$ receptor antagonists methysergide, methiothepin, ketanserin, and granisetron. In this model, SB-205800 showed both a time and dose-dependent reduction in the response to 5-HT with an $ID_{50}$ of ~10 µg/kg 75 minutes after oral dosing.

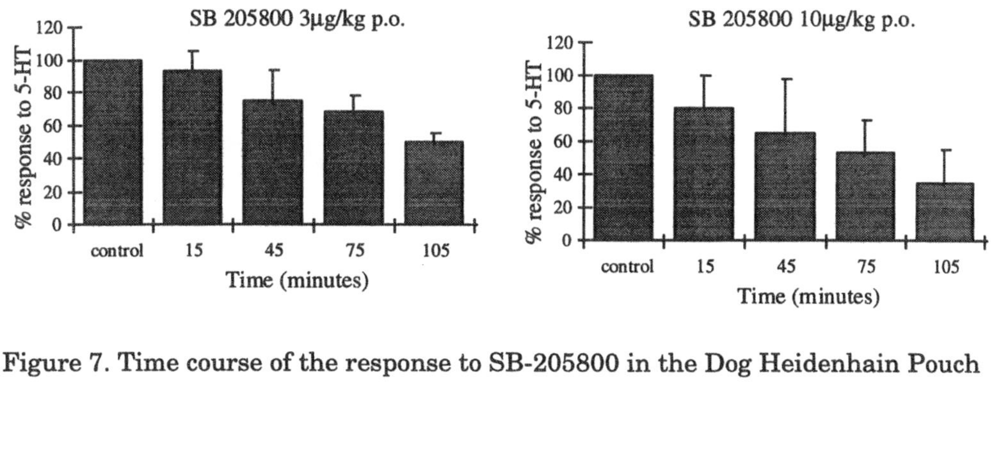


Figure 7. Time course of the response to SB-205800 in the Dog Heidenhain Pouch

## 4. IDENTIFICATION OF OXAZINOINDOLE 5-HT$_4$ ANTAGONISTS

In addition to the benzoate/benzamide series of 5-HT$_4$ receptor antagonists, we were pursuing a parallel programme of heteroaromatics based on the non-selective antagonist, tropisetron. By introducing the N-butylpiperidinylmethyl side chain we again achieved a greater than 1000 fold increase in 5-HT$_4$ receptor antagonist potency with a concomitant loss in 5-HT$_3$ receptor affinity consistent with the known intolerance of bulky N-alkyl groups at this receptor. Simple SAR on this series showed that there was a general intolerance of substitution in the 4, 5 and 6 positions of the indole, that 1-alkyl groups could be tolerated, but with no particular advantage. However, 2-methoxy substitution gave an order of magnitude increase in potency, SB-206922.

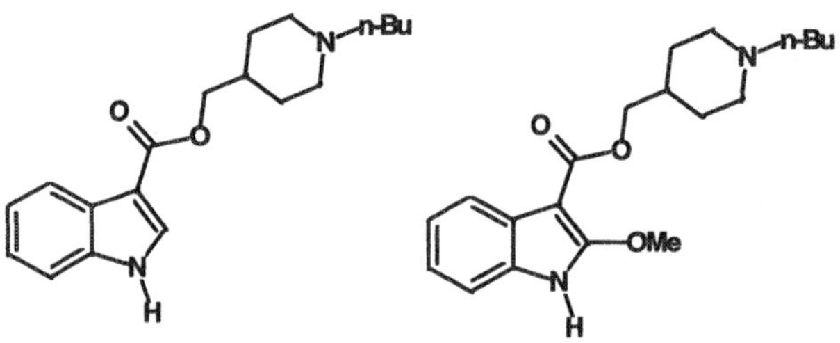


SB-204139 pIC$_{50}$ 9.0 SB-206922 pIC$_{50}$ 10.0

Figure 8. Structures of SB-204139 and SB-206922.

The metabolic and chemical stability of the methoxy group was a concern and we therefore investigated combining the tolerance of substitution in the 1-position with the advantages of the 2-alkoxy substituent by cyclisation in a similar manner to the benzodioxans. The cyclic analogues should be more stable to O-dealkylation. Investigation of ring size showed that the 5, 6 and 7 membered ring sizes were potent, but that potency was optimised again with the 6-membered ring.

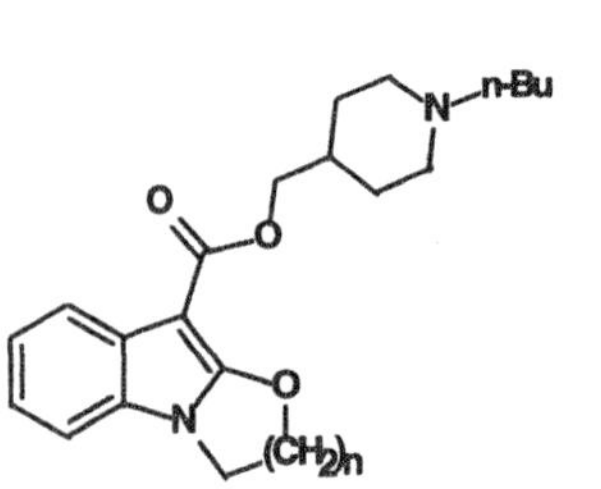


| n | pIC$_{50}$ |
|---|---|
| 1 | 10.0 |
| 2 | 10.6 |
| 3 | 9.8 |

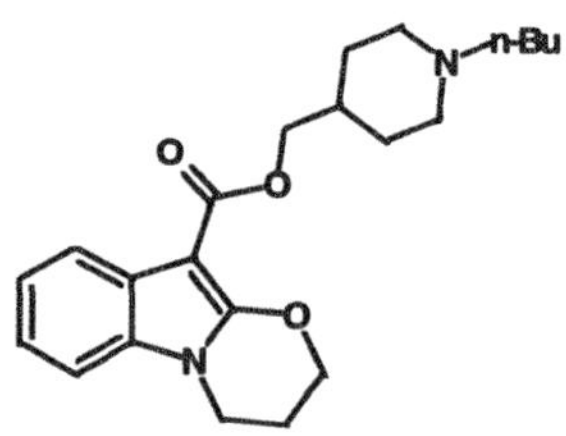


SB-207058: pKi 10.1

Figure 9. Effect of ring size on 5-HT$_4$ receptor potency and the structure of SB-207058

Thus the ester, SB-207058, with a $pIC_{50}$ of 10.6 had sufficient potency to consider investigation of the equivalent amide. Once again, conversion to the amide, SB-207266, gave a near 10 fold reduction in 5-$HT_4$ receptor affinity, but still retained a greater than 1,000 fold selectivity over the other receptors in our cross-screening package and had the potential for good oral bioavailability [12].

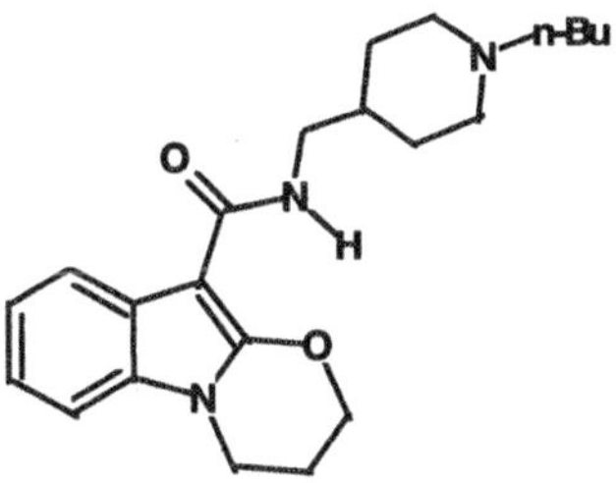


SB-207266: pKi 9.3
pA$_2$ 10.6

| Receptor | Affinity pKi |
|---|---|
| 5-$HT_{1A}$ | <5 |
| 5-$HT_{1D\alpha}$ | <5 |
| 5-$HT_{1D\beta}$ | <5 |
| 5-$HT_{1E}$ | <5 |
| 5-$HT_{2A}$ | <6 |
| 5-$HT_{2C}$ | <6 |
| 5-$HT_3$ | <6 |
| 5-$HT_4$ | 9.3 |
| $D_2$ | <6 |
| $D_3$ | <5 |

Figure 10. Structure of SB-207266 and selectivity data

Further characterisation in the guinea pig isolated colon model showed that at low concentrations, SB-207266 was a competitive antagonist of the 5-HT-evoked contractions with a $pA_2$ of 10.6. At higher concentrations, however, there appeared to be a non-competitive effect which we subsequently found was due to a slow recovery of the tissue preparation (Figure 11) [13]. In our selectivity package SB-207266 was also over 1000 fold selective and this has been confirmed over a wider range of receptors, enzymes and ion channels.

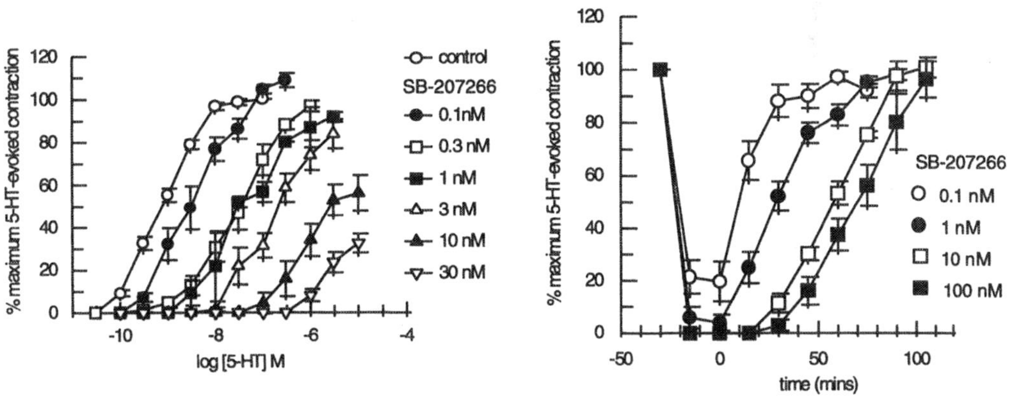


Figure 11. Dose response, and onset and recovery of the 5-$HT_4$ receptor antagonistic effects of SB-207266 in the guinea pig distal colon longitudinal muscle-myenteric plexus preparation.

To date, activity of the 5-$HT_4$ receptor antagonists had only been demonstrated in non-human tissue and animal models. To increase our confidence that SB-207266 would be effective in man, we investigated the compound in two models of 5-$HT_4$ receptor-mediated activity in human isolated tissue (Figure 14) [14].

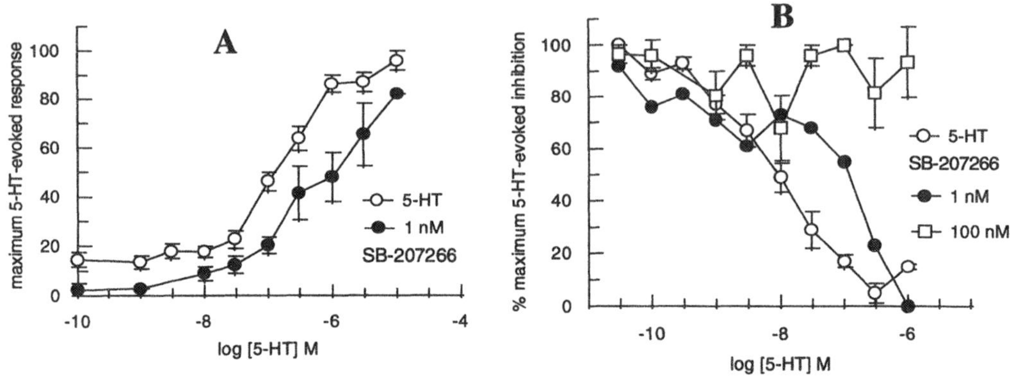


Figure 14. Antagonism by SB-207266 of **A**: 5-HT-evoked contractions of human terminal ileum circular muscle and **B**: 5-HT-evoked inhibition of electrically-evoked contractions of human colon circular muscle.

In the human terminal ileum, SB-207266 caused a reduction in the 5-HT-evoked excitatory response with an estimated $pK_b$ of 9.8 (Figure 14A). In a second model, now using electrically stimulated human colon circular muscle, SB-207266 blocked the 5-HT-evoked inhibition of stimulation with an estimated $pK_b$ of 10.7 (Figure 14**B**). Thus in two *in vitro* models of human 5-$HT_4$ receptor function, SB-207266 acted as an antagonist with potency consistent with its animal *in vitro* pharmacology.

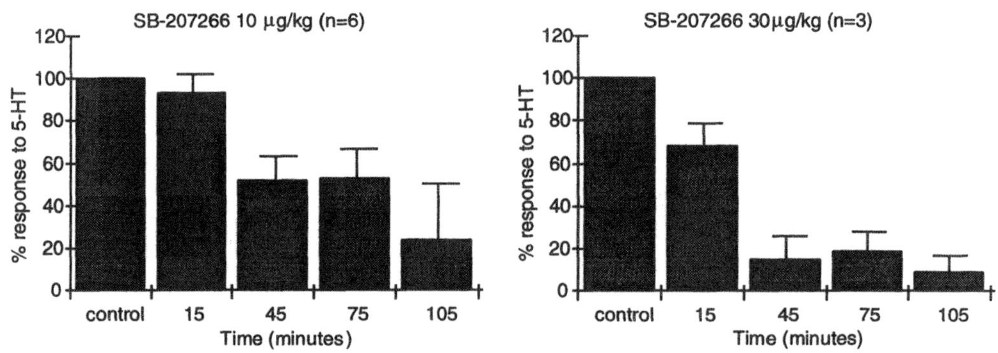


Figure 12. Time course of the response to SB-207266 p.o. in the dog Heidenhain pouch pharmacodynamic model.

*In vivo*, in our pharmacodynamic model of the Heidenhain pouch dog, at both 10 and 30μg/kg p.o., SB-207266 reversibly inhibited the effects of 5-HT with a time of onset between 15 and 45 minutes and lasting for the entire duration of the experiment (Figure 12). At a time of 75 min. post dose SB-207266 had an $ID_{50}$ of 9.6 μg/kg p.o. [13]

Thus we had two highly selective, highly potent 5-$HT_4$ receptor antagonists, both of which were potential development candidates. SB-207266 was finally chosen for a number of reasons including a marginal higher potency, structurally less related to the benzamides and so was less likely to produce unexpected partial agonist activity, and the synthesis was more amenable to scale-up. SB-207266 could be prepared by a 3-step procedure in an overall yield of 36% (Figure 13).

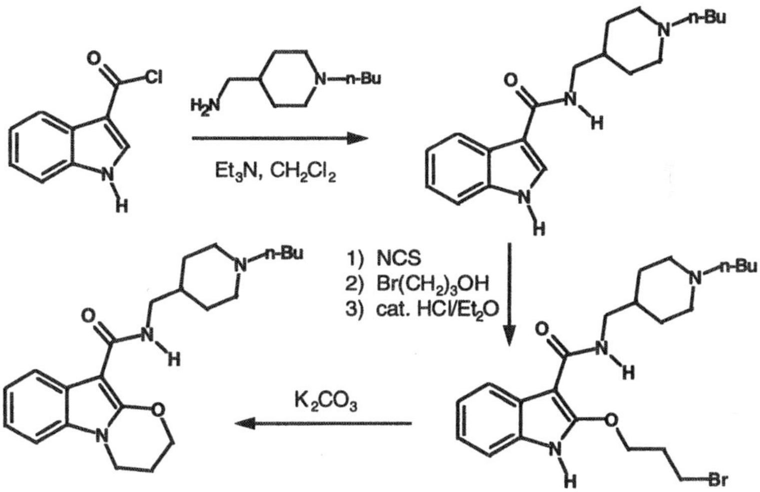


SB-207266 overall yield 36%

Figure 13. Synthesis of SB-207266

Reaction of 3-indoloyl chloride with the aminomethylpiperidine gives the intermediate amide. Reaction with N-chlorosuccinimide is believed to initially give an unstable 3-chloro adduct which is susceptible to nucleophilic attack at the 2-position. Addition of 3-bromopropanol and adjustment of the pH with HCl gives, in a one-pot procedure, the 2-(3-bromopropoxy) amide. Cyclisation is then achieved by treatment with potassium carbonate in acetone.

## 5. CONCLUSION

In conclusion, starting from a screening exercise of previously investigated 5-$HT_4$ receptor partial agonists, we identified BRL-49332 as a candidate for further exploitation. Optimisation of the aromatic for both potency and antagonist function led to the identification of the benzodioxan ester SB-204070. This ester proved to be inactive orally, however conversion to the amide SB-205800 gave the first orally active, potent and selective 5 $HT_4$ receptor antagonist. In parallel, using the side chain of BRL-49332, we identified the oxazinoindole amide SB-207266 which satisfied all our potency, functional and selectivity criteria for entry to development. SB-207266 is currently being evaluated in Phase II clinical trials for the treatment of irritable bowel syndrome.

## ACKNOWLEDGEMENTS

The authors thank all their colleagues, far too numerous to mention here, for their contributions to the work described. Their contributions have, and will be, properly acknowledged on the original scientific articles.

## REFERENCES

1. K. M. Dhasmana, Y.N. Zhu, S.L. Cruz and C.M. Villalón, Life Sciences, 53 (1993) 1651.
2. M.R. Briejer, L.M.A. Akkermans and J.A.J. Schuurkes, Pharmacol. Revs., 47 (1995) 631.
3. G.J. Sanger, Neurogastroenterol. Mot., 8 (1996) 319.
4. L.M. Gaster and F.D. King, Med. Res. Revs., 17 (1997) 163.
5. F.D. King, M.S. Hadley, K.T. Joiner, R.T. Martin, G.J. Sanger, D.M. Smith, G.E. Smith, P. Smith, D.H. Turner and E.A. Watts, J. Med. Chem., 36 (1993) 683.
6. J. Bockaert, M. Sedden and A. Dumius, Mol. Pharmacol., 37 (1990) 408.
7. K.A. Wardle and G.J. Sanger, Br. J. Pharmacol., 110 (1993) 1593.
8. K.-H. Bucheit, R. Gamse and H.-J. Pfannkuche, N-S. Arch Pharmacol.,345 (1992) 387.
9. L.M. Gaster, A.J. Jennings, G.F. Joiner, F.D. King, K.R. Mulholland, S.K. Rahman, S. Starr, P.A. Wyman, K.A. Wardle, E.S. Ellis and G.J. Sanger, J. Med. Chem., 36 (1993) 4121.
10. S.E. Banner, M.I. Smith, D. Bywater, L.M. Gaster and G.J. Sanger, Eur J Pharmacol. 308 (1996) 181.
11. A.M. Brown, T.J. Young, T.L. Patch, C.W. Cheung, A.J. Kaumann, L.M. Gaster and F.D. King, Br. J. Pharmacol., 110 (1993) 10P.
12. L.M. Gaster, G,F, Joiner, F.D. King, P.A. Wyman, J.M. Sutton, S. Bingham, E.S. Ellis, G.J. Sanger and K.A. Wardle, J. Med. Chem., 38 (1995) 4760.
13. K.A. Wardle, S. Bingham, E.S. Ellis, L.M. Gaster, B. Rushant, M.I. Smith and G.J. Sanger, Br J Pharmacol., 118 (1996) 665.

K.A. Wardle and G.J. Sanger, Br J Pharmacol., 120 (1997) 192P.

H. van der Goot, (Editor)
TRENDS IN DRUG RESEARCH II



# The impact of robotics and novel assay technologies on lead discovery processes

John G. Houston

Lead Discovery Unit, GlaxoWellcome Medicines Research Centre, Stevenage, UK

## 1. INTRODUCTION

Drug discovery is one of the most competitive activities within the pharmaceutical industry. Quality drugs can have huge commercial potential in the multi-million, sometimes billion dollar markets that are available worldwide.

The search for novel therapeutic molecules has continued unabated over the last century highlighted by moments of serendipity and inspired chemical intuition However, as we move into the new millennium the increasingly competitive nature of the pharma industry has meant that many companies do not want their future success dependent on chance discoveries or to be beaten at the line by a competitor company. Serendipity must be turned into certainty in this race to find new therapeutic molecules and increasingly the way forward for many companies has been to invest heavily in novel drug discovery technologies to help them in this goal.

This paper describes some of the approaches taken by GlaxoWellcome to accelerate the drug discovery process.

## 2. DRUG DISCOVERY AND DEVELOPMENT PROCESS

The basic premise of drug discovery is that modulation of certain molecular targets by drug intervention will have a positive effect on the outcome of disease. Years of basic research can take place before these candidate molecular targets are identified and validated. Even then it is not always certain how some of these targets are implicated in the disease state, and sometimes the true mechanism of action is found years after the initial molecule has been found.

The process of drug discovery and development through to a product being launched on the market is an extremely complicated and multifaceted process. It is also a very time consuming and incredibly expensive process with high attrition rates for molecules moving through the pipeline.
The industry average from discovery to launch over the last 10 years has been around 15 years.

Many difficult hurdles have to be overcome before an early lead molecule can be regarded as a therapeutic drug (Fig. 1).

| | ***Years*** | |
|---|---|---|
| **Understand Disease** | ? | **Basic research** |
| **Select Mechanism/Target** | 1 | |
| **Design Primary Screen** | 2 | **Compound discovery phase** |
| **Screen/Identify Hit** | 3 | **Highthroughput screening and rational design** |
| **Convert Hit to Lead** | 4 | |
| **Convert Lead to Development Candidate** | 5<br>6 | **Pre-Clinical testing and candidate selection** |
| **Clinical Trials (I, II, III)** | 7 | **Patents registered** |
| **Registration/Licensing** | 13 | **Marketing, advertising, Surveillance** |
| **Product Launch** | 14 | |

Figure 1. **Lead Discovery and Drug Development Process**

Early lead molecules need to be novel, easily synthesized and chemically tractable with some significant potency against the target of choice. To turn this compound into a quality molecule worthy of entry into a drug optimization program, the desired physico chemical characteristics have to be designed in. Ideally, the compound should be nontoxic with no drug-drug interactions, be potent in vitro and in vivo, have a good half life and usually be orally bioavailable.

The next stage in the process are the clinical trials when the molecule first enters man and when development costs start to escalate.
The trials will show any obvious toxicity and safety problems, then move on to showing pilot efficacy and safety dose ranges before moving onto the final phase (III) when major efficacy is shown.

Getting a candidate molecule to this stage can take up to 10 years and for every one molecule that makes it hundreds of thousands do not. So, why do we need to accelerate the drug discovery process, shouldn't we just improve it?

C.Cimarusti (BMS) at a recent drug discovery conference (IBC, Boston 96) quoted the average cost of development of a new drug to be approximately £225 million. One can see why the the major players in the pharma industry need R&D budgets of over £1 billion per year to sustain their competitive edge.

With this in mind any process impact that reduces the discovery and development time of a drug and maximises its patent life can therefore have huge rewards. A billion pounds a year "blockbuster" drug can earn £2.7 million a day. If ever the old adage "time is money" were true it is in the pharma industry.

In fact the stated goals of several large drug companies, including Glaxo Wellcome, is to reduce time to market by 40%. This has been added to increases in overall productivity goals with Glaxo Wellcome setting the target of three significant new medicines per annum by the year 2000.

To achieve these goals major improvements in various key stages of the drug discovery and development process were needed within most companies.

Four years ago at Glaxo a review of the lead discovery process took place to look at ways of radically re-designing how we found molecules. The goal was to significantly increase productivity, diversity and improve quality whilst cutting the discovery timescales from 3 or 4 years to 3 or 4 months. So both accelerating the process and improving it was essential.

How was this challenge tackled?

## 3. LEAD DISCOVERY: RATIONAL DESIGN OR RANDOM SCREENING

A model scenario that most drug companies would like to aspire to is to have the ability to design drugs rationally using detailed structural knowledge of the molecular target. By use of NMR spectroscopy, protein x-ray crystallography and impressive computer aided molecular modeling tools a molecule is constructed that "fits" exactly with the protein target. This is a painstakingly, slow process and can take many years to produce the candidate molecule. This method of drug design has so far been unable to meet the increasing requirements of today's expanding drug development pipelines. These tools are extremely useful in the latter stages of drug optimization, and adaptation of existing drugs, natural enzyme substrates and receptor agonists/antagonists. However, pharma companies need alternative approaches for finding chemistry start points for drug development programmes. Therefore the major weapon in the armoury of most large pharmaceutical organisations for finding novel chemical entities over the last decade has been the use of rapid, high throughput screening (HTS) systems.

Over this period we have witnessed an explosion in drug discovery technology driven largely by the need for "random" compound screening.

High throughput 'random' screening requires a cohesive matrix of different functions and groups within the research company to work together to produce the final output of a drug candidate (Fig. 2).

Random screening is the process by which large numbers of samples (usually random compound collections from company files and natural product extracts) are tested against a wide range of target assays e.g. enzymes/receptors that have previously been identified or implicated in a disease state.

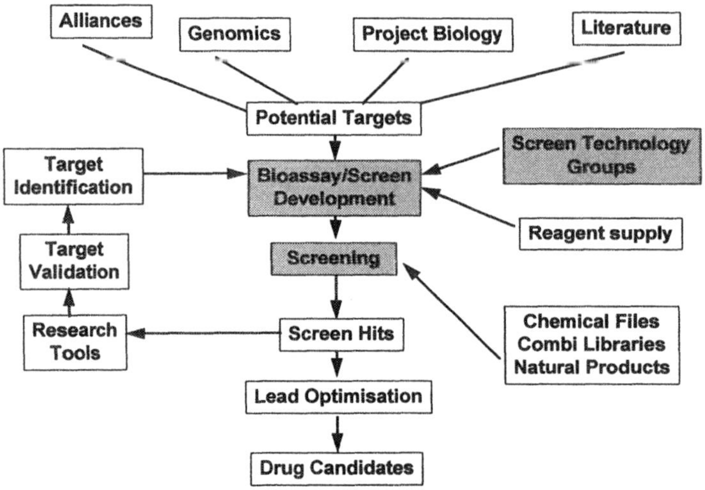


Figure 2. **Random Screening Process**

An example of a typical screening assay would be in the search for an enzyme inhibitor such as an ACE inhibitor. Artificial substances can be made that change colour when they come into contact with an active enzyme, and if one of the compounds in the screen stops this colour change it is probable that this compound will be an inhibitor of the enzyme. Similar assays using a range of different design formats can be used to detect the binding of compounds to receptors, using radiolabelled, fluorescent or coloured markers.

Biologically active compounds from screening or screen 'hits' are regarded as potential drug candidates and are used as the starting point for lead optimisation medicinal chemistry programs where the structure/activity relationships studies are carried out to find the best candidate for full drug development. These initial lead series have Ki's usually <100μM and show clear opportunities for chemical improvement. The process of improving the lead series involves driving down the Ki (<500nM), establishing a structural knowledge base and helping to provide some chemical/biological rationale for the SAR studies.

The next stage in optimisation is improving the biological properties of the lead i.e. tox, pk. Hopefully, keeping the Ki and maximising structural diversity. Sometimes these screen hits are not used in lead optimisation programs but are immediately fed back into exploratory research projects were the compounds are used as 'tools' to aid target validation and/or identification.

In the mid-eighties, Glaxo's screening programs focused primarily on infectious diseases. Assay designs were simple and the major source of samples for screening came from natural product extracts. By the end of the eighties, drug discovery approaches had moved on to include enzyme targets, receptor ligand interactions and mammalian whole cell assays. At this time the screening process was a labour intensive, manual process with each target taking approximately 2/3 years to be screened against a suitable number of samples (approximately 200,000) to ensure molecules of sufficient interest being found. The majority of these samples were crude natural product extracts that needed a significant amount of downstream processing (structure elucidation) once a screen hit was identified.

Limited data capture and analysis systems were available and almost all of the process from sample weighing, compound dissolution, transferring to test plates and screening itself was a manual process.

As the screening programme was extended to include complex molecular targets associated with mammalian diseases the total number of targets requiring screening increased rapidly as did the number of samples available for testing. By the early 90's it was clear that new screening technologies and robotics were needed to improve the whole process and take the companies into the new millennium as a world leader in drug discovery.

A root and branch review of all the components of the drug discovery process was initiated (Fig 3).

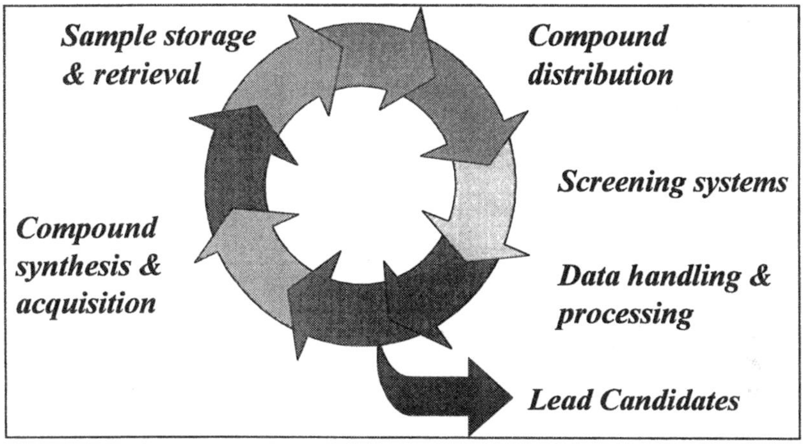


Figure 3. **Drug Discovery Components**

## 4. IMPROVING THE SCREENING PROCESS

The complete process from compound acquisition and *de novo* synthesis, through to storage and distribution, and finally HTS, data analysis and hit identification all needed to be overhauled and re-constructed. The science of lead discovery had to be established firmly in the minds of researchers to improve the number of valid, therapeutically orientated targets moving through to HTS. Rapid assay to screen development needed to be implemented with a specific resource identified with strong screen development skills.

Sample diversity also had to be improved enormously with more emphasis on combinatorial chemistry approaches. Cost effective approaches to high capacity screening had to be integrated with rapid hits to leads conversion.

What was needed was a screening revolution by rapid evolution to fully automated sample synthesis and supply systems as well as robotic high throughput screening and automated data analysis.

## 5. COMBINATORIAL CHEMISTRY AND SAMPLE SUPPLY ROBOTICS

Sample diversity is generally agreed to be the key component for successful screening and lead discovery. The typical collections of samples available to most pharma companies include company compound collections, new synthesis, combinatorial libraries, commercial and academic sources, compound collection sharing and natural products.

Past experience has shown that natural product extracts from bacterial, fungal, plant and marine sources have been extremely useful for generating diverse structures that have eventually reached the market as drugs e.g. Taxol (from Taxus brevifolia), Lanoxin (from Digitalis lanata), Larium (from the bark of the cinchona tree) to name but a few.

However, finding a chemical start point that is amenable to easy synthesis and simple modifications can prove difficult with a natural product structure. The whole process of identification and structure elucidation is now seen as increasingly time consuming and wasteful of resource. Alternative approaches, have of course, included testing the company files which represents the history of the various chemical programs over many years. These files can be diverse but are more likely to show large clusters of similar compounds such as steroids or betalactams reflecting the research focus of the company over the years.

The recent explosion in combinatorial chemistry technology has enabled the creation of very large peptide, oligonucleotide and carbohydrate libraries containing millions of compounds. This technology is likely to mature over the next 5-10 years as various companies, large and small find even better ways of applying the technique to making large libraries of small molecular weight compounds.

The techniques involves producing large chemical libraries composed of diverse structural entities built around a common molecular scaffold. This results in a very large, randomised collection of compounds as opposed to well characterised, discrete compounds. The potential of this approach is clearly in the size and diversity of compounds that could be created. Also, the ability to react quickly to optimise any screen hit using the same combinational approach and the speed to make very large primary libraries for screening. Systems are already in place in many companies that have automated this approach. For example Affymax's Encoded Synthetic Library (ESL) synthesiser which can generate approximately 50K member libraries in 48 hours. Also, the commercially available ChemTech system which is capable of generating large libraries of around 3000 compounds in 36 hours.

All of this automation in the production of libraries means that the numbers and variety of samples available for screening has increased enormously.

This also means that storage of samples and making them easily and rapidly available for screening has become more of an issue.

Again Glaxo Wellcome, like several other companies has invested heavily in automating sample storage and retrieval systems. At Glaxo Wellcome a fully automated compound store (ACS) that can retrieve 15,000 compounds per day has been operational at the Stevenage research site for approx. 2 years. The acquisition of Wellcome also gave the new company access to the Haystack liquid sample storage and retrieval system, which has an automated weighing and dispensing station. Eventually, this system will be replaced by a fully automated liquid store (ALS) with the capacity to retrieve 250,000 samples per day and hold up to 3 million samples.

A plethora of liquid handling robots will make replica copies of screen plates from master stocks for distribution around the GW research sites.

## 6. SCREENING ROBOTICS AND NOVEL ASSAY TECHNOLOGIES

As part of the strategy to fully automate the screening process Glaxo acquired a fully integrated turnkey robotic screening system from Robocon GesmbH, Austria (Fig 4), RoboLab 9600. The system was custom designed to handle a variety of assay types based on the 96 well microtitre plate such as ligand binding assays, ELISA's and a range of enzyme assays with radiochemical, colourmetric and luminescent readouts.

Figure 4. **RoboLab 9600**

The system consisted of a CRS A255 articulated robotic arm mounted on a 6 metre linear track surrounded by a range of instruments such as dispensers, barcode readers, incubator, harvestor, shakers, plate hotels, plate readers and a Topcount scintillation counter.

Pre-dispensed plates for testing arrive from the sample supply store on a daily basis and the robot runs constantly for up to 12 days until all samples have been tested (approximately 200,000 assay points or 15K assay points per day. Operationally, the successful usage of the system was analysed over 2 years of running (Fig 5). The data assumes that the total capacity of the system would reflect the robolab running 24 hours per day, 7 days per week all year. This is unrealistic but the analysis shows that the robot usage is now up at nearly 80% of capacity. The 20% downtime is accounted for by public holidays, weekends and development time on the robot.
Overall, introduction of this first screening system dramatically reduced the time taken to screen samples against a target of interest, improved the quality of the data generated, enhanced safety in the workplace and allowed staff skills to change - many becoming expert in robot usage.

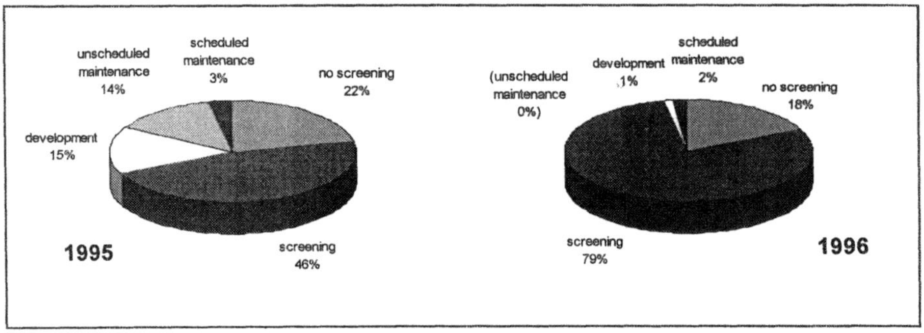


Figure 5 **R1 screening performance 1996 vs. 1995 (Banks etal, 1996)**

However, by the end of 95 this one system was only carrying out 25% of our overall screening campaigns, the rest still heavily reliant on manual or semi-automated systems. So, building on the success of R1 a second robotic system was commissioned R2.

An extensive user requirement specification was written and various tenders accepted prior to choosing Thurnall Plc., Manchester, UK to build a new robot system.

This new system consists of 2 robotic cells one for use of radioisotopic assay ('hot') the other for non-isotopic screens ('cold') (Fig 6).

Again these systems were based on linear track robots surrounded by various instrument modules. There were many new features added to R2 to improve on R1 including better liquid handling systems, more plate readers and counters and plate storage features. The other major difference to R1 is that R2 is configured to operate with 96 and 384 well plates offering a considerable boost to the overall capacity of the system.

Both of the robotic cells that constitute R2 have double the capacity of R1. The 384 model could have up to 4 times the capacity of R1.

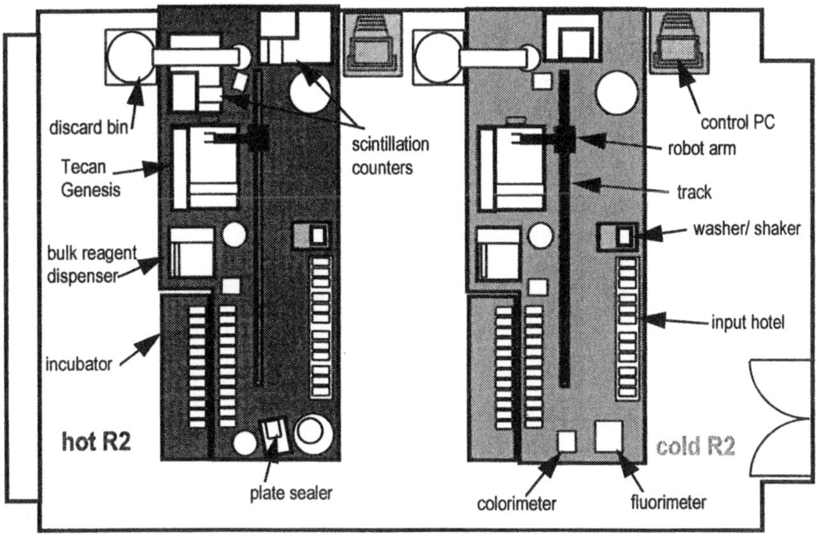


Figure 6 **Schematic layout of hot and cold R2 (Banks etal, 1996)**

Overall, the combination of R1 and R2 systems allows primary screening at the rate of up to 500,000 assay points per week.

By the end of 97 we expect that 75% of the primary screening will be on these fully automated systems.

Other robotic systems have also had to be added now the primary screening "bottle neck" has been released. Using Biomek and Tecan systems for downstream processing of screen hits is now essential to cope with the flux of compounds emerging from the screening programmes.

Alongside this heavy investment in screening hardware has come new developments in assay design and bioassay technology (Fig 7). This has resulted in focusing on a range of generic assay technologies that are amenable to automation.

**Moving from Isotopic to non-isotopic**

**Isotopic**

- **Scintillation proximity assay (SPA and cytostar)**

**Non-isotopic**

- **Homogenous time resolved flourescence (hTRF)**
- **Electrochemiluminescence (ECL)**
- **Fluorescence correlation spectrometry (FCS)**
- **Fluorescence Resonance Energy Transfer (FRET)**
- **Multiple Receptor Assay Systems**

Figure 7 **Novel Assay Designs**

Because of the variety of samples generated for screening, assays need to be robust with minimum susceptibility to interference that causes false positives and false negatives. Rapid adaptation and flexibility of the assay formats to a variety of complex molecular targets is an essential feature of any assay technology suitable for automation.

One such assay technology which has been used extensively at Glaxo Wellcome on the screening robotics has been scintillation proximity assay (SPA) from Amersham International (Houston J.G., 1996)

This technology is based on plastic beads or microspheres that contain a light generating molecule called a fluorophore. The microspheres can be treated in many ways allowing binding of a variety of biological reagents such as receptors. When radiolabelled ligand binds to the receptor on the bead, particles released by radioactive decay excite the fluorophore which then emits a light signal which is easily measured. Unbound radiolabelled ligand generates no signal.

This is a simple, robust, homogenous assay system that can generally apply to a range of molecular targets, perfect for screen automation.

Over 50% of the HT screens run during 1995 at Glaxo were SPA formats.

One major issue for SPA format assays is the fact that they are all radioisotope based assays (iodine, tritium) and the generation of radioactive waste from these assays can easily outstrip normal permitted lab levels.

With this in mind and also with a look to the future Glaxo Wellcome has also been developing and acquiring non-isotopic, generic assay formats such as homogenous time resolved fluorescence and fluorescence resonance energy transfer (FRET).

These technologies offer a variety of approaches when trying to design screening systems and are essential as assays become miniaturised to the level where radioisotope assays are unfeasible.

## 7. THE FUTURE

What will the future bring?

Consolidating on the advantages of integrated automated systems will almost definitely include incorporation of miniatursied screening formats. Breaking away from the 96 well plate format into 384, 864 and 3456 well plates. All keeping the same footprint as the 96 well plate, to ensure that they are coupled with robotic systems plate handling.

The challenges for miniaturisation will be in keeping generic assay formats as the volumes in the assays are reduced below 1 µl. More precise liquid handling systems based on inkjet technologies, highly sensitive reporter substrates and image analysis stations probably replacing plate readers.

The advantage of miniaturisation are primarily: higher capacity, faster screening, reagent savings, lower costs. (Table 1)

Impact of Miniaturisation on screening

Example: 3 million samples to be tested.

Table 1

| Plate Format | No. of plates tested for 10 targets | No. of plates tested for 100 targets |
|---|---|---|
| 96 | 375,000 | 3,750,000 |
| 384 | 80,000 | 800,000 |
| 864 | 35,000 | 350,000 |
| 3456 | 900 | 9,000 |

One can see from Table 1, assuming an average of £20 per plate screened, the savings to be gained by miniaturising could be enormous.

Another cost saving strategy is to combine miniaturisation with a compound pooling strategy with 10, 20, or 30 compounds being tested per well.

The driving force for miniaturisation will be cost effectiveness, the ability to carry out many more tests at lower cost. How far will we go in miniaturisation depends on the limits of liquid dispensing and detection technology and the biology of the assay itself

Some technologists have visions of Ultra-HTS formats in the nanoliter to picoliter range with up to 500,000 assay points per day. Even taking the bioassay and the chemistry onto microchips building integrated, miniaturised combichem and Ultra-HTS systems that could be run in the lab by a bench scientist.

Use of new technologies such as combinatorial chemistry, robotic high-throughput screening, along with advances in medical genetics and the application of bioinformatics will be the main drive behind increases in productivity within Glaxo Wellcome.

## References

1. C. Cimarusti, "Integration of Discovery and Development Phases: A Paradigm for Project Acceleration". IBC Conference on Drug Discovery Technology, Boston (1996)

2. M.N. Banks, S. Fogarty, M. Valler, K. Mills, S. O'Brien, A. Binnie and J.G. Houston, Automated High throughput screening to aid the drug hunter at Glaxo Wellcome, proceedings of the International Symposium on Laboratory Automation and Robotics (1996), p220-232

3. J.G. Houston, Partnerships Power Drug Discovery, Pharmaceutical Executive, 16, 62-66 (1996)

## Acknowledgements

I would like to thank all members of the Lead Discovery Unit for their work described in this paper

H. van der Goot, (Editor)
TRENDS IN DRUG RESEARCH II



# New animal models in target discovery.

R.S. Oosting*, K.L. Stark**, R. Hen**, and G.J.M. van Scharrenburg*

*Department of Biotechnology, Solvay Duphar B.V. P.O. Box 900, 1380 DA Weesp, The Netherlands.
**Center for Neurobiology & Behavior, Columbia University, 722 West 168th Street, New York, NY 10032.

## 1. INTRODUCTION

Pressure on the pharmaceutical industry has been constantly increasing. Since the late 1980s it has become increasingly clear that if a pharmaceutical company wants to survive, it must develop effective drugs with novel modes of action, preferably against diseases that are only poorly treated today. Knowledge of the etiology of a particular disease, and the biochemical pathways and genes involved in the disease are therefore crucial.

There is hope that current developments in genomic research, like large scale sequencing (1), positional cloning (2), and expression profiling (3), will lead us to new therapeutic targets. "Genomics" should thus provide a great number of potential targets. However, a major problem is how to discriminate useful targets from non-useful targets. Knowledge of the physiological or pathological function of the proposed target *in vivo* is therefore necessary. Although genetic studies with lower experimental organims such as yeast, *C. elegans*, and *Drosophila melanogaster* have provided insight into many fundamental properties of various molecules, these organisms are not very suitable for studying complex human behaviors and diseases. Therefore, studies with these organisms may have only limited value in the process of therapeutic target discovery. In this review we will focus on "genetic pharmacology" approaches in mice and rats. Several approaches will be discussed: 1) gene transfer, 2) antisense knockouts, and 3) gene knockouts. Since our interest lies primarily in the central nervous system, the examples cited in the text will be primarily from this field.

## 2. GENE TRANSFER

One way to study the physiological and/or pathological function of a protein *in vivo* is simply to overexpress it. By introducing new genetic information into the germ line of an organism, "transgenic" animals can be produced (4). This is now possible in a number of different animals, including mice, rats, cattle, sheep, and pigs. The first step in making a transgenic animal is to micro-inject DNA containing the gene of interest into the pronuclei of fertilized oocytes. These manipulated oocytes are then transferred to the oviduct of foster mothers where they are allowed to develop to term. If the DNA construct integrates stably into the genome, then it is likely that it will be inherited by most, if not all, of the developing animal's cells, including the germ cells. The DNA construct must contain the complete coding sequence of the gene under the control of promoter/enhancer sequences. These transcriptional regulatory elements can be constitutively active, active only at certain developmental stages, active in an inducible way, ubiquitously active, or active in a tissue/cell-type specific way. Table 1 lists some examples of transgenic (and knockout) animals which have been produced in the CNS field and which have particularly interesting phenotypes.

## 3. KNOCKOUTS

One of the best ways to establish the *in vivo* function of a protein is to observe the consequences that occur when that protein is no longer produced, or is produced at reduced levels. The antisense and gene knockout strategies described below are two strategies by which this is possible.

### 3.1 Antisense knockouts

Antisense oligonucleotides inhibit gene expression by hybridizing to the coding (sense) sequences of a specific messenger RNA (mRNA) target by Watson-Crick base pairing. The formed DNA-RNA or RNA-RNA duplexes are then recognized by cellular enzymes (e.g. RNase-H) which degrade the mRNAs, ultimately leading to a decrease in the cellular levels of the protein.

So far, the most use of antisense oligonucleotides *in vivo* has been in the central nervous system. This may be for the following reasons: 1.) the CNS is a closed system and therefore injected oligonucleotides remain localized, and 2.) nerve cells may effectively take up oligonucleotides without a delivery system. The dopamine-2 receptor (31,32), 5-HT6 receptor (33), the neuropeptide Y-Y1 receptor (34), and the V1 vasopressin receptor (35) are among the proteins in the CNS that have been successfully targeted with the *in vivo* antisense oligonucleotide approach. In the above mentioned experiments, antisense inhibition exhibited a functional outcome similar to that resulting from the use of receptor antagonists.

Unfortunately, the interpretation of antisense experiments is hampered by the high degree of non-specificity of oligonucleotides. Oligonucleotides show sequence-specific and non-specific binding to several kinds of small molecules and proteins. Therefore, it may be difficult to separate true antisense effects from the non-specific biological effects of the antisense oligonucleotides (36). The antisense method is aimed at lowering target mRNA levels. However in many antisense studies, this has not been found (for instance ref. 35). In this respect it may be worth mentioning our own experiences with the *in vivo* use of antisense oligonucleotides. We tried to block

Table 1. *Knockout and transgenic animal models for psychiatric and neurological disorders.*

| Disorder | Transgene or Knockout | Gene | Ref. |
|---|---|---|---|
| Anxiety | T | CRF (corticoid releasing factor) | 5 |
| | KO | Ca/calmodulin kinase II, α subunit | 6 |
| | KO | pre-pro enkephalin | 7 |
| Addiction | KO | 5-HT1B (serotonin 1B receptor) | 8 |
| | KO | Dopamine transporter | 9 |
| | KO | D1R (Dopamine 1 receptor) | 10 |
| | KO | protein kinase C-γ isozyme | 11 |
| Aggression | KO | Ca/calmodulin kinase II, α subunit | 6 |
| | KO | monoamine oxidase A | 12 |
| | KO | pre-pro enkephalin | 7 |
| | KO | nNOS (neuronal nitric oxide synthase) | 13 |
| | KO | 5-HT1B | 14 |
| Alzheimer's Dis. | T | β-amyloid precusor protein | 15 |
| Epilepsy | KO | 5-HT2C | 16 |
| | KO | t-PA (tissue plasminogen activator) | 17 |
| | KO | TNF-receptor (tumor necrosis factor) | 18 |
| | KO | Editing-deficient GluR-B allele | 19 |
| | KO | OTX-1 (orthodenticle homolog 1) | 20 |
| | KO | Ca/calmodulin kinase II, α subunit | 21 |
| | KO | IP3-R1 (type 1 inositol 1,4,5, triphosphate rec.) | 22 |
| Ischemia/Stroke | KO | TNF-receptor | 18 |
| | KO | nNOS | 13 |
| Memory/Learning | KO | CREB (cAMP-response element binding prot.) | 23 |
| | KO | mGLuR1 | 24 |
| | KO | NMDA receptor ε1 subunit | 25 |
| | KO | Ca/calmodulin kinase II, α subunit | 26 |
| | T | Ca/calmodulin kinase II | 27 |
| Eating disorders | KO | 5-HT2C | 16 |
| | KO | Neuropeptide Y | 28 |
| Parkinson's Disease | KO | D2R | 29 |
| | KO | Nurr1 | 30 |
| Schizophrenia | KO | DA transporter | 9 |

the expression of the D2 receptor and also the 5HT-1A receptor in rats by intracerebroventricular injection of antisense oligonucleotides. Our purpose was to to repeat published data (D2: ref. 31 and 5HT-1A: ref. 37). In the 5HT-1A experiment we saw no effect with the oligonucleotides, and in the D2 experiment the only observed effect was with a random control oligonucleotide. Additionally, we know from others that there is a high degree of non-reproducibility in *in vivo* experiments with antisense oligonucleotides. We therefore believe that the antisense approach is not yet a reliable method for therapeutic target discovery.

### 3.2 Gene knockouts

The technology of gene targeting has proven to be extremely powerful in various fields of biological research. Through gene targeting, mutant mice lacking a particular gene and thus a particular protein, can be generated. By analyzing such mutant mice, both expected and unexpected functions of various proteins have been discovered. Proteins that were once thought to be important have appeared to be redundant or even superfluous (38), and for others, unknown physiological functions or properties have been found (e.g. refs 13 and 16). Table 1 lists some knockout (and transgenic) animals in the CNS field with interesting phenotypes. This table is meant only as an illustration and is in no way exhaustive.

Gene knockout technology relies on the fact that blastocyst derived embryonic stem (ES) cells can be grown *in vitro* and modified by transfection. In particular, it is possible to replace a given gene by means of homologous recombination with a fragment of the same gene that bears a mutation. Such engineered ES cells can be injected into recipient blastocysts and contribute to the formation of a chimeric mouse. Transmission of the mutation to the germ line will result in heterozygous mice which can be further bred to generate mice that are homozygous for the desired mutation.

### 3.3 Constitutive versus inducible and tissue-specific knockout strategies

A major problem which hampers the interpretation of experiments done with knockout mice is that any altered behavior or phenotype may be due to a developmental abnormality caused by the absence of the protein during development. In the brain, many developmental abnormalities will be difficult to detect, such as the alteration of certain projections. Nonetheless, these changes may have a large phenotypic impact. An additional disadvantage of the current knockout strategy is the inability to provide region specific control of the gene disruption. In other words, with the current strategy the targeted protein is absent in all the areas that normally express this protein.

To circumvent the above mentioned problems, we are currently working on inducible and tissue-specific knockout strategies.

### 3.4 Inducible knockouts

An inducible knockout in an adult animal will represent the acute loss of the targeted protein as opposed to a chronic loss as seen in classic constitutive knockouts. The degree to which developmental abnormalities are responsible for the phenotypic changes observed in classical constitutive knockouts can therefore be seen by comparing the two types of knockouts. An inducible knockout strategy will therefore allow us to evaluate the contribution of developmental compensations. Of course, compensatory changes may still take place in such an adult animal since related systems may try to adapt the animal to its new conditions. Results that can be obtained with an inducible knockout will be similar to those that can be obtained with a highly specific antagonist.

In our laboratory, we are using an inducible system that was originally developed by Gossen and Bujard (39). It consists of a CMV minimal promoter (essentially a TATA box) placed downstream of 7 operator sequences (tet-o) of the E.Coli Tn10 tetracycline resistance operon and a fusion protein consisting of the Tet repressor and the herpes simplex virus protein 16. Binding of this fusion protein, which is inhibited by tetracycline, to the tet-o sequences (trans)activates transcription from the minimal promoter. Hence, it was named the tetracycline transactivator (tTA). Figure 1 shows a schematic diagram of the tTA/tet-o system.

In transgenic mice this system has proven to give a highly regulated temporal and, depending on the promoter used, spatial control of transcription. In these experiments, mice transgenic for tTA under the control of a viral or tissue-specific promoter were crossed with animals transgenic for certain genes under the control of the tet-o/minimal promoter (26,40,41). The resulting double mutants showed tetracycline-sensitive expression of the transgene.

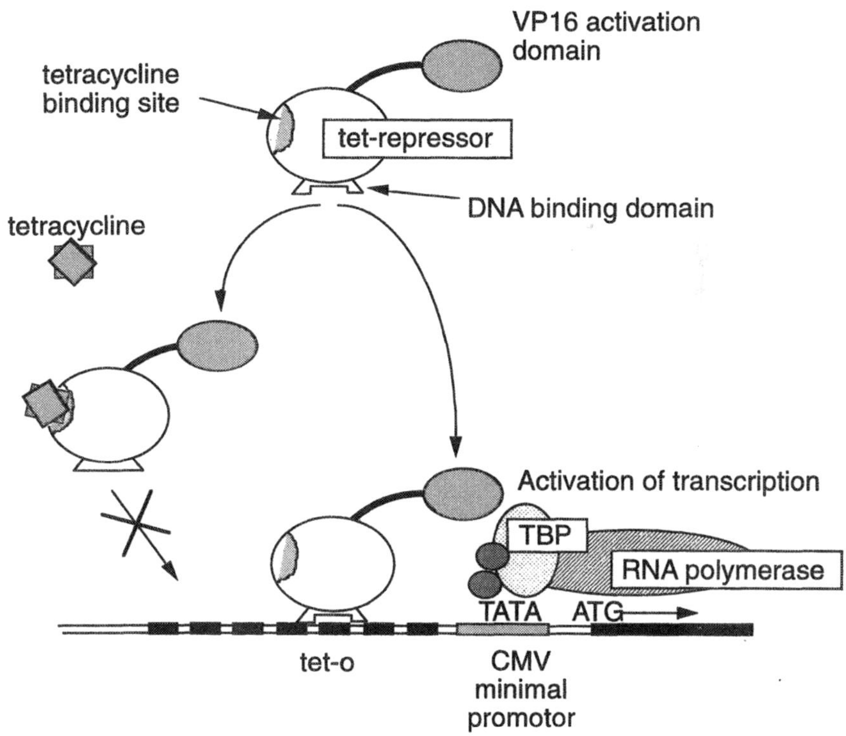


*Figure 1. Schematic diagram of the working mechanism of the tTA/tet-o system.*

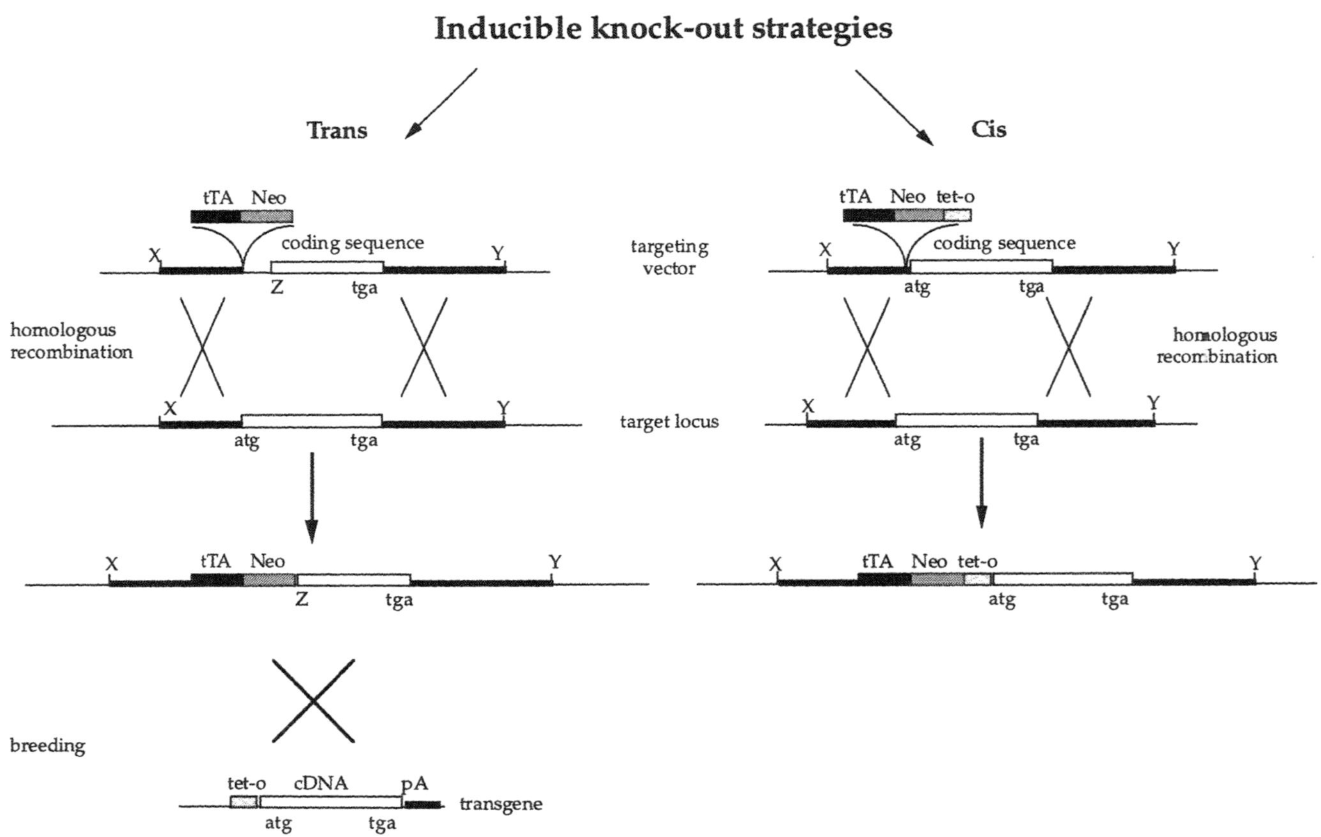


Figure 2. *Inducible Knockout strategies (X, Y, and Z are unique restriction sites).*

In our inducible knockout strategy, tTA is knocked into the targeted gene at the site of the initiating ATG. In this way, the expression of tTA will be under the control of the promoter of the targeted gene such that its spatial and temporal expression should reflect that of the endogenous gene. In addition to knocking in tTA, we are also deleting part of the endogenous gene at the same time, thus ensuring that the resulting animal is a knockout for that locus. By crossing this animal with a transgenic animal containing the targeted cDNA under the control of the tet-o /minimal promoter, we should generate animals in which the level of expression can be regulated by tetracycline (see figure 2 for a schematic diagram of this "trans" approach).

A drawback of the above described approach is that 2 different (one knockout and one transgenic) mouse lines are required. In addition, since the integration of the tet-o containing construct occurs randomly, the integration site may affect the inducibility. Therefore, several founder lines must be established and crossed with tTA bearing knockout animals in order to obtain animals with the desired regulation of transcription. We reasoned that it may be possible to put tTA and the tet-o/minimal promoter in the same construct and still get tetracycline sensitive transcription. A potential problem of this approach is that several promoters are located very close together and that they may affect each others activity. We therefore made several reporter constructs to test this "cis" approach. We used the CMV promoter to drive the transcription of tTA and used β-galactosidase as a reporter gene. Figure 3a shows a diagram of the different plasmids we have used. These plasmids were transfected by $Ca_3(PO_4)_2$ into COS-7 cells which were grown either in the presence or absence of tetracycline (1 μg/ml). The β-galactosidase activity was then assayed after 36 hours. Construct 1 contains: the CMV promoter, tTA, pGK-neo-pA, tet-o/minimal promoter/β-galactosidase. As controls, constructs 2 and 3 were made to test the effects of the pGK promoter in the construct and to investigate the "leakiness" of the cis strategy. Additionally, the plasmid RSV-βgal was used as a control. On this plasmid the RSV promoter drives the expression of β-galactosidase. In the experiments with construct 3, the β-galactosidase activity was very low. This suggests that the "leakiness" of the "cis" system is minimal. With both constructs 1 and 2 the expression of β-galactosidase was sensitive to tetracycline. Additionally, the expression of β-galactosidase in the control plasmid RSV-βgal was not effected by tetracycline, thus demonstrating that at this concentration, tetracycline is not toxic to the cells. The results are shown in figure 3b. This figure also shows the results of an experiment in which a plasmid containing the CMV promoter and tTA was cotransfected with a plasmid containing tet-o and β-galactosidase. The ratio of β-galactosidase activity in the absence of tetracycline over the activity in the presence of this antibiotic were 14 $\pm$ 3 (n = 3), 9 $\pm$ 1 (n = 2) and 7 $\pm$ 0.3 (n = 2) for constructs 1,2, and the cotransfection, respectively. It is clear from these experiments that the presence of the phosphoglycerate kinase promoter (pGK) upstream of the tet-o sequences does not affect the tetracycline sensitivity of the "cis" system.

We concluded from these experiments that the above described "cis" approach is feasible and we are currently in the process of making knockout animals in this manner (see fig 2). After completion of the above described experiments, a paper was published in which the "cis" approach was used to generate transgenic mice (42). The expression of the transgene luciferase in these animals could be completely and reversibly repressed by tetracycline.

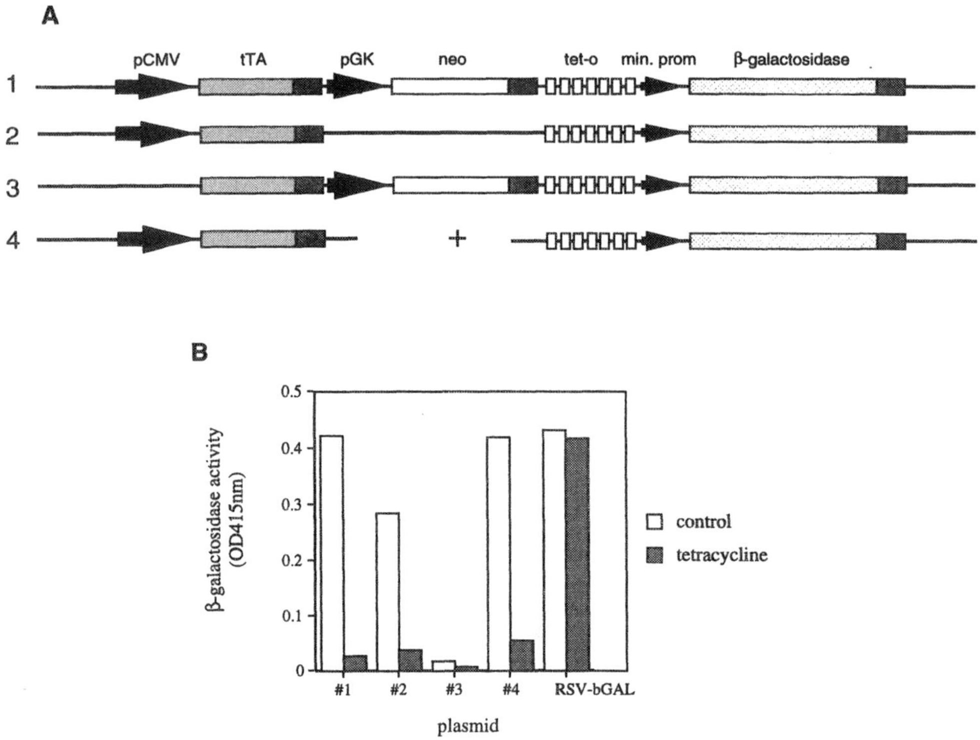


*Figure 3. (A) Schematic diagrams of the tTA/tet-o reporter constructs. (B) Effects of tetracycline on the beta-galactosidase activity in transiently transfected COS-7 cells. Results of a representative experiment are shown.*

### 3.4 Tissue/cell type restricted gene knockouts

In order to assess the contribution of certain proteins to the phenotype of an animal, it would be advantageous to be able to manipulate not only the time of their expression, as is possible with the tetracycline inducible system, but also the location of their expression. One way to gain tissue/cell type specific control would be to inject tetracycline stereotaxically into the brain of an animal that has the expression of the gene of interest under the control of tetracycline, as was discussed earlier. An alternative approach, that is possible if a knockout animal of the gene of interest is available, is to re-express the gene in the knockout background only in particular structures. This second approach, the tissue-specific "rescue" is possible if a tissue-specific promoter is available, and is currently being persued for the 5HT-1B receptor by one of us (K.L.S).

A third way to generate tissue/cell type-restricted gene knockouts makes use of the bacteriophage P1-derived Cre/LoxP recombination system. Cre is a DNA recombinase which recognizes a particular short sequence: the loxP site. Two loxP sites, as far apart as 200 kb apart (43), can be recognized by the Cre recombinase, which will excise all the DNA between the two loxP sites when recombining them. To make a tissue-restricted knockout the following two animals must be crossed: 1) a mouse that harbors the gene/exon to be knocked out between loxP sites and 2) a mouse that is transgenic for the Cre recombinase under the control of a tissue/cell type specific promoter. The loxP sites must be inserted into the genome of embryonic stem cells by homologous recombination. In this way Tsien et al (44) produced a mouse in which the NMDA-R1 gene is only deleted in the CA1 pyramidal cells of the hippocampus.

## 4 CONCLUDING REMARKS

The different genomic animal models, as discussed here, have each there advantages and disadvantages.

The main advantage of the antisense knockout approach is that results can be obtained within several weeks. However, the present inability to convincingly discriminate between antisense and nonantisense effects makes this approach an as yet unreliable method for target discovery.

The main disadvantage in the transgenic and gene knockout approaches is that it takes at least a year to obtain sufficient animals for testing. However, this disadvantage is easily outweighted by the advantages of these methods for therapeutic target discovery. Constitutive knockout mice have provided a wealth of information about the physiological and pathological functions of a large number of proteins. Additionaly, these mice are able to serve as good models for genetic diseases. With the recent development of inducible and tissue-specific knockouts (and transgenics), we now predict that knockout animals will become even more widely used. These mice offer the researcher the opportunity to obtain results comparable to those which would be obtained through classical pharmacological techniques using specific antagonists.

## REFERENCES

1. M.D. Adams, J.M. Kelley, J.D. Gocayne, M. Dubnick, M.H. Polymeropoulos, H. Xiao, C.R. Merril, A. Wu, B. Olde, R.F. Moreno, A.R. Kerlavage, W.R. McCombie, and J.C. Venter, Science 252, 1991, 1651-1656.
2. E.S. Lander, and N.J. Schork, Science 265, 1994, 2037-2048.
3. V.E. Velculescu, L. Zhang, B. Vogelstein, and K.W. Kinzler, Science 270, 1995, 484-487.
4. R. Jaenisch, Science 240, 1988, 1468-1474.

5. M.P. Stenzel-Poore, S.C. Heinrichs, S. Rivest, G.F. Koob, and W.W. Vale, J. Neuroscience 14, 1994, 2579-2584.
6. C. Chen, D.G. Rainnie, R.W. Greene, and S. Tonegawa, Science 266, 1994, 291-294.
7. M. König, A.M. Zimmer, H. Steiner, P.V. Holmes, J.N. Crawley, M.J. Brownstein, and A. Zimmer, Nature 383, 1996, 535-538.
8. J.C. Crabbe, T.J. Phillips, D.J. Feller, R. Hen, C.D. Wenger, C.N. Lessov, and G.L. Schafer, Nature Genetics 14, 1996, 98-101.
9. B. Giros, M. Jaber, S.R. Jones, R.M. Wightman, and M.G. Caron., Nature 379, 1996, 606-612
10. M. Xu, X-T. Hu, D.C. Cooper, R. Moratalla, A.M. Graybiel, F.J. White, and S. Tonegawa, Cell 79, 1994, 945-955.
11. R.A. Harris, S.J. McQuilkin, R. Paylor, A. Abeliovich, S. Tonegawa, and J.M. Wehner, P.N.A.S. USA 92, 1995,3658-3662.
12. O. Cases, I. Seif, J. Grimsby, P. Gaspar, K. Chen, S. Pournin, U. Müller, M. Aguet, C. Babinet, J. Chen Shih, and E. De Maeyer, Science 268, 1995, 1763-1766.
13. R.J. Nelson, G.E. Demas, P.L. Huang, M.C. Fishman, V.L. Dawson, T.M. Dawson, and S.H. Snyder, Nature 378, 1995, 383-386
14. S. Ramboz, F. Saudou, D.A. Amara, C. Belzung, L. Segu, R. Misslin, M-C. Buhot, and R. Hen, Behavioural brain research 73, 1996, 305-312.
15. K. Hsiao, P. Chapman, S. Nilsen, C. Eckman, Y. Harigaya, S. Younkin, F. Yang, and G. Cole, Science 274, 1996, 99-102.
16. L.H. Tecott, L.M. Sun, S.F. Akana, A.M. Strack, D.H. Lowenstein, M.F. Dallman, and D. Julius, Nature 374, 1995, 546
17. S.E. Tsirka, A. Gualandris, D.G. Amaral, and S. Strickland, Nature 377, 1995, 340-344.
18. A.J. Bruce, W. Boling, M.S. Kindy, J. Peschon, P.J. Kraemer, M.K. Carpenter, F.W. Holtsberg, and M.P. Mattson, Nature Medicine 2, 1996, 788-794.
19. R. Brusa, F. Zimmermann, D-S. Koh, D. Feldmeyer, P. Gass, P.H. Seeburg, and R. Sprengel, Science 270, 1995, 1677-1680.
20. D. Acampora, S. Mazan, V. Avantaggiato, P. Barone, F. Tuorto, Y. Lallemand, P. Brulet, and A. Simeone, Nature Genetics 14, 1996, 218-222.
21. L.S. Butler, A.J. Silva, A. Abeliovich, Y. Watanabe, S. Tonegawa, and J.O. McNamara, P.N.A.S. USA 92, 1995, 6852-6855.
22. M. Matsumoto, T. Nakagawa, T. Inoue, E. Nagata, K. Tanaka, H. Takano, O. Minowa, J. Kuno, S. Sakakibara, M. Yamada, H. Yoneshima, A. Miyawaki, Y. Fukuuchi, T. Furuichi, H. Okano, K. Mikoshiba, and T. Noda, Nature 379, 1996, 168-171.
23. R. Bourtchuladze, B. Frenguelli, J. Blendy, D. Cioffi, G. Schutz, and A.J. Silva, Cell 79, 1994, 59-68.
24. A. Aiba, C. Chen, K. Herrup, C. Rosenmund, C.F. Stevens, and S. Tonegawa, Cell 79, 1994, 365-375.
25. K. Sakimura, T. Kutsuwada, I. Ito, T. Manabe, C. Takayama, E. Kushiya, T. Yagi, S. Aizawa, Y. Inoue, H. Sugiyama, and M. Mishina, Nature 373, 1995, 151-155.
26. M. Mayford, M.E. Bach, Y-Y. Huang, L. Wang, R.D. Hawkins, and E.R. Kandel, Science 274, 1996, 1678-1683.

27. A.J. Silva, C.F. Stevens, S. Tonegawa, and Y. Wang, Science 257, 1992, 201-211
28. J.C. Erickson, K.E. Clegg, and R.D. Palmiter, Nature 381, 1996, 415-418.
29. J.H. Baik, R. Picetti, A.Saiardi, G. Thiriet, A. Dierich, A. Depaulis, M. Le-Meur, and E. Borrelli, Nature 377, 1995, 424-428.
30. R.H. Zetterström, L. Solomin, L. Jansson, B.J. Hoffer, L. Olson, and T. Perlmann., Science 276, 1997, 248-250.
31. M. Zhang and I. Creese, Neuroscience Letters, 161, 1993, 223-226.
32. L-W. Zhou, S-P. Zhang, Z.H. Qin and B. Weiss, J. Pharmacol. Therapeutics 268, 1994, 1015-1023
33. A. Bourson, E. Borroni, R.H. Austin, F.J. Monsma, and A.J. Sleight, J. Pharmacol. Therapeutics 274, 1995, 173-180.
34. C. Wahlestedt, E.M. Pich, G.F. Koob, F. Yee, and M. Heilig, Science 259, 1993, 528-531.
35. R. Landgraf, R. Gerstberger, A. Montkowski, J.C. Probst, C.T. Wotjak, F. Holsboer, and M. Engelmann, J. Neuroscience 15, 1995, 4250-4258
36. R.W. Wagner, Nature 372, 1994, 333-335.
37. R.T. Windh, R. De La Garza II, E.J. Mah, M.L. Thomas, and K.A. Cunningham, Society for Neuroscience abstracts 20, 1994, 1545.
38. H.P. Erickson, J. Cell Biol. 120, (1993), 1079-1081
39. M. Gossen, and H. Bujard, Proc. Natl. Acad. Sci. U.S.A. 89, (1992), 5547-5551
40. P.A. Furth, L. St. Onge, H. Böger, P. Gruss, M. Gossen, A. Kistner, H. Bujard, and L. Hennighausen, Proc. Natl. Acad. Sci. U.S.A. 91, (1994), 9302-9306.
41. R.S. Passman, and G.I. Fishman, J. Clin. Invest. 94, (1994), 2421-2425.
42. N. Schultze, Y. Burki, Y. Lang, U. Certa, and H. Bluethmann, Nature Biotechnology 14, 1996, 499-503
43. Z.W. Li, G. Stark, J. Gotz, T. Rulicke, U. Muller, and C. Weissmann, P.N.A.S. USA 93, 1996, 6158-6162.
44. J.Z. Tsien, P.T. Huerta, and S. Tonegawa, Cell 87, 1996, 1327-1338.

H. van der Goot, (Editor)
TRENDS IN DRUG RESEARCH II



# Hot Topics
# Introduction

P. Angeli

Dipartimento di Scienze Chimiche, Università degli Studi di Camerino,
Via S. Agostino 1, 62032 Camerino, Italy

This section, being dedicated to hot topics in Medicinal Chemistry, is by its nature heterogeneous; as a consequence my introduction will be divided into three parts according to the topics of the lectures.

## 1. INVERSE AGONISM

A starting date could be 1993 when Hoyer and Boddeke reported in TiPS [1] the possibility that a single receptor in a homogeneous cell population may couple differently to a variety of effector systems: this means that, in principle, a ligand can behave as an antagonist, a partial agonist, or even as a full agonist at the same receptor. On the other hand, classical receptor theory says that some ligands behave as agonists, partial agonists, or antagonists, depending on the amount of receptor reserve. Variations in intrinsic activity and potency can be observed with a ligand even in a homogeneous population of receptors when the drug acts as a silent antagonist or as a full agonist. Of course, this makes it difficult to establish whether different receptor subtypes are involved.

In 1994 Terry Kenakin in the same journal [2] proposed the definition of negative efficacy related to antagonists which possess the ability to destabilize spontaneous complexes of receptors and G-proteins. In this way, the pharmacological dictionary acquired new definitions such as positive and negative efficacy in addition to neutral antagonists, that is ligands which are presumed to be devoid of efficacy. Again in TiPS, Graeme Milligan and colleagues in 1995 [3] described the two-state model of receptor activation in which receptors exist in equilibrium between the inactive state R and the active state R*: agonists stabilize the active state that interacts productively with a G-protein, while neutral antagonists do not alter the equilibrium. Receptor ligands which preferentially stabilize the inactive state R reducing the basal regulation of the effector system, are defined as inverse agonists. Many examples of inverse agonism have been suggested in δ-opioid, $\beta_2$- and $\alpha_2$-adrenergic, 5-HT, benzodiazepine and dopamine $D_2$

receptors, and an intense debate on the role of inverse agonists and on their therapeutic potentials is ongoing at the moment.

## 2. APOPTOSIS

There are two different types of cell death: a traumatic one (necrosis) and a programmed physiological one (apoptosis). The therapeutic induction of apoptosis allows the rapid killing of a large number of effector cells, without the inconveniences of necrosis, which releases all cellular contents into the surrounding tissue, thereby causing chronic inflammatory disorders. Apoptosis is a biological process that helps to keep the right number of cells in the organism. Some cell genes play a prominent part in the modulation of the apoptotic process: among them, the protooncogene BCL-2, which behaves as a trap for free radicals, and P53 (tumor suppressor gene), a cell sensory which plays a major role in the protection of cells against DNA damage.

It is possible to promote the apoptotic process *in vitro* in cell lines with biological (cytokines), physical (ionizing and ultraviolet radiations), and chemical agents, or through the withdrawal of growth factors. Oxidative events are important in apoptosis: mild oxidative stress can induce it in some cells. Protein kinase C (PKC) is one of the known regulators of cellular apoptosis since its inhibition or activation can both induce and suppress this process. The regulatory role of this kinase in cell death thus suggests the design of new drugs active as PKC activators or inhibitors [4].

The study of the mechanisms of apoptosis, and knowledge and regulation of them, can be an important aid in ageing-bound pathologies (in many neurodegenerative disorders the neuronal loss is associated with an excessive aliquot of apoptosis), in many infectious diseases such as AIDS (anomalous propensity to apoptosis of T $CD4^+$ lymphocytes), in cancer (apoptosis can be considered one of the main antitumoral mechanisms a cell possesses; anticancer drugs in therapy presumably work by means of apoptosis induction), in asthma and chronic inflammation.

## 3. BACTERIAL RESISTANCE

Many years have gone by since the first descriptions of penicillin-resistant pneumococci introduced the phenomenon of drug resistance, the reason why the treatment of infections and cancer with chemotherapeutics is frequently not efficacious. Changes in the point of attack of chemotherapeutics, a reduction of their penetration into the intracellular compartment where they are not allowed to stay in a sufficient concentration, and an enzymatic inactivation are some of the mechanisms that neutralize the effect of the chemotherapic drug.

We can distinguish between two types of drug resistance: intrinsic and acquired. The first relates to the failure of many microorganisms and tumors to respond to initial chemotherapy, while the second occurs when a microorganism or a tumor initially responds, but later relapses and appears to be strongly resistant to the original treatment. Multidrug resistance is a particular case of acquired drug resistance, first evidenced in tumor chemotherapy, but now found acting also in microorganisms. It results from overexpression of transmembrane glycoproteins such as the glycoprotein P-170 that functions as an ATP-dependent pump, clearing cells from the chemotherapeutic drug [5].

The strategy for overcoming this problem from a medicinal chemistry standpoint is the design of drugs able to reverse multidrug resistance, thereby avoiding the specific inactivation mechanisms.

## REFERENCES

1. D. Hoyer and H. W. G. M. Boddeke, TiPS 14 (1993) 270.
2. T. Kenakin, TiPS 15 (1994) 408.
3. G. Milligan, R. A. Bond and M. Lee, TiPS 16 (1995) 10.
4. D. J. McConkey and S. Orrenius, TiCB 4 (1994) 370.
5. F. Gualtieri, La Chimica e l'Industria 78 (1996) 1233.

H. van der Goot, (Editor)
TRENDS IN DRUG RESEARCH II



# Inverse agonism at G protein-coupled receptors. Studies with wild type and mutated adrenergic and opioid receptors

G. Milligan

Molecular Pharmacology Group, Division of Biochemistry and Molecular Biology, Institute of Biomedical and Life Sciences, University of Glasgow, Glasgow G12 8QQ, Scotland, United Kingdom

## 1. THE NATURE OF INVERSE AGONISM

The traditional view of second messenger regulation produced via seven transmembrane element G protein-coupled receptors (GPCRs) was that the receptor was quiescent until the agonist binding site became occupied by a suitable ligand. This binding was then communicated to the appropriate G protein by a conformational change in the structure of the receptor transmembrane helices. This resulted in guanine nucleotide exchange on the G protein $\alpha$ subunit and subsequent regulation of effector enzymes by either the GTP bound G protein $\alpha$ subunit or the G protein $\beta\gamma$ complex.

The concept of entirely quiescent GPCRs was not particularly attractive both because there is known to be 'basal' levels of generation of second messengers and because the application of molecular genetic approaches to the field indicated that increasing levels of expression of GPCRs frequently resulted in increasing levels of apparently ligand-independent second messenger regulation. Furthermore, if the binding of agonist promotes the adoption of an active conformational state of the GPCR then, thermodynamically, this should also be anticipated to occur with some level of probability in the absence of ligand even if, in analogy with certain ligand-gated ion channels, this probability may be very low in the absence of agonist.

If a GPCR is not entirely quiescent and agonists favour stabilisation (and thus the probability of maintenance) of an active state then it must be anticipated that ligands which preferentially favour adoption and maintenance of the inactive state will result in a reduction in basal second messenger regulation. Such ligands are termed inverse agonists. Because the majority of GPCRs are designed to produce a biochemical or membrane potential alteration, and thus a physiological effect, in response to the presence of an agonist ligand, the resting state of the GPCRs usually heavily favours the inactive conformation. As such, it can be difficult in many circumstances to discriminate inverse agonists from neutral antagonists (ligands which bind to but do not alter the equilibrium between active and inactive conformations of a GPCR). This is particularly true in systems in which GPCR expression levels are low. For this reason much of the basic characterisation of inverse agonist ligands has been performed in transfected cell systems expressing relatively high levels of a GPCR or by examining GPCRs which harbour mutations which result in varying degrees of increased agonist-independent (constitutive) activity. Herein I will describe strategies we have used to measure inverse agonist function at adrenergic and opioid receptors.

## 2. THE δ OPIOID RECEPTOR

The first clearly characterised demonstration of inverse agonism at a GPCR was the capacity of ICI174864 to cause a reduction in basal high affinity GTPase activity in membranes of the neuroblastoma x glioma hybrid cell line, NG108-15, which endogenously expresses the δ-opioid receptor (Costa and Herz, 1989). Following expression of this receptor in Rat-1 fibroblasts we were able to isolate a clone, D2, expressing some 6 pmol/mg membrane protein of this receptor. In membranes of these cells agonist ligands, such as the synthetic enkephalin [D-ala$^2$]-leucine enkephalin (DADLE), increased high affinity GTPase activity and the binding of [$^{35}$S]GTPγS. Both of these effects were prevented by prior treatment of the cells with pertussis toxin (Mullaney *et al.* 1996) demonstrating the stimulation to represent activation of $G_i$-like G proteins. Direct visualisation of agonist activation of $G_i$-like G proteins expressed by these cells was achieved by use of an assay in which cholera toxin is able to catalyse [$^{32}$P]ADP-ribosylation only of copies of $G_i\alpha$ which have previously been activated (Milligan *et al.* 1991). These are then detected by autoradiography following resolution of membrane proteins by SDS-PAGE. Because a clear signal of activated $G_i\alpha$ could be observed in this assay without addition of opioid agonist and because pertussis toxin-pretreatment of clone D2 cells reduced the basal high affinity GTPase activity in membranes of these cells to a value similar to that observed in membranes of untransfected Rat-1 cells (Mullaney *et al.* 1996) we assessed whether ICI174864 would show inverse agonist characteristics in clone D2 cells. ICI174864 reduced both basal high affinity GTPase activity and [$^{35}$S]GTPγS binding in a concentration-dependent manner (Mullaney *et al.* 1996). Furthermore it eliminated the agonist-independent cholera toxin-catalysed [$^{32}$P]ADP-ribosylation of $G_i$ (Mullaney *et al.* 1996). These effects of ICI174864 were not observed in membranes of untransfected cells. In the same way that the stimulatory effects of DADLE on high affinity GTPase were prevented by pertussis toxin pretreatment of the cells, pertussis toxin treatment also eliminated the inhibitory effect of ICI174864 (Mullaney *et al.* 1996). A central concern in many studies of inverse agonism is that the observed effects actually represent a competition between an antagonist ligand and levels of endogenous agonist present in the preparation. The δ opioid receptor ligand H-Tyr-Tic[$CH_2$-NH]Phe-Phe-OH (TIPP[ψ]) was unable to regulate the function of the δ-opioid receptor either positively or negatively and is thus a neutral antagonist. This mitigated against the possibility that ICI174864 was simply competing for the binding site with an unidentified agonist. Furthermore, co-addition of (TIPP[ψ]) prevented the effects of both DADLE and ICI174864 (Mullaney *et al.* 1996).

One major criticism of many studies on inverse agonism is that although the effects are observed in membrane preparations equivalent effects are not easily observed on whole cells. We addressed this issue in clone D2 cells using two distinct strategies. In the first of these we performed whole cell adenylyl cyclase assays following prelabelling of the cells with [$^3$H]adenine to allow this to be incorporated into the cellular ATP pool. Basal adenylyl cyclase activity was low in these cells and was little altered by the addition of either DADLE or ICI174864. However, addition of a range of concentrations of forskolin or cholera toxin to cause a degree of activation of the adenylyl cyclase population allowed both inhibition of this activity by DADLE and further stimulation by ICI174864 above that produced by these agents (Mercouris *et al.* 1997). Again, this effect of ICI174864 was prevented by pertussis toxin pretreatment of the cells. This appears to imply that a tonic δ–opioid receptor-$G_i$ activation is present in these cells in the absence of opioid agonist. However, pertussis toxin treatment alone did not mimic the capacity of ICI174864 to conditionally further stimulate forskolin- amplified adenylyl cyclase activity. These results imply an element of control of inverse agonist function which requires further analysis.

The Cytosensor™ microphysiometer records alteration in cellular metabolic activity as changes in the rate of extrusion of proteins and thus of extracellular pH. An increase in proton output in response to DADLE and a decrease in response to ICI174864 was observed in clone D2 cells (Mercouris *et al.* 1997).

## 3. ADRENOCEPTORS

The use of so called constitutively active mutant (CAM) forms of GPCRs potentially offers a novel and convenient means to examine and screen for inverse agonist ligands. This is a reflection that if they represent a conformational state of the receptor closer to that adopted upon binding of agonist it should be easier to detect the function of inverse agonists. Stable expression of a CAM version of the human $\beta_2$-adrenoceptor (Samama *et al.* 1993) in NG108-15 cells results in a large increase in basal adenylyl cyclase activity, which is related to the level of expression of the receptor (MacEwan and Milligan, 1996). This activity can be further stimulated by the addition of isoprenaline (MacEwan and Milligan, 1996) but as adenylyl cyclase is the limiting element for stimulatory regulation of adenylyl cyclase activity in these cells (MacEwan *et al.* 1996, Milligan, 1996) then the apparent maximal effect of isoprenaline at the CAM$\beta_2$-adrenoceptor is actually less than that produced following transfection of the wild type $\beta_2$-adrenoceptor when presented as "fold stimulation". The β-adrenoceptor ligands betaxolol and sotalol act as inverse agonists at the CAM$\beta_2$-adrenoceptor as assessed by their ability to inhibit this CAM receptor-induced basal adenylyl cyclase activity (MacEwan and Milligan, 1996). Maintenance of NG108-15 cells expressing the CAM$\beta_2$-adrenoceptor in the presence of either betaxolol or sotalol resulted in a strong upregulation of levels of the receptor as measured in subsequent [$^3$H]dihydroalprenolol binding studies (MacEwan and Milligan, 1996). Although detectable, the capacity of these agents to cause upregulation of the wild type

$\beta_2$-adrenoceptor expressed in NG108-15 cells was less impressive. Upregulation of levels of the CAM$\beta_2$-adrenoceptor was not observed by exposure to all $\beta$-adrenoceptor ligands. For example, alprenolol, which failed to inhibit basal adenylyl cyclase activity, and thus acted as an antagonist ligand also failed to cause an upregulation of the CAM$\beta_2$-adrenoceptor (MacEwan and Milligan, 1996). A similar set of observations has recently been reported by Smits *et al.* (1996) following stable expression of the wild type $H_2$ histamine receptor in CHO cells. Sustained treatment with inverse agonists but not with neutral antagonists resulted in an upregulation of levels of the receptor. However, recent studies in which $\beta$-ligands were released from an osmotic minipump in mice which express transgenically the CAM-$\beta_2$-adrenoceptor in a cardiac specific manner resulted in massive (up to 50 fold) upregulation of this receptor whether the ligand employed was an inverse agonist or neutral antagonist (Samama *et al* 1997)

The $EC_{50}$s for betaxolol-induced upregulation of the CAM$\beta_2$-adrenoceptor and inhibition of basal adenylyl cyclase in transfected NG108-15 cells were very similar and close to the $K_i$ for this ligand again indicating that the functional effects required occupancy of the receptor. Transient transfection of COS-7 cells with CAM$\beta_2$-adrenoceptor cDNA also provided a convenient means to detect the inverse agonist properties of betaxolol and sotalol at this receptor. The maintained presence of these ligands during the period of transient expression also resulted in a large upregulation of levels of the CAM$\beta_2$-adrenoceptor (Sautel and Milligan, unpublished observations).

It was possible that the ability of betaxolol to cause upregulation of the CAM$\beta_2$-adrenoceptor reflected the inhibition of basal cAMP production caused by this ligand as destabilisation of $\beta_2$-adrenoceptor mRNA has been noted as a prominent contributory mechanism to agonist-induced receptor downregulation (Hadcock and Malbon, 1991, Collins *et al.* 1992). To assess this question we treated NG108-15 cells expressing the CAM$\beta_2$-adrenoceptor with the adenylyl cyclase P-site inhibitor (Johnson *et al.* 1989) SQ 22536 (Desaubry *et al.* 1996). This agent (at 100 $\mu$M) was able to cause some 75% inhibition of adenylyl cyclase activity but despite this it was unable to mimic the effect of betaxolol. Likewise, we addressed whether maintaining cellular levels of cAMP would prevent the effect of betaxolol. Treatment of the cells with dibutyryl cAMP, a cell permeant and phosphodiesterase resistant analogue of cAMP, along with betaxolol also failed to prevent the receptor upregulation caused by the $\beta$-receptor ligand. We also assessed whether betaxolol treatment would cause a substantive alteration in steady state levels of mRNA encoding the CAM $\beta_2$-adrenoceptor. However, over a range of concentrations and times of exposure to betaxolol no consistent or large scale changes were observed in northern blots.

CAM mutations of the $\alpha_{1B}$-adrenoceptor have also been produced. Following stable expression in Rat-1 fibroblasts a CAM mutant of the hamster $\alpha_{1B}$-adrenoceptor is upregulated by sustained exposure to a variety of ligands traditionally viewed as $\alpha_1$-adrenoceptor antagonists (Lee *et al.* 1997). Such an event was not produced by equivalent treatments of cells expressing the wild type $\alpha_{1B}$-adrenoceptor. Each of the ligands used could be shown to function as an inverse agonist at the CAM $\alpha_{1B}$-adrenoceptor due to their capacity to lower basal phospholipase C and/or phospholipase D activities and following upregulation of the CAM $\alpha_{1B}$-adrenoceptor by sustained exposure to phentolamine, the agonist phenylephrine caused greater maximal stimulation of both phospholipase C and phospholipase D without alteration in agonist $EC_{50}$ (Lee *et al.* 1997). One limitation in these experiments was that we failed to identify a compound which acted as a neutral antagonist at the CAM $\alpha_{1B}$-adrenoceptor. By contrast, the wild type $\alpha_{1B}$-adrenoceptor displayed such limited agonist-independent constitutive activity that it was not possible to examine if the ligands tested had the capacity to inhibit basal signal transduction from this receptor, although as noted above they were not able to cause significant regulation of levels of the wild type receptor.

An issue in the development of inverse agonists and their comparison with neutral antagonists for clinical use is the possibility that inverse agonists might be more likely to cause receptor upregulation and thus following sudden removal might result in a greater degree of supersensitivity. Based on the available results to date this issue might be dependent upon the identity of the specific receptor targeted. Inverse agonism can clearly be detected at a variety of wild type receptors but as noted above the degree of upregulation of receptor levels produced by such ligands has tended to be rather limited in comparison to that achieved when examining CAM variants of these receptors. The capacity of antagonist drugs to cause upregulation of receptors in physiological settings in well known but may reflect a competition with and blockade of receptor downregulation produced by circulating levels of endogenous agonists.

## REFERENCES

1. Collins, S., Caron M.G. and Lefkowitz, R.J. (1992) Trends Biochem. Sci. 17, 37-39.
2. Costa, T. and Herz, A. (1989) Proc. Natl. Acad. Sci. USA 86, 7321-7325.
3. Desaubry, L., Shoshani I. and Johnson R.A. (1996) J. Biol. Chem. 271, 2380-2382.
4. Hadcock, J.R. and Malbon C.C. (1991) Trends Neurosci. 14, 242-247.
5. Johnson, R.A., Yeung, S.M.H., Stubner, D., Bushfield, M. and Shoshani, I. (1989) Mol. Pharmacol. 35, 681-688.
6. Lee, T.W., Cotecchia, S. and Milligan, G. (1997) Biochem. J. (in press).
7. MacEwan, D.J. and Milligan, G. (1996) Mol. Pharmacol. 50, 1479-1486.
8. MacEwan, D.J., Kim, G.D. and Milligan, G. (1996) Biochem. J. 318, 1033-1039.
9. Merkouris, M., Mullaney, I., Georgoussi, Z. and Milligan, G. (1997) J. Neurochem. (in press).
10. Milligan, G. (1996) Cell Signalling 8, 87-96.
11. Milligan, G., Carr, C., Gould, G.W., Mullaney, I. and Lavan, B.E. (1991) J. Biol. Chem. 266, 6447-6455.
12. Mullaney, I., Carr, I.C. and Milligan, G. (1996) Biochem. J. 315, 227-234.

13. Samama, P., Cotecchia, S., Costa, T. and Lefkowitz, R. J. (1993) J. Biol. Chem. 268, 4625-4636.
14. Samama, P, Bond, R.A., Rockman, H.A., Milano, C.A. and Lefkowitz, R.J. (1997) Proc. Natl. Acad. Sci.USA 94, 137-141.
15. Smit, M.J., Leurs, R., Alewijnse, A.E., Blauw, J., Van Nieuw Amerongen, G.P., Van De Vrede, Y., Roovers, E. and Timmerman, H. (1996) Proc. Natl. Acad. Sci. 93, 6802-6807.

H. van der Goot, (Editor)
TRENDS IN DRUG RESEARCH II



# Life and Death Decisions: Medicinal Chemistry Approaches to Apoptosis

Alan P. Kozikowski

Drug Discovery Laboratory, Georgetown University Medical Center, Institute for Cognitive and Computational Sciences, 3970 Reservoir Road N. W., Washington, DC 20007-2197

## 1. INTRODUCTION

**Apoptosis** derives from a Greek word that means to fall off, as leaves from a tree. It is a process in which the cell shrinks, and eventually the apoptotic cell bodies that are produced are phagocytosed. Apoptosis is characterized by preservation of membrane integrity, cytoplasmic and nuclear condensation, reduction in cellular volume, plasma membrane bleb formation, and nuclear fragmentation. During apoptosis the morphological changes are often accompanied by internucleosomal cleavage of genomic DNA; the breakdown of the DNA occurs in discrete fragments of 180-200 base pairs, which appear as a DNA ladder in agarose gel electrophoresis. Apoptosis differs from necrosis in that the latter involves cell death that is accompanied by inflammatory processes. Thus in apoptosis cells can die one at a time, among a group of healthy cells, whereas in necrosis a whole group of cells may die at the same time. Death by necrosis is generally associated with a traumatic injury, ischemia, chemical exposure, and the like. Apoptosis is often confused or mixed up with programmed cell death (PCD). The latter term originated to describe cell death occurring at normal stages of the developmental program and that is initiated by some physiological trigger. In some cases PCD of cells shares morphological changes that are similar to those of apoptosis, but this is not always so [1].

PCD was probably first realized to take place back in ancient times, by individuals who followed the developmental progression of tadpoles to frogs, where during the process of metamorphosis frogs were observed to lose their tails. Later, in the landmark research of Rita Levi-Montalcini and Victor Hamburger this same process was found to play an important role in the development of the central nervous system, for at discrete stages of development both neuronal and glial cells undergo a highly orchestrated disappearance [2].

It is now believed that among other factors, apoptosis occurs when a cell suffers some form of physiological imbalance. This imbalance leads to improper cross-talk among signal transducing systems, for example, a rise in intracellular calcium levels may be initiated in the absence of another required signal, and that consequently the cell is forced into apoptotic death [3].

In some cases apoptosis is desirable, whereas in other cases, it is not. Thus, diseases can be associated with either an undesired inhibition of apoptosis, or in other cases with an undesired increase in apoptosis. Cases in point are cancer and Alzheimer's disease. In cancer, including breast and prostate cancer, autoimmune disorders, and viral infections, apoptosis is inhibited through different mechanisms. In AIDS, in neurodegenerative conditions including Alzheimer's disease, Parkinson's disease, and amyotrophic lateral sclerosis, in ischemic injury including myocardial infarction, stroke, and reperfusion injury, and in toxin-induced liver diseases, an improper increase in apoptosis takes place.

A number of agents are known to act as inhibitors of apoptosis. These include physiological inhibitors like growth factors (NGF, BDNF, CNTF), zinc, estrogens, androgens, neutral amino acids, viral genes like Cowpox virus crmA, and pharmacological agents such as cysteine protease inhibitors (ICE inhibitors), calpain inhibitors, tumor promoters (PMA), and free radical inhibitors (TMPO). On the other hand, the TNF family members including TNF, Fas ligand, neurotransmitters like glutamate and dopamine, calcium, glucocorticoids, damage-related inducers like heat shock, viral infection, oncogenes (myc), tumor suppressers (p53), oxidants, free radicals, nutrient deprivation, chemotherapeutics (cisplatin, bleomycin, nitrogen mustards), gamma radiation, UV radiation, and toxins (including MPTP and beta-amyloid) act as inducers of apoptosis. In view of the various agents that act as inducers or inhibitors of apoptosis, the medicinal chemist is left only to his or her imagination to invent agents that might counter (or even in some cases assist) the activators or block the inhibitors of apoptosis.

## 2. CPP32

While a complete pathway for cell death has not been defined, it is very clear from a number of studies that the CED-3/ICE family of proteases plays an essential role in apoptosis. The CED-3 homolog, CPP32, also known as apopain, plays a definitive role in cell death, for it is involved in the destruction of certain key homeostatic and repair enzymes at the onset of apoptosis. CPP32 degrades proteins by effecting cleavage of functional domains comprised of the $(P_4)$Asp-X-X-Asp$(P_1)$ motifs. Targets for destruction by CPP32 include poly(ADP-ribose) polymerase (PARP, which is an enzyme involved in DNA repair), the 460,000 $M_r$ catalytic subunit of the DNA-dependent kinase essential for double-strand break repair, the 70000 $M_r$ small nuclear ribonucleoprotein necessary for mRNA splicing, and other enzymes including protein kinase C delta [4]. Studies have revealed that levels of CPP32 are elevated during apoptosis, and that its removal from apoptotic cellular extracts can lead to a reduction in apoptosis in an *in vitro* type of assay. Activation of one protease can, in fact, lead to the processing and activation of the same or other proteases to lead to an amplification of the protease cascade. ICE can, for example, process CPP32 to its active form, but the reverse is not true.

Of particular interest to our own research is the structural work that has been done on apopain. Researchers at the Merck Research Laboratories have recently been able to determine the three-dimensional structure of apopain in complex with the tetrapeptide aldehyde inhibitor, Ac-DEVD-CHO [5]. This inhibitor has been shown to have a 49-fold preference for CPP32 versus ICE (interleukin-1β converting enzyme, which belongs to a second phylogenic subfamily of proteases, and whose role in apoptosis is less secure). In contrast, the tetrapeptide aldehyde Ac-YVAD-CHO is far more selective for ICE ($K_i$ = 0.76 nM) than for CPP32 ($K_i$ = 10,000 nM).

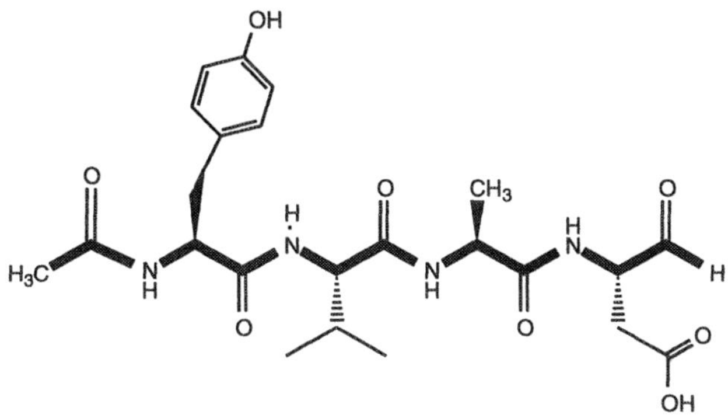


**Figure 1**. Ac-YVAD-CHO, selective inhibitor of ICE

The x-ray work reveals that the tertiary and quaternary structure of these cysteine proteases are similar. In both proteases two heterodimers associate to form a tetramer, a four-chain assembly with two-fold rotational symmetry. A diagram of the tetrapeptide aldehyde inhibitor bound to CPP32 is provided below. As is clear from this diagram, Cys 285 engages in

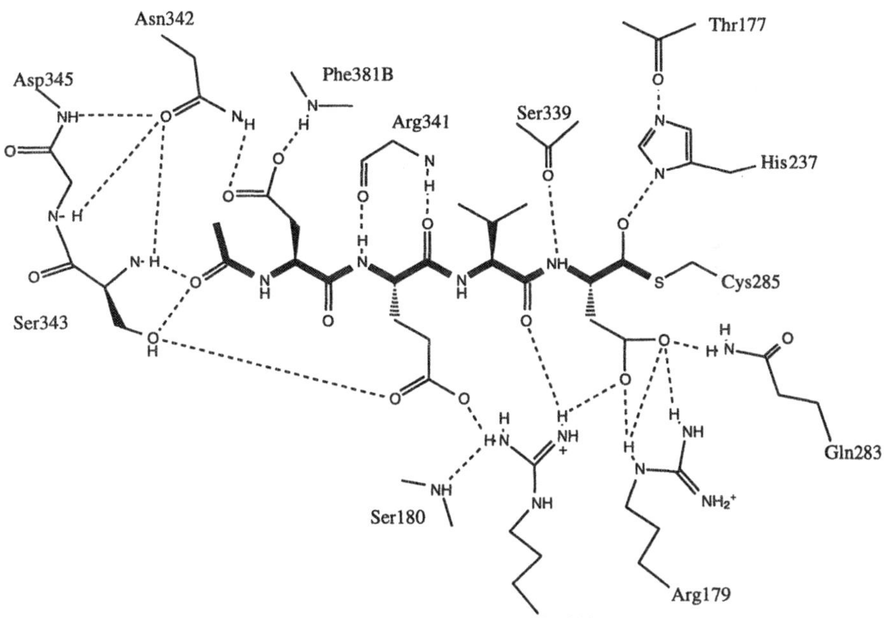


**Figure 2.** Hydrogen bonds and other polar interactions between the bound tetrapeptide aldehyde inhibitor and CPP32/apopain (adapted from reference 5).

thiohemiacetal formation with the aldehyde group, thus mimicking the transition state for amide bond hydrolysis. The binding pockets of ICE and apopain are similar, with the exception particularly of the $S_4$ subsite which appears to be responsible for the differences in substrate specificities. As expected from the preference for a $P_4$ Tyr in the ICE inhibitors, the corresponding subsite in the ICE enzyme corresponds to a large, shallow hydrophobic depression, while the $P_4$ site in apopain is narrow and intimately engulfs the $P_4$ aspartate residue side chain. Time dependent irreversible inhibitors of ICE have been designed that are acyloxymethylketones, and which react by displacement of the carboxylate group by the active site cysteine 285 [4].

In order to discover compounds that may serve as cell permeable and selective inhibitors of these cysteine proteases, and in particular CPP32, we have chosen to investigate approaches based upon database screening methods in which a pharmacophore query is set up employing various elements of the x-ray structural information of the tetrapeptide inhibitor Ac-DEVD-CHO in complex with CPP32. Using a fluorescent assay that depends upon cleavage of the CPP32-like substrate, Ac-DEVD-AMC (AMC = aminomethylcoumarin), we have been able to screen a large number of compounds in a single day. Several interesting compounds were identified by this process, and the results will be reported.

## 3. PKC

As mentioned above, PKC is also a target for CPP32. It has been shown in a variety of systems that PKC may be involved in the negative regulation of cell death. For example, phorbol esters, which bind to the regulatory site of PKC, have been shown to protect thymocytes from apoptosis initiated by T cell activation, glucocorticoids, cAMP, etc. Phorbol esters have also been shown to directly block calcium induced internucleosomal DNA fragmentation in isolated thymocyte nuclei, which would suggest that PKC mediates its effects within the nucleus. Concanavalin A, a T-cell mitogen which exerts its effects through PKC, has been shown to result in a sustained increase of intracellular $Ca^{2+}$ in immature thymocytes, thereby triggering cell proliferation. When the concanavalin A treatment is repeated in the presence of the non-selective PKC inhibitor H-7 or sphingosine, the cells undergo apoptosis instead [6]. This would suggest that PKC provides some sort of protective mechanism that diverts the $Ca^{2+}$ signal from the cell death pathway to a cell proliferation pathway. Also, inhibitors of PKC induce apoptosis in neoplastic cell lines. The actual nature of the role that PKC plays in apoptosis is, however, a cloudy one, for other cases can be cited in which PKC activation induces apoptosis. For example, stimulation of PKC-β by phorbol esters leads to death in myeloid leukemia cells. Moreover, others have reported that phorbol esters stimulate cAMP-induced thymocyte death [6]. It is clear that more work is needed, and that part of the problem relating to the link between apoptosis and PKC may derive from the relative non-selectivity of the compounds used in such assays, differences in the isozyme levels of various cell lines, and differences in the pathways governing the stability and turnover of these isozymes.

In order to sort out some of these issues we have been looking at the design of new PKC activators that may be more isozyme selective. PKC is a ubiquitous signal transducing enzyme system that plays a marked role in diverse cellular processes including not only apoptosis, but the regulation of ion channels, neurotransmitter release, growth and differentiation, and neuronal plasticity as well. This complex enzyme system serves to mediate numerous signals originating from the induction of lipid hydrolysis. The discovery of isozyme-selective activators and inhibitors (modulators) of PKC is crucial to both dissecting the role of the individual isozymes in physiological and pathophysiological processes and to manipulating these pathways. The discovery of such small molecule modulators of PKC is still in the early stages. It has been our objective to synthesize analogues of the tumor promoting substance lyngbyatoxin (or more precisely, its simpler congener indolactam V or ILV) and to assess the ability of these new materials to modulate the various isozymes of PKC as well as to examine the effects of such agents on cellular growth and differentiation.

Specifically, it is our aim to learn more about the nature of the structure-activity relationships that govern isozyme-selective binding, activation, inhibition, translocation, and/or down-regulation of PKC by analogues of indolactam (ILV) and to discover isozyme selective modulators of this multifunctional kinase which function through the phorbol ester binding site. Biological investigations have revealed that there exist at least eleven isozymes of protein kinase C bearing closely related amino acid sequences [7]. The isozymes exhibit distinct patterns of tissue specific expression and intracellular localization. Complete details of the functional significance of these different isozymes and their substrate specificities remain to be elucidated; however, recent work reveals that the "distinct isozymes of PKC can exert different effects on growth control and malignant transformation in the same cell type"[8].

In view of the fact that bryostatin-1, a non-promoting PKC activator, is in clinical evaluation as an anti-cancer agent, it is likely that isozyme selective activators of PKC that are not tumor promoters will find use in treating neoplastic conditions, perhaps by initiating apoptosis. The recent x-ray structural determination of the complex between phorbol 13-acetate and the CRD2 activator-binding domain of protein kinase Cδ, [9] coupled with the insights gained from modeling the interactions of diverse ligands to PKC, provides unique opportunities for the discovery of PKC modulators with novel properties.

Teleocidin was first isolated from the mycelia of *Streptomyces mediocidicus* as a mixture of highly toxic compounds by Sakai *et al.*[10]. The structure of one of these metabolites was assigned by Hirata as shown by formula **1**. Lyngbyatoxin A was first isolated from the lipid extract of a Hawaiian shallow-water variety of blue green alga, *Lyngbya majuscula* Gomont [11]. The structure of this compound is closely related to that of the teleocidin family. The lyngbyatoxin series can be obtained together with the teleocidin B group from *Streptomyces mediocidicus* [12]. Therefore, they were also named as teleocidin A-1 (**2a**, which is identical to lyngbyatoxin A) and A-2 (**2b**) by Sakai. Indolactam V (**3**, ILV), which contains the basic ring structure of the teleocidins, is the simplest member of the family, and is produced in large quantities by actinomycetes strain NA34-17 [13,14]. On the basis of both solution NMR studies and molecular mechanics calculations, it has been reported that the indolactam portion (indolactam V, **3**) of the teleocidins and lyngbyatoxins can exist in two conformational states, the sofa-like conformation containing a trans amide bond and the twist-like conformation containing a cis amide bond. At equilibrium, the ratio of twist/sofa was 2.8; the twist form of ILV represents the biologically active conformation as is discussed further below [15,16].

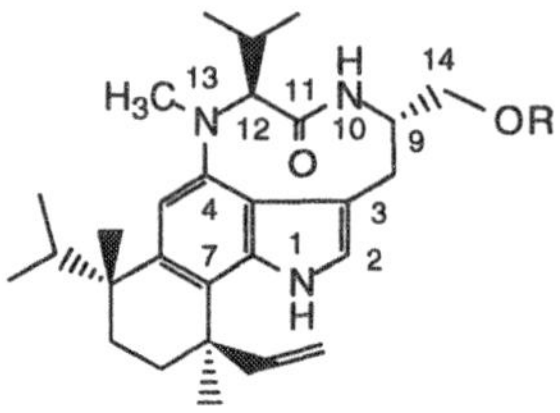

**1**, R = H: Teleocidin B-4 or Olivoretin D

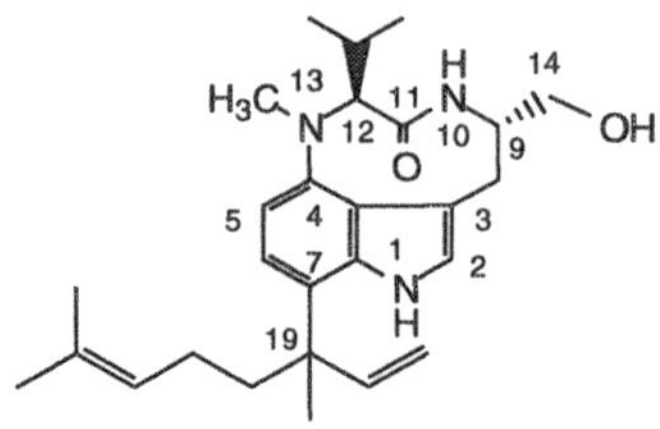

**2a**, 19 R: Teleocidin A-1; (or Lyngbyatoxin A)
**2b**, 19 S: Teleocidin A-2

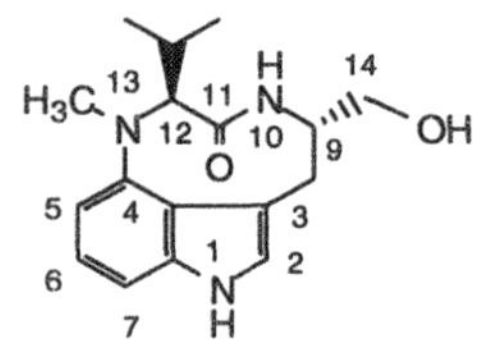

**3**, Indolactam V (ILV)

Based upon our previous synthetic efforts aimed at identifying analogues of ILV that were synthetically more accessible, we were the first group to report routes to certain benzolactam analogues of ILV in which the pyrrole ring was absent [17]. Previously, we had found that the 9-membered benzolactam **4** was not active as a PKC activator, presumably because of its inability to adopt the twist conformation. In continuation of this theme, we now describe a novel synthetic approach to the 8-membered benzolactams **5a** and **5b**. These compounds, unlike **4**, are forced to adopt the twist conformation as a consequence of the smaller ring size constraining the amide group to adopt solely the cis stereochemistry. Endo *et al.* have previously reported the synthesis of some related compounds,[16] however, our modeling results suggested that the location of a hydrophobic side chain at position 8 rather than 9 might confer improved isozyme selectivity to the compounds.

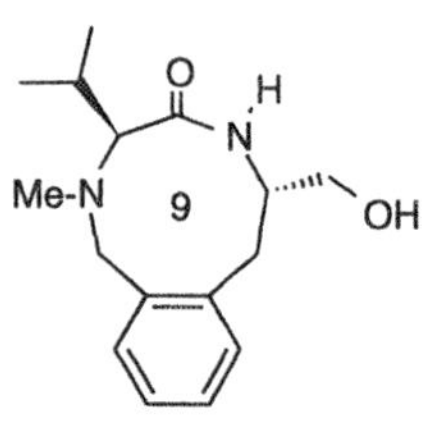

**4**, 9-membered ring benzolactam

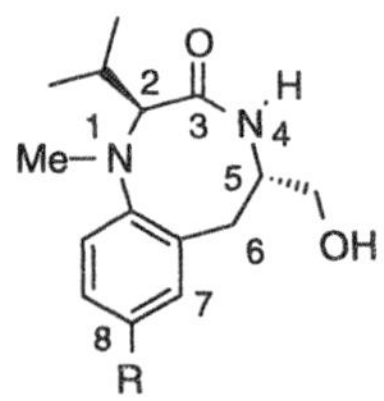

**5**, R = $-C{\equiv}C(CH_2)_7CH_3$

**6**, R = $-(CH_2)_9CH_3$

In order to build a proper foundation for achieving our aims, we decided to fully explore the possible binding modes of teleocidin, ILV, and the 8-membered ring benzolactam to PKC using molecular modeling methods. Teleocidin, ILV, and the benzolactam are known to directly compete with phorbol ester binding to PKC. Our recent mutagenesis studies have also clearly shown that a number of residues, such as Pro 11, Leu 21, and Gln 27, are important to the binding of both phorbol esters and ILV to PKC [18]. These results all suggest that ILV, teleocidin, the 8-membered ring benzolactam, and the phorbol esters should occupy the same binding region or that their binding regions at least partially overlap.

This binding model allows us to fully appreciate the role that each functional group in the 8-membered ring benzolactam plays in its binding to PKC (Figure 3). The primary hydroxyl group at C-5, acting as a hydrogen bond donor, forms a hydrogen bond with the carbonyl group of residue Leu 21, while this same group, acting as a hydrogen bond acceptor, forms a hydrogen bond with the amido group of Thr 12 (Figure 4). Thus, this primary hydroxyl group is functionally equivalent to the hydroxyl group at C-20 in phorbol 13-acetate. The carbonyl group at position 3 forms a hydrogen bond with the amido group of Gly 23, which is equivalent to the carbonyl group at position 3 in phorbol 13-acetate. The amido group at position 4 forms a hydrogen bond with the carbonyl group of Leu 21. It is noteworthy that this hydrogen bond was not found in the hydrogen bond network formed between phorbol 13-acetate and PKC, and is thus unique to the 8-membered ring benzolactam, ILV, and teleocidin. On the other hand, the hydrogen bond formed between the hydroxyl group at C-4 in phorbol 13-acetate and the carbonyl group of Gly 23 in PKC is absent in the hydrogen bond network of the benzolactam and PKC system. All four hydrogen bonds formed between the 8-membered ring benzolactam and PKC have optimal geometries for strong hydrogen bond formation.

Isozyme Studies.

An enantiospecific route to these eight-membered benzolactam analogues bearing a 10-carbon chain that uses L-phenylalanine as the starting material has been developed which is described in a recent paper [17]. The two new benzolactams **5** and **6** have been tested for isozyme

**Table 1. $K_i$ values ± SEM for the inhibition of [$^3$H]PDBU binding by the compounds tested.**

| | α | β | γ | δ | ε |
|---|---|---|---|---|---|
| **Compound** | | | | | |
| **5** | 14.7 ± 1.3 | 17.4 ± 2.2 | 40.7 ± 8.9 | 122 ± 22 | 142 ± 3 |
| **6** | 46.6 ± 8.0 | 58.2 ± 12.6 | 145 ± 25 | 185 ± 30 | 187 ± 22 |
| **ILV*** | 11.0 | 6.1 | 19.4 | 8.2 | 21.9 |
| ***n*-octyl-ILV*** | 0.5 | 0.4 | 1.2 | 0.8 | 1.0 |

*Data taken from reference 19.

selectivity by investigating their ability to displace [$^3$H]PDBU binding to recombinant PKC isozymes expressed in the baculovirus system[19]. These data are presented below along with comparison data for ILV and *n*-octyl-ILV. As is apparent, the acetylenic derivative **5** shows a rather nice level of selectivity, with the compound showing a higher affinity for the α and β isozymes, in comparison to γ, δ, and ε. The greatest selectivity of **5** is the approximately ten-fold difference in affinity between PKC α and ε. In contrast, *n*-octyl-ILV shows a pairwise selectivity of only about two for α vs. ε, and at best a three-fold selectivity for β vs. γ. As is also apparent, the electronic effect of the acetylenic group of **5** plays some role in its isozyme selectivity, for its saturated alkyl counterpart **6** shows poorer selectivity with only a four-fold difference in the $K_i$ for α versus ε.

<u>Cell proliferation assay and PKC Downregulation</u>.

The acetylenic benzolactam **5** and ILV were tested for antiproliferative activity against breast carcinoma cell lines MCF-7 and MDA-MB-231 [20]. Exposure of MCF-7 and MDA-MB-231 cells to **5** for four days resulted in $IC_{50}$ values of 20 and 30 μM, respectively, whereas ILV was inactive.

PKC isozyme levels were determined in MCF-7 and MDA-MB-231 cells exposed to **5** for 24 hours. In MCF-7 cells, PKCβ and PKCε were virtually eliminated, PKCδ was reduced to a lesser extent, and PKCζ was unchanged. MDA-MB-231 cells exhibited a similar reduction in PKCβ, whereas PKCδ and ζ were slightly reduced, while PKCα and PKCε remained unchanged. These results indicate that **5**, while not completely selective, preferentially downregulates PKCβ in both cell lines. The varying degree of selectivity of **5** for other PKC isozymes was to some degree cell type specific. This result was not completely unexpected, since different tumors would be expected to exhibit different PKC isozyme patterns as well as different pathways governing their stability and turnover. The acetylenic benzolactam is presently being tested in animal tumor models and shows promising results.

## 4. SPHINGOSINE PATHWAY

While the PKC signaling pathway is well studied, perhaps less is known regarding the importance of second messengers derived from another major class of membrane lipids, the sphingolipids. All sphingolipids contain as their backbone long chain sphingoid bases of which sphingosine is the most prominent, an amide-linked fatty acid (to form the residue known as ceramide which is embedded in the lipid bilayer) and a polar head group (hydroxyl for ceramide, phosphoryl choline for sphingomyelin, or carbohydrate residues for glycosphingolipids) [21,22]. There is now considerable information that the breakdown products of membrane sphingolipids also play a fundamental role in the transmission of external signals that impinge upon cells into internal signals through the activation of certain intracellular enzymes. Much like the glycerophospholipid cycle, it has now been shown that a sphingolipid metabolic cycle also operates by the generation of its own intermediates possessing second messenger functions. These intermediates include ceramide, sphingosine, and sphingosine-1-phosphate or SPP.

**Sphingolipid Metabolic Pathway Illustrating the Formation of the Intracellular Second Messengers, Sphingosine, Spingosine-1-phosphate, and Ceramide.**

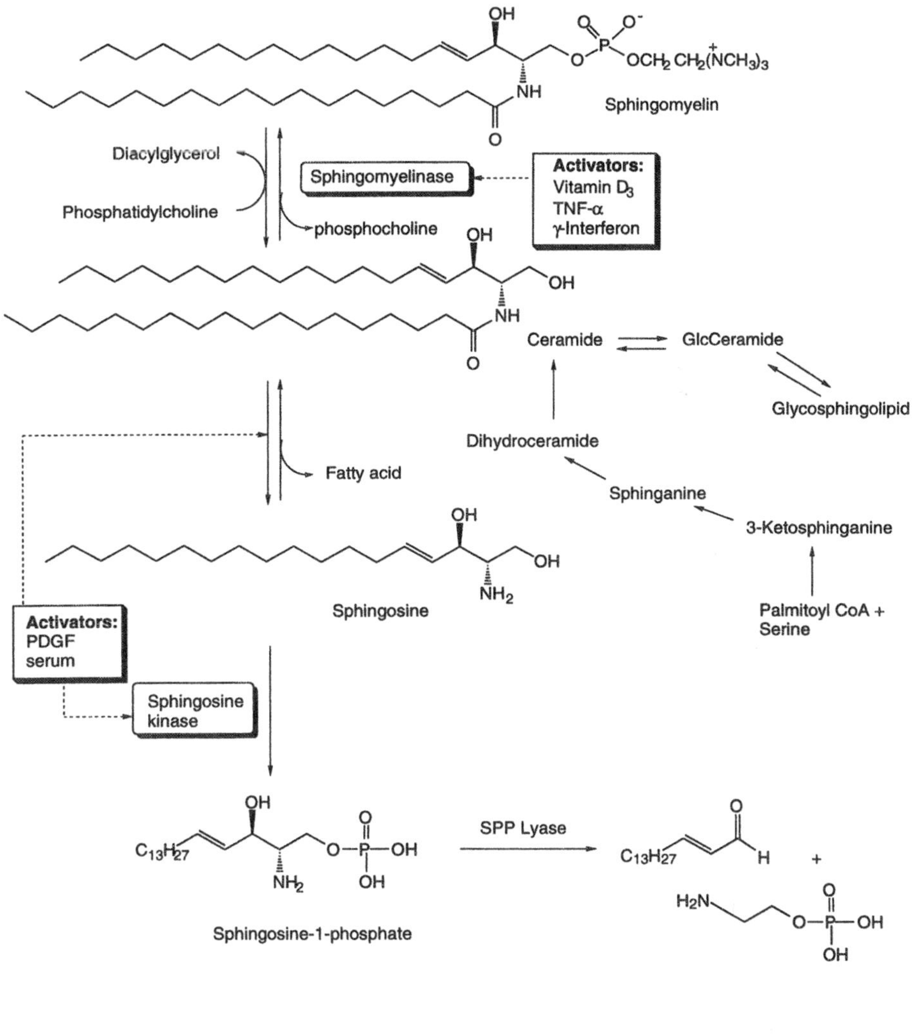


Ceramide has been reported to induce apoptosis and cell arrest in several mammalian cell lines, whereas sphingosine and its phosphorylated product SPP generally provide positive mitogenic signals. SPP has also been shown to block the appearance of the intranucleosomal DNA fragmentation associated with apoptosis. Several cytokines including TNF-α, vitamin $D_3$, γ-interferon, IL-1, and nerve growth factor stimulate the activity of sphingomyelinase [21,22].

This enzyme catalyzes the breakdown of sphingomyelin to form ceramide. Ceramide in turn plays a previously unrecognized role in cellular differentiation and apoptosis. While the cellular targets that mediate the action of ceramide are not fully known, likely candidates include a ceramide stimulated kinase and a ceramide-stimulated phosphatase. Recently, Mizushima *et al.* have shown that a cell permeable ceramide derivative can induce cleavage and activation of CPP32, but not ICE [23]. Ceramide-induced apoptosis of Jurkat cells was also shown to be blocked by the CPP32-tetrapeptide aldehyde inhibitor DEVD-CHO, whereas Jurkat cells with defective CPP32 activation were resistant to ceramide induced apoptosis. These results would accordingly suggest that the ceramide-sphingomyelin pathway must function upstream of CPP32, but that this protease is required for ceramide-induce apoptosis. Moreover, ceramide-induced apoptosis has been reported to be blocked by the PKC activator TPA through a mechanism relating to the translocation of PKC $\delta$ and $\varepsilon$ from the membrane to cytosol; the TPA translocates these isozymes to the membrane, thus blocking their ceramide-induced translocation to cytosol [24]. Additionally, other evidence shows that PKC activators may block ceramide induced apoptosis by stimulating sphingosine kinase, thus increasing levels of SPP that overcome the ceramide [25]. While many details still remain to be worked out regarding the sphingolipid signaling cascade, both SPP and ceramide should be viewed as possible starting points in the design of agents useful in regulating diverse apoptotic events.

While this brief article cannot give adequate coverage to all the recent biological breakthroughs in the current understanding of apoptosis, it should be clear that this event is a complicated process, a process that involves the interplay of a multitude of cellular signaling systems. Its very complexity extends the possibility, however, to the "creative" medicinal chemist to invent the new drug therapies of tomorrow based upon interventions targeted to various stages in the death cascade.

## ACKNOWLEDGMENTS

I wish to thank Professors Shaomeng Wang, Robert Glazer, Alexander Yakovlev, and Sarah Spiegel for their collaborative efforts in the area of apoptosis research.

## REFERENCES

1. N. J. Holbrook, G. R. Martin, and R. A. Lockshin, Eds., Cellular Aging and Cell Death, Wiley-Liss, New York, 1996.
2. W. Bursch, F. Oberhammer, and R. Schulte-Hermann, Trends Pharmacol. Sci. 13 (1992) 245.
3. A. Levitzki, Eur. J. Biochem. 226 (1994) 1.
4. D. K. Miller, Annual Reports in Medicinal Chemistry 26 (1996) 249.
5. J. Rotonda, D. W. Nicholson, K. M. Fazil, M. Gallant, Y. Gareau, M. Labelle, E. P. Peterson, D. M. Rasper, R. Ruel, J. P. Vaillancourt, N. A. Thornberry, and J. W. Becker, Nature Structural Biology 3 (1996) 619.
6. R. A. Schwartzman and J. A. Cidlowski, Endocrine Reviews 14 (1993) 133.
7. A. Basu, Pharmacol. Ther. 59 (1993) 257.
8. S. Stabel and P. J. Parker, Pharmacol. Ther. 51 (1991) 71.

9. G. Zhang, M. G. Kazanietz, P. M. Blumberg, and J. H. Hurley, Cell 81 (1995) 917.
10. M. Takashima and H. Sakai, Bull. Agr. Chem. Soc. Japan 24 (1960) 647.
11. J. H. Cardellina II, F.-J. Marner, and R. E. Moore, Science 204 (1979) 1193.
12. S. Sakai, Y. Hitotsuyanagi, N. Aimi, H. Fujiki, M. Suganuma, T. Sugimura, Y. Endo, and K. Shudo, Tetrahedron Lett. 27 (1986) 5219.
13. T. Sugimura, H. Fujiki, M. Mori, M. Nakayasu, M. Terada, K. Umezawa, and R. E. Moore, Carcinogenesis 7 (1982) 69.
14. K. Irie and K. Koshimizu, Memoirs of the College of Agriculture, Kyoto University, No. 132, 1988.
15. Y. Endo, M. Hasegawa, A. Itai, K. Shudo, M. Tori, Y. Asakawa, and S. Sakai, Tetrahedron Lett. 26 (1985) 1069.
16. Y. Endo, M. Ohno, M. Hirano, A. Itai, and K. Shudo, J. Am. Chem. Soc. 118 (1996) 1841.
17. A. P. Kozikowski, D. Ma, Y.-P. Pang, P. Shum, V. Likic, P. K. Mishra, S. Macura, A. Basu, J. S. Lazo, and R. G. Ball, J. Am. Chem. Soc. 115 (1993) 3957; A. P. Kozikowski, S. Wang, D. Ma, J. Yao, S. Ahmad, R. I. Glazer, K. Bogi, P. Acs, S. Modarres, N. E. Lewin, and P. M. Blumberg, J. Med. Chem. (1997) in press.
18. M. G. Kazanietz, W. Wang, G. W. A. Milne, N. E. Lewin, H. L. Liu, P. M. Blumberg, J. Biol. Chem. 270 (1995) 21852.
19. M. G. Kazanietz, L. B. Areces, A. Bahador, H. Mischak, J. Goodnight, F. Mushinski, P. M. Blumberg, Mol. Pharmacol. 44 (1993) 298.
20. G. Yu, S. Ahmad, A. Aquine, C. R. Fairchild, J. B. Trepel, S. Ohno, K. Suzuki, T. Tsuruo, K. H. Cowan, and R. I. Glazer, Cancer Commun. 3 (1991) 181.
21. R. Kolesnick and D. W. Golde, Cell 77 (1994) 325.
22. Y. A. Hannun, J. Biol. Chem. 269 (1994) 3125.
23. N. Mizushima, R. Koike, H. Kohsaka, Y. Kushi, S. Handa, H. Yagita, and N. Miyasaka, FEBS Lett. 395 (1996) 267.
24. H. Sawai, T. Okazaki, Y. Takeda, M. Tashima, H. Sawada, M. Okuma, S. Kishi, H. Umehara, and N. Domae, J. Biol. Chem. 272 (1996) 2452.
25. O. Cuvillier, G. Pirianov, B. Kleuser, P. G. Vanek, O. A. Coso, J. S. Gutkind, and S. Spiegel, Nature 381 (1996) 800.

# Antibiotic resistance - a view from the pharmaceutical industry

R. Bax

Director and Vice-President, Anti-Infective Therapeutic Unit,
Clinical Research and Development
SmithKline Beecham Pharmaceuticals, New Frontiers Science Park (South), Third Avenue, Harlow, Essex, CM19 5AW, England

The worldwide market for pharmaceuticals was 280 billion US dollars in 1995. The world antibacterial market that year was $27 billion of which 40% was for cephalosporins, 20% penicillins, 12% quinolones and 8% for macrolides. Of the 20 leading pharmaceuticals sales in 1996 there are three antibiotic products. Ciprofloxacin ranked seven overall, with sales of $1.13 billion, followed by clarithromycin with sales of around $1 billion and, number twelve, Augmentin, with sales of $990 million. The growth in these three products over the year averaged 17%.

Within the eight leading markets for antibacterials, there is a difference in usage of penicillins, cephalosporins, quinolones, tetracyclines and macrolides. For example, Spain and UK are very high users of penicillins, whereas France and Italy are very low users of cephalosporins. Japan, the second largest market after the US, is atypical with high shares of prescriptions within the cephalosporin and quinolone area, and very low usage of tetracylines and macrolides.

SmithKline Beecham currently has a leadership position in antibacterials by value with sales in excess of £1.8 billion per year. Clearly leadership would imply breakthrough products. However, within the antibacterial area, there have been no new major distinctly different antibiotics with novel modes of action since the launch of nalidixic acid in 1960. Since then, all the new oral and injectible antibiotics which have been developed have been variations on existing antibiotic nuclei such as the β-lactam and quinolones antibiotics.

Antimicrobial resistance was first discovered with the introduction of benzylpenicillin. It was first noted in 1940 in Esterichia coli. In 1944 the first penicillin resistant Staphylococcus aureus was noted in the US. In the late 1950s Shigella became much more resistant in Japan to tetracyclines which was related to direct tetracycline [1]. In the 1960s the rise of multi resistant Staphylococcus aureus including methecillin and penicillin resistance [2]. The 1960s, '70s and '80s saw the rise of multiresistant Gram negative bacteria particularly to the pencillins and cephalosporins. It was in the 1970s that both the gonococcus and Haemophillus influenzae acquired resistance factors via the tem enzyme which destroyed the currently available penicillin and cephalosporin antibiotics [3].

The 1980s saw a general rise of Gram negative resistance especially to agents such as the aminoglycosides [4] and to the cephalosporins [5]. This continued into the 1990s, especially within hospitals. A rise in resistance, particularly multiresistance, across classes of antibiotics from a whole range of bacteria found in the community included the multi drug resistant Streptococcus pneumoniae [6], group A Streptococci. Moraxella catarrhalis [7] acquired almost universally a betalactamase and therefore became almost totally resistant to amoxycillin. Set against this within the hospital has been the rise and plateauing out of methicillin resistant Staphylococci and the increase of resistance in, for example, acinetobacter and enterobacter species to virtually all third generation cephalosporins, Enterococcus faecalis and facecalis in hospital. In 1997 from Japan the first reports of vancomycin resistant Staph. aureus have been noted.

We now have a situation where the prevalence of groups of genes determining resistance are now commonplace both in the hospital and also increasingly in the community. A good example of the spread of resistant determinants is the multi resistant tuberculosis. It is well known that the health costs, morbidity and mortality of multiresistant tuberculosis both in the hospital and in the community is very large [8]. There is an increase of methicillin resistant staphylococci in the community and day care centres [9]. Pseudomem branous colitis is a nosocomial (hospital acquired infection) caused by Clos being spread under the influence of antibiotic use, is also on the increase. Within the community we have the widespread dissemination of resistant genes to penicillins, cephalosporins, macrolides, quinalones and tetracyclines.

In hospitals, increase in Gram negative resistance has increased to a whole range of antibiotics. We have Carbapenem resistance in Pseudomonas aeruginosa, even occasionally in Esherishia coli and the increasing importance of Stenotrophomonas multophilia and many others.

The WHO report in 1996 [10] stated that in the contest for supremacy, microbes are sprinting ahead; the gap between their ability to mutate into drug resistant strains and man's ability to counter them is widening fast.

Resistance can be divided into two areas: innate resistance where resistance has arisen with the need for the antibiotic to develop resistance and that acquired through either antibiotic use or transmission of these resistant strains within the same species or between different species of bacteria. Reasons for this are complex and include demongraphic population changes and increases in vulnerable patients. Social changes often increase the reservoirs and the carriers of resistance. Travel, increased use of daycare centres where children spread resistance genes [11]. A major determinant is antibiotic use, although the exact relationship between antibiotic use and the magnitude and rate of development of resistance is poorly understood and even more poorly quantified.

Antibiotics have been used since the 1930s when Prontocil, a sulphonamide, was first launched for things such as lobar pneumonia. Antibiotics are perceived as of great value and, in common with many other products, their use has increased expectations. The market is extremely large and very mature. With little actual demonstratable differentiation between products especially between patient outcomes treated with different classes. The 1930s to 1980s were really a challenge in identifying and developing new antibiotics to combat infectious diseases and bacterial resistance. It was generally accepted that this challenge was largely met by the pharmaceutical industry and that most infectious diseases were amenable to treatment.

However, partly because of the total absence of the introduction of totally new classes of antibiotics since 1960 and the social changes and economic changes, along with the increased use of antibiotics, the whole area of antibiotics and resistance has become incredibly complex and of increasing importance. Antimicrobial resistance is increasing not only in hospital but also in the community, and it is occurring across all classes of antibiotics. The magnitude and rate of change and affect on patients' outcomes are poorly defined and incompletely understood. Currently there is an approach in the community use of antibiotics that patients present with symptoms there is a presumption of bacterial infection, there is an assumption that initial antibiotic treatment is useful and antibiotics are prescribed based on a whole variety of reasons including initial presentation and patient desire.

The roles taken up by the major players of this antibiotic symphony cause the problems. Providers treat individual patients, patients expect antibiotics, the systems or the payers provide funding, and drug developers discover, develop, provide product and promote heavily, leading to increasing antibiotic use. There needs to be a change in view of all these major players that perhaps providers should recognise the community value of antimicrobial, patients should acknowledge the value of initial non-drug approach, the payers and governments funding health care should participate in tailoring resource use to the community health outcomes, and the pharmaceutical industry should consolidate, evaluate, research and deliver knowledge about appropriate care reactive to community outcomes rather than individual patient outcomes.

There are four challenges within the antibiotic area. Highly successful therapies have fundamentally altered the nature of challenge. More rational but less intuitive choices have led to very complex decisions. There is much more available data than there ever was but less useful knowledge in determining efficacy of treatment in individual patient and community situations.

Challenge 1
Infectious disease, morbidity and mortality has decreased although there is a suggestion that this is now increasing. Progress was in large part due to antibiotic use. Because of the rise of the market in the other disease areas such as in the ulcer treatments, the relatively slow growth of the antibiotic market, lack of ability to differentiate and the complete paucity of new antibiotics, some pharmaceutical companies have halted antibiotic discovery and development. However, problems are emerging and the demand for and use of antibiotics is increasing. This will increase bacterial resistance and there is a suggestion, as shown by drug resistant tuberculosis hospital acquired infections, and the rise of resistance within the community pathogens, that there may become an inability to treat even common infections.

Challenge 2
Many antibiotics have well-defined but overlapping antimicrobial effectiveness. However, it is very difficult to devise clear, scientifically appropriate studies to show differences between antibiotics. This has led to a questionable link between the use of antibiotics and improved patient outcome. The most appropriate choice is not clear even in common diseases and the clinical and societal implications of bacterial resistance are unclear.

Challenge 3
More available data but less useful knowledge. There is an enormous increase in research, clinical and community information, including databases. We have microbial genetics, we have

microbial surveillance, knowledge of prescribing patterns, knowledge of healthcare utilisation, knowledge of demographic changes and knowledge of morbidity. However, databases are isolated, the episodes that occur are unconnected and often the patient outcomes we are most interested in are not documented.

Challenge 4
Increase in tension between individual patient desires and society's priorities. Prescriber and indiscriminate patient driven use of antibiotics is believed to be linked to increasing bacterial resistance. However, fears of the andromeda strain are rising. The potential threat to society of individual medical treatment decisions is more finely drawn in antibiotics use than any other field of medicine. There is a big and growing pressing need for the development of new antimicrobial agents with new modes of action which will not be inherently resistant. There are major new targets in antimicrobial development which are looking at the mediators of cell division, genes associated with gene expression, potentiators of antibiotics through a multiplicity of means including, of course, the tried and tested betalactamase inhibition and also looking at the genes responsible for pathogens virulence and spread.

The strategic goals for improved antimicrobial use include the optimal use of prophylatic, empiric and therapeutic prescribing in hospitals and in the community. We need to develop processes for detecting, reporting and preventing the transmission of antimicrobial resistant organisms. Before we can go anywhere in actually managing the issue of improved antimicrobial use we have to define optimum use. We need pure and applied research into definition of optimum use. There has to be a priority order of either benefit effects for the patient or for the community at large. There needs to be a comprehensive, multi-disciplinary approach to antimicrobial use where all concerned, including the patients, the doctors, the organisations responsible for the delivery of healthcare and the pharmaceutical industry are involved. We need to continue to develop the old classes of antimicrobials and their usage. We need to develop totally new classes of agents and hopefully bacterial genomics are of prime interest. When these products become available is an open question. It is only by these means that we will be able to combat the threat of multi resistant bacteria in our communities and on the effects of us as individuals after the year 2000 [12].

## REFERENCES


1. Finland K. Occurrence of serious bacterial infections since the introduction of antibacterial agents. JAMA 170 (1959) 2188-2197.
2. Barker M. Methicillin resistant staphylococci. J. Clin. Pathol. 14 (.1961) 385-393.
3. Medeiros A. Evolution and dissemination of beta-lactamases accelerated by generations of β-lactam antibiotics. Clin. Infect. Dis. 21 (suppl 1) (1997) s19-45.
4. Dorbusch K. European study group of antibiotic resistance. Eur. J. Clin. Micribiol. 6 (1987) 378-385.
5. Knothe H., Shah P., Krcmery V., Antal M., Mitsuhashi S. Transferable resistance to cefotaxime, cefoxitin, cefamandole and cefuroxime in clinical isolates of Klebsiella pneumonia and Senatia marcescens. Infection 11 (1983) 315-317.
6. Baquero F., Martinez-Beltian J. & Loza E. A review of antibiotic resistance patterns of Streptococcus pnewmoniae in Europe. J.A.C. 28, Suppl C (1991) 31-38.
7. Berk S. L., Kalbfleich et al. Antibiotic susceptability patterns of community acquired

respiratory isolates of Moraxella catarrhalis in Western Europe and in the USA. J.A.C. 38, Suppl A (1996) 85-96.
8. Raviglione M. C., Snider D. E., Kochi A. Global epidemiology of tuberculosis: morbidity and mortality of a worldwide epidemic. JAMA 273 (1995) 220-226.
9. Sarovoltz L. D., Pohlod D. J., Arking L.M. Community-acquired methicillin resistant Staphylococcus aureus infections; a new source for nosocomial outbreaks. Ann. Intern. Med. 97 (1987) 325-329.
10. The World Health Report 1996. Report of the Director General.
11. Report of the ASM Task Force on Antibiotic Resistance 1995. ASM Publications p.8-9.
12. Bax R. Antibiotic Resistance. Clin. Infect. Dis. 24 (Suppl 1) (1997) s151-153.

# AUTHOR INDEX

## SUBJECT INDEX

FSC
www.fsc.org
MIX
Papier | Fördert
gute Waldnutzung
FSC® C083411

Kolibri 360 GmbH
Ferdinand-Jühlke-Straße 7
99095 Erfurt, Deutschland
produktsicherheit@kolibri360.de